PETROLEUM GEOCHEMISTRY AND EXPLORATION IN THE AFRO-ASIAN REGION

BALKEMA – Proceedings and Monographs
in Engineering, Water and Earth Sciences

PROCEEDINGS OF THE 6TH AAAPG INTERNATIONAL CONFERENCE, BEIJING, CHINA, 12–14 OCTOBER 2004

Petroleum Geochemistry and Exploration in the Afro-Asian Region

Editors

Digang Liang
Key Lab of Petroleum Geochemistry, PetroChina, Beijing, China

Darui Wang
The Editorial Board of Petroleum Exploration & Development, Beijing, China

Zhenxi Li
Key Lab of Gas Geochemistry, Institute of Geology & Geophysics, Chinese Academy of Sciences, Lanzhou, China

CRC Press
Taylor & Francis Group
Boca Raton London New York

CRC Press is an imprint of the
Taylor & Francis Group, an **informa** business
A BALKEMA BOOK

Front cover photo: courtesy of Dr. Jianping Chen

Published by:
CRC Press/Balkema
P.O. Box 447, 2300 AK Leiden, The Netherlands
e-mail: Pub.NL@taylorandfrancis.com
www.crcpress.com – www.taylorandfrancis.com

ISBN-13: 978-0-415-44083-7 (hbk)

Typeset by Charon Tec Ltd (A Macmillan Company), Chennai, India

Petroleum Geochemistry and Exploration in the Afro-Asian Region –
Digang Liang, Darui Wang & Zhenxi Li (eds)
© 2008 Taylor & Francis Group, London, ISBN 978-0-415-44083-7

Table of Contents

Preface ix

Opening Speech xi

Scientific Committee of AAAPG 2004 xiii

Beijing Declaration xi

By-Laws of AAAPG xvii

Executive Council of AAAPG xix

1. Petroleum geochemistry and exploration in the Afro-Asian Frontier Basins

Developments in petroleum geochemistry in the last decade 3
K. Chandra & A.K. Sinha

Characteristics and genesis of the western slope thick oil in Songliao Basin, China 11
Z. Wang, C. Zou, G. Xu, X. Wu, X. Sun & J. Wang

Evaluation of petroleum source rocks of Bengal Basin, India 21
S. Pahari, I.V.S.V. Prasad, A. Banerjee & M. Varshney

Accumulation history and features of natural gases in the Kuqa foreland basin 27
M. Zhao, Z. Wang, S. Zhang, Q. Wang, Y. Song, S. Liu & S. Qin

Biomarker indications as evidence for organofacies variations within the Abu Gabra formation of the Muglad basin, Sudan 37
O. Ibrahim Mohamed Fadul & A. Mahgoub Mohamed

Accumulation periods of hydrocarbon and its distribution in Dongying Sag, Jiyang Depression, Eastern China 47
G. Zhu, S. Zhang, J. Dai, Y. Liang, Q. Jin & L. Zhang

Correlation between contents of TOC and strontium distribution within source rocks of Shahejie Formation in Dongying Sag, Eastern China 59
G. Zhu, S. Zhang, J. Dai, Q. Jin & G. Wang

Comparison of Proterozoic basins of India with similar basins of the world: Implications to hydrocarbon resource prospects 69
C. Vishnu Vardhan, S.K. Panicker & B. Kumar

Hydrocarbon-related diagenetic zones and its implication for the Yingnan 2 gas pool, Eastern depression of Tarim Basin 75
Y. Li, N. Zhong & Z. Zhang

2. Source rock evaluation, migration and reservoir geochemistry

Structural change of extracted lignin by white rot and brown rot fungi and thermal alteration of the lignin after the microbial treatments 87
K. Toh & T. Murae

Application of reservoir geochemistry in Jiuxi Depression 95
Y. Hu, J. Da, D. Zhang, K. Cheng & Y. Xong

Solid residue of thermal simulation experiment on *Laminaria* and its petroleum geological significance 103
W. Liu, X. Lu, J. Qin, K. Hu, Z. Li, L. Bian & M. Hu

Algorithm for production allocation in multilayer commingled production: Research and application 113
Y. Wang, S. Lu, W. Fang & J. Zhang

Oil-source rock correlation with multivariate statistical analysis 119
Z. Li

Interaction between basalts and oil source rocks in rift basins: CO_2 generation 127
Q. Jin, L. Xu, Y. Zeng & C. Wan

3. Molecular and isotopic geochemistry

Surface geochemical exploration practice for oils and gases in China 137
C. Liu, K. Zhao & C. Wu

Maximising the yield of bound aliphatic biomarkers via a convenient two-stage hydropyrolysis procedure 145
W. Meredith, C.E. Snape, C. Uguna & G.D. Love

Comparison of hydrocarbon generation simulation of source rocks in close and open system: A case of oil shale from Lower Jurassic in Qaidam Basin, China 153
J. Qin, Z. Li & B. Liu

Carbon isotopic signature of CO_2 in Arthit gas field, northern Malay basin, the Gulf of Thailand 161
S. Pisutha-Arnond, A. Sirimongkolkitti & V. Pisutha-Arnond

4. Basin modeling, novel geochemical approaches and petroleum strategies

The sedimentary model of $Ng_{(1+2)}$ in Gudao Oilfield and its controls on the remaining oil distribution 169
X. Wang, J. Zhong, L. Jiang & P. Jia

Organic and trace elements geochemistry of western Niger delta oils 177
A.A. Mustafa & J.I. Nwachukwu

Predictive models of the formation volume factor and density of the saturated crude oil 187
H. Xue, S. Lu, B. Wang, X. Liu & X. Fu

The application of rock pyrolysis technique to geochemistry evaluation of reservoirs in China 195
L. Wu & Z. Zhang

Study of fault sealing by geochemical techniques: Fault sealing evaluation in petroleum migration and reservoir continuity 201
D. Hou, J. Zhang, Z. Long, J. Zhu, Y. Tang, X. Xu & B. Huang

Biodegraded characteristics of ground soil for crude oil contaminants 209
X. Liu, C. Li, J. Li, C. Dai, J. Wen & Y. Li

Application of fluid inclusion to understand oil charging in Palogue Field, Melut Basin, Sudan 213
D. Cheng, L. Dou, Z. Li & Y. Li

Geochemical monitoring technique in reservoir performance 219
X. Jin, Q. Shi, R. Lin & D. Zhu

The formation and recognition of thermo-evaporative condensate oil and gas 225
J. Shen, S. Lu, M. Wang & W. Fang

Characteristics of crude oil components for different current recoveries 235
X. Liu, J. Li, C. Dai, C. Li, J. Feng & J. Wen

Author Index 241

Petroleum Geochemistry and Exploration in the Afro-Asian Region –
Digang Liang, Darui Wang & Zhenxi Li (eds)
© 2008 Taylor & Francis Group, London, ISBN 978-0-415-44083-7

Preface

The Afro-Asian Association of Petroleum Geochemists (AAAPG) successfully held the 6th International Conference on Petroleum Geochemistry and Exploration in the Afro-Asian Region during October 12–14 in Beijing. The conference was organized by China National Petroleum Corporation (CNPC) and co-organized by Chinese Academy of Sciences (CAS). 233 delegates from 23 countries of Asia, Africa, Europe, North America and Australia participated in the conference and this conference was remarked as a splendid meeting for the petroleum geochemistry domain in the Afro-Asian region.

The Afro-Asian Association of Petroleum Geochemists (AAAPG) is an international academic organization in the Afro-Asian region. Since its foundation, the association has held six international academic conferences, the first was convened in Dehradun, India in 1985, the second in Beijing, China in 1988, the third in Melbourne, Australia in 1992, the fourth in Arusha, Tanzania in 1996, the fifth in New Delhi, India in 2000 and the sixth in Beijing, China again. These past conferences have played an important role in enhancing academic exchanging between the Afro-Asian petroleum geochemists and their colleagues in other continents as well as in strengthening our mutual friendship and cooperation.

The sixth international conference (AAAPG-2004, Beijing) received 210 abstracts and during the conference 21 talks were given at the plenary meeting, 92 talks were addressed at the parallel sessions and 56 posters were demonstrated. Many famous geochemists delivered highlight talks at the plenary meeting, they were Mr. K. Chandra (India), Chairman of the Fifth Executive Council of AAAPG; Prof. Ping'an Peng (China), Vice Chairman of the Fifth Executive Council of AAAPG; Dr. C.M. Ekweozor (Nigeria), Vice Chairman of the Fifth Executive Council of AAAPG; Prof. Shuichang Zhang (China), Member of the Fifth Executive Council of AAAPG; Dr. K. Kagya (Tanzania), Member of the Fifth Executive Council of AAAPG; Dr. E. Subroto (Indonesia), Member of the Fifth Executive Council of AAAPG and some world-known scholars including Prof. D.W. Welte (Germany), Prof. J.M. Moldowan (U.S.A.), Dr. L.R. Snowdon (Canada), Dr. S. George (Australia), Dr. W.G. Dow (U.S.A.), Dr. L.H. Nilsen (Denmark) and Dr. Maowen Li (Canada).

The talks and papers presented to this conference covered varieties of research fields including (1) geochemistry on exploration in newly prospecting areas and on the formation of giant oil/gas fields; (2) petroleum system; (3) basin modeling; (4) dynamics of hydrocarbon generation and carbon isotopes; (5) geochemistry on source rocks and oil-source correlation; (6) geochemistry of natural gases and the biogenic gas; (7) oil/gas expulsion and primary migration; (8) molecular geochemistry and biomarkers; (9) organic petrology; and (10) the relationship between present or paleo- geothermal fields and hydrocarbons. These topics, on the whole, are involved with nearly all the hotspots and frontiers of current petroleum geochemistry researches in the world and have mirrored the newest development of petroleum geochemistry in the Afro-Asian region.

The papers and talks presented at this conference have drawn the following conclusions:

1. Petroleum geochemistry has played a non-substitutable important role in determining effective source rocks, classifying petroleum systems, delineating geneses of conventional and unconventional oils and gases, tracing back the dynamic process of petroleum accumulations, guiding the final determination of prospecting target areas and reducing the exploration risk in a series of new and significant oil-gas discoveries in the Afro-Asian region.
2. Basin modeling and petroleum system classification that are fully applying geochemical data are the two developing orientations to enhance the petroleum exploration effects.
3. Molecular geochemistry and biomarkers have made significant progresses in their applications in petroleum exploration, which are symbolized in three aspects by the identification and application of age-indicating biomarkers, the application of diamondoids in appraising the cracking level of

crude oils, and the application of the compound-specific isotope analysis of biomarkers and the compound-specific isotope analysis of diamondoids (CSIAB and CSIAD) in oil-source correlation as well as in quantitative identification of source-commingled oils.

4. The dynamics of hydrocarbon generation, the dynamics of carbon isotopic fractionation and the analysis of liquid historical recordings (inclusions) are the three significant approaches to reconstruct the history of oil and gas accumulations.

Beijing Declaration consentaneously approved of at the conference fully affirmed the important role of geochemistry research in oil-gas exploration, appealed to strengthening international exchanging and cooperation between Afro-Asian petroleum geochemists and their colleagues over the world, and to working hard so as to make more contributions to the petroleum exploration and economical development in the Afro-Asian region. We wish the publication of the present proceedings play a role in exchanging research achievements, enhancing mutual understanding and strengthening our friendship.

It was approved at the AAAPG-2004 international conference that the 7th International Conference on Petroleum Geochemistry and Exploration in the Afro-Asian Region will convene in Abuja, the capital of Nigeria. Dr. C.M. Ekweozor, Vice Chairman of the Sixth Executive Council of AAAPG will serve as the Chairman of the Organizing Committee of this conference. We congratulate beforehand the coming conference a great success and present the proceedings as a gift to the 7th AAAPG international conference.

The proceedings include 29 research papers presented at the AAAPG-2004 international conference. It should be noted that some of the outstanding presentations addressed at the conference have been submitted to Organic Geochemistry and they, much to our regret, can't be included in the present proceedings.

At last, we sincerely thank Taylor & Francis Group for hard and persistent working for the publication of the proceedings.

Executive Council of AAAPG
AAAPG-2004 International Conference Proceedings
Chief Editors: Digang Liang
Darui Wang
Zhenxi Li

2006.12. Beijing

Petroleum Geochemistry and Exploration in the Afro-Asian Region –
Digang Liang, Darui Wang & Zhenxi Li (eds)

The 6th Conference on Petroleum Geochemistry and Exploration in Afro-Asian Regions – Opening Speech

Dr. Chengzao Jia, Conference Chairman
(Oct. 12, Beijing)

Good Morning, Ladies and Gentlemen:

"The 6th Conference on Petroleum Geochemistry and Exploration in the Afro-Asian Regions" co-organized by the China National Petroleum Corporation and the Chinese Academy of Sciences has its grand opening here in the golden Autumn Beijing! Firstly, please allow me in the name of the Conference Chairman and on behalf of Dr. Tao Wang, the Honorary Conference Chairman and the executives of China National Petroleum Corporation, extend our warm welcome to all the petroleum geochemists from 23 countries, and express our warm welcome to Dr. Chandra, Chairman of the AAAPG Executive Council, and Prof. Pu Fan, the Honorary Chairman of the AAAPG Executive Council!

It is all known that petroleum geology, petroleum geophysics and petroleum geochemistry are the three pillars for petroleum exploration. And in the field of petroleum geochemistry, four major progresses have been made in the past more than 30 years, these progresses include the theory of hydrocarbon generation by organic matter thermal maturation, the oil generation theory by kerogen thermal degradation, the application of biomarkers in petroleum exploration, and development of the concept of petroleum system. These progresses lead to significant changes in terms of thoughts and methodology in petroleum exploration. Today, we are very pleased to see that our invited leading specialists in these fields have joined us and will have their presentations in these fields at this conference.

We are also very happy to see that the petroleum geochemists in the Afro-Asian region have made their important contributions to oil/gas exploration, and Chinese petroleum geochemists, through their study on non-marine petroleum generation theory and their practices in petroleum exploration, have contributed to the development of the petroleum geochemistry science as well.

Under the leadership of the AAAPG Executive Council, the AAAPG conference is held every four years, and since 1985, AAAPG conference has been held successively in India, China, Australia, and Tanzania. All these past conferences got a complete success, and played its active role in large oil/gas discoveries in the Afro-Asian region. We expect that this AAAPG conference today will again present the fruitful research achievements by the African and Asian petroleum geochemists in the past recent four years, and our delegates will be able to learn each other and exchange ideas and experience, making this conference another complete success.

As of today, we have received 210 abstracts from 23 countries and regions. 21 presentations in three morning plenary sessions and 92 presentations in three afternoon four parallel sessions are scheduled at the conference. One hour poster session each day is also arranged. We strongly suggest your active involvement and discussion on these presentations and hope all of you will have a most rewarding three-day meeting.

Ladies and Gentlemen, I would like to say again: You are warmly welcomed in China, and you are warmly welcomed in Beijing! The Chinese petroleum geochemists treasure this great gathering and they are eager to learn from you and have as many discussions as they can possibly have.

Finally, wish the 6th AAAPG conference a complete success! Wish all of you will develop friendship and have new gains of knowledge! And wish each and every one of you to have a pleasant stay and an everlasting memory in Beijing!

Now, I announce, the 6th Conference on Petroleum Geochemistry and Exploration in the Afro-Asian Region opens!

Petroleum Geochemistry and Exploration in the Afro-Asian Region –
Digang Liang, Darui Wang & Zhenxi Li (eds)
© 2008 Taylor & Francis Group, London, ISBN 978-0-415-44083-7

Scientific Committee of AAAPG 2004

Honorary Chairman of AAAPG-2004
Prof. Tao Wang (Senior Vice-Chairman of World Petroleum Congress)

Chairman of AAAPG-2004
Dr. Chengzao Jia (Academician of CAS, Chief Geologist of PetroChina, Director of RIPED)

Vice-Chairman of AAAPG-2004
Prof. Jiamo Fu (Academician of CAS)

Mr. Kuldeep Chandra (Chairman of the Executive Council of AAAPG, India)
Prof. Pu Fan (Honorary Chairman of the Executive Council of AAAPG, China)
Prof. Jiamo Fu (Member of the Executive Council of AAAPG, Academician of CAS, China)
Prof. Digang Liang (Vice-Chairman of the Executive Council of AAAPG, China)
Prof. Ping'an Peng (Vice-Chairman of the Executive Council of AAAPG, China)
Dr. C. M. Ekweozor (Vice-Chairman of the Executive Council of AAAPG, Nigeria)
Prof. T. Murae (Vice-Chairman of the Executive Council of AAAPG, Japan)
Prof. Difan Huang (Member of the Executive Council of AAAPG, China)
Dr. E. A. Subroto (Member of the Executive Council of AAAPG, Indonesia)
Mr. U. Nwaezeapu (Member of the Executive Council of AAAPG, Nigeria)
Dr. M. Kagya (Member of the Executive Council of AAAPG, Tanzania)
Prof. S. O. Akande (Member of the Executive Council of AAAPG, Nigeria)
Prof. J. M. Moldowan (Dept. of Geological & Environmental Sciences, Stanford Uni., USA)
Dr. L. R. Snowdon (Geological Survey of Canada, Canada)
Dr. W. G. Dow (Consultant, The Woodlands, Texas, USA)
Prof. D. H. Welte (Gesellschaft für Integrierte Explorationsysteme mbH, Jülich, Germany)
Dr. S. C. George (CSIRO Petroleum, Australia)
Mr. Y. B. Sinha (Oil & Natural Gas Corporation Ltd., India)
Dr. L. H. Nielsen (Geological Survey of Denmark & Greenland, Denmark)
Dr. A. D. Carr (Advanced Geochemical Systems Ltd., U.K.)
Dr. Ulrich Ritter (Sintef Petroleum Research, Norway)
Dr. V. A. Kashirtsev (Institute of Oil & Gas Problems, Siberian Division of the RAS, Russia)
Mr. Omer I. M. Fadul (Greater Nile Petroleum Operation Company, Sudan)
Prof. S.P. Bindra (International Energy Foundation, Libya)

Petroleum Geochemistry and Exploration in the Afro-Asian Region –
Digang Liang, Darui Wang & Zhenxi Li (eds)
© 2008 Taylor & Francis Group, London, ISBN 978-0-415-44083-7

Beijing Declaration

The 6th AAAPG International Conference on Petroleum Geochemistry and Exploration in the Afro-Asian Region was convened in Beijing from October 12–14, 2004. More than 200 delegates from 23 countries and regions participated in the conference. Out of 212 abstracts, 169 were selected for oral or poster presentations. This voluminous contribution documents the important role that petroleum geochemistry has played in hydrocarbon exploration in general and in the Afro-Asian region in particular.

Since the Fifth AAAPG International Conference held in India in 2000, important petroleum discoveries have been made in Asia and Africa. In Asia, these include the Kashagan giant oilfield in Caspian Sea (Kazakhstan), three giant oilfields around Bushehr Port in southern Iran, five large oilfields including PL19-3 in the Bohai Gulf of China, four giant gas or oil fields in Tarim and Ordos basins, China. Other important discoveries have been made in Rajasthan, a giant gas field in deep-waters in a Godavari offshore basin, and large oil fields in offshore Orissa and Bombay offshore, and the Vasai East in India, in South Sumatra Basin, Indonesia, in offshore NW Australia and in offshore Malaysia and Viet Nam. In Africa, over 100 million tons of new oil reserves have been established, in the Eastern Suez Gulf of Egypt, Central-Western Libya, Agomi, Bonga and Erha fields in Nigeria offshore, in four abyssal blocks in Angola and in the Abhubezi region of South Africa. It is important to note that four of the five most significant petroleum discoveries in the world since January 2000 have been made in Asia and Africa.

The participants of the AAAPG – 2004 International Conference unanimously agree upon the following declaration:

1. The Asia-Pacific region and Africa occupy nearly half of the total land area in the world and possess large areas of continental shelf. As this region is rich in hydrocarbon resources, the potential for new discoveries in the region continues to be good to high. Towards ensuring energy security in the region in the back drop of high volatility in crude oil prices, the Conference considers the urgent need to encourage systemic petroleum systems studies and exploration modeling integrating geochemical and geophysical data in the reference frame of geology, sequence stratigraphy and basin evolutionary histories.
2. The growing knowledge on petroleum geochemistry of petroleum systems of different types of the sedimentary basins of the region brought out in this Conference is going to help formulate improved exploration alternatives and that such knowledge should be encouraged to grow through providing thrust to research and development by way of dynamic industry-academia relationship.
3. The significant discoveries in hydrocarbon exploration of the Asian-African region are closely related to petroleum geochemical research. For exploration in the frontier regions in the superimposed basins, foreland basins and deep-water basins, geochemical research remains one of the most important tools for hydrocarbon resource assessment and appraisal. The Asian-African region contains abundant hydrocarbon resources derived from lacustrine and deltaic facies. Advances in geochemical study have provided theoretical foundations for the hydrocarbon exploration related to coal-derived hydrocarbons, hydrocarbons in the deep earth, biodegraded oil, heavy oil, immature oil, and shallow biogas. The understanding of the mechanisms and timing for hydrocarbon accumulations can potentially serve as a means for high-resolution basin modeling. Therefore, it can be reasonably expected that the petroleum geochemical research in the Asian-African region will face more opportunities and challenges, and hence make more significant contributions to more giant oil and gas discoveries in the foreseeable future. We are confident that important breakthroughs are possible in fields such as geochemistry of reservoir formation and constrained basin modeling.
4. Strengthening international academic exchange and cooperation is definitely one of the important approaches for pushing forward the development of petroleum geochemical research in the Asian-African region. The AAAPG Conference, which has been held once every four years, serves as an

important forum for the academic exchange among petroleum geochemists not only in this region but also over the world and as a good opportunity for us to enhance the friendship and to learn more from geochemists in Europe and America. As one of the three major organizations in organic geochemistry in the world, AAAPG will continue to hold such a periodic international conference and to expand its international influence. We believe AAAPG will turn into a link, a bridge and a core organization for the petroleum geochemists in the Asian-African region to conduct academic exchange. The Association shall be happy to see initiation of seminars and workshops in different member countries of AAAPG intervening the major International AAAPG Conference and make the AAAPG event more vibrant.

Our fellow petroleum geochemists from the Afro-Asian region and our colleagues from other parts of the world, for the sake of the development of petroleum geochemistry, for the hydrocarbon exploration and the economic development in the Asian-African region, for the global prosperity and peace, let us work harder and make more significant contributions.

This declaration is hereby jointly made by
The participants of
The AAAPG-2004 International Conference
In Beijing, China
On October 14, 2004

*Petroleum Geochemistry and Exploration in the Afro-Asian Region –
Digang Liang, Darui Wang & Zhenxi Li (eds)
© 2008 Taylor & Francis Group, London, ISBN 978-0-415-44083-7*

The Association of Afro-Asian Petroleum Geochemists (AAAPG)

By-Laws

Article I The Name

The official name of this Association shall be the "Association of Afro-Asian Petroleum Geochemists", hereinafter be referred to as AAAPG.

Article II Purposes

The purposes of AAAPG are to advance the science of geochemistry, especially as it relates to the petroleum and natural gas; to promote the geochemical technology of exploration for petroleum and natural gas; to issue some aperiodic communication brochures, to plan, organize and administrate a triennial or quadrennial academic meeting which is called "Conference on Petroleum Geochemistry and Exploration in the Afro-Asian Region".

Article III Membership

Section 1
Any person/organization in the Afro-Asian Region, who endorses the Bylaws of the AAAPG and is active in organic geochemistry, petroleum/natural gas geochemistry and geochemical exploration, and other disciplines concerning petroleum and gas geochemistry, is eligible to vote, hold elective office, serve on the Council, and enjoy the privilege of obtaining the publication of AAAPG in special price.

Section 2
All the members should be honest, sincere and guided by high-standard professional morality and professional responsibility.

Article IV Administration

Section 1
There shall be an Executive Council for AAAPG. Members of this Council shall be nominated either by the chairmen or by other members of the Executive Council on the basis of geographical consideration.

Section 2
Members of the Executive Council are liable for organizing petroleum geochemists of their own country to actively participate in the AAAPG conference and other academic events.

Section 3
The Executive Council is usually composed of the Chairman, Vice Chairmen, Honorary Chairmen, and members. The chairman of the Organizing Committee of the previous conference will automatically be the Chairman of Executive Council without other election or nomination, consequently the past Chairman of the Executive Council will be the Honorary Chairman of the Council automatically.

Section 4
The Chairman of the Executive Council can nominate and organize a Special Committee on geographic basis for the purpose of solving the problems concerning the AAAPG within such a region.

Section 5
The Secretaries of AAAPG shall be delegated by the Executive Council, under the direct leadership of the Chairman, to conduct the general affairs of the Association.

Article V Meeting

Section 1
There shall be a regularly triennial or quadrennial academic meeting, named "Conference on the Petroleum Geochemistry and Exploration in the Afro-Asian Region".

Section 2
There shall be a regularly meeting of the Executive Council during the triennial or quadrennial academic meeting. This meeting shall be for the purpose of planning the next Conference (time, place, organizer) and for any other business concerned that may arise.

Article VI Amendments to the By-Laws

Any amendment to this By-Laws should be discussed, endorsed and declared by the Executive Council when adopted.

Petroleum Geochemistry and Exploration in the Afro-Asian Region – Digang Liang, Darui Wang & Zhenxi Li (eds)
© 2008 Taylor & Francis Group, London, ISBN 978-0-415-44083-7

Executive Council of AAAPG

Honorary Chairmen
Prof. Pu Fan, Lanzhou Branch of CAS, **China**
Mr. Kuldeep Chandra, KDMIPE, ONGC Ltd, **India**

Chairman
Prof. Digang Liang, Key Lab of Petroleum Geochemistry, PetroChina, **China**

Vice Chairmen
Dr. C.M. Ekweozor, Getamme Geochem, **Nigeria**
Dr. A.K. Balyan, ONGC Ltd., **India**
Dr. Simon George, CSIRO Petroleum, **Australia**
Prof. Alexey E. Kontrorovich, Institute of Petroleum Geology SB RAS, **Russia**
Dr. Meshack L. N. Kagya, Tanzania Petroleum Development Corporation, **Tanzania**
Dr. Eddy Subroto, Department of Geology, Institute of Technology, Bandung (ITB), **Indonesia**
Prof. Ping'an Peng, State Key Lab of Organic Geochemistry, **China**

Members
Mr. Abdul Jalil Muhamad, Petronas Research & Scientific Services, **Malaysia**
Dr. Abdulla Elhenshir, Petroleum Research Center, **Lybya**
Mr. Ahmed Mahgoub, Section Head Basin Studies, **Sudan**
Dr. Akinsehinwa Akinlua, Department of Chemistry, Obafemi Awolowo University, **Nigeria**
Mr. A.K. Sinha, Geochemistry Division, KDMIPE, ONGC Ltd, **India**
Dr. Alla Nemchenko-Rovenskaya, Vernadskiy Institute, Russian Academy of Science, **Russia**
Dr. Anil Garg, Regional Training Institute, **India**
Dr. Arie Nissenbaum, Dept. of Environmental Sciences & Energy Research, Weizmann Institute of Science, **Israel**
Mr. A.U. Nwaezeapu, Overt Ventures LTD, **Nigeria**
Dr. Baleshwar Kumar, Stable Isotope & Suface Geochemical Prospecting Lab, National Geophysical Research Institute, **India**
Prof. Baojia Huang, Exploration Division, Western South China Sea Ltd., CNOOC, **China**
Dr. Ben G. K. van Aarssen, Centre for Applied Organic Geochemistry, **Australia**
Dr. Chidi Eneogwe, ExxonMobil Mobil Producing Nigeria, Geochemistry Laboratory, **Nigeria**
Dr. Chih-Hsien Sun, Exploration & Development Research Institute, Chinese Petroleum Corporation, Taiwan 360, **China**
Dr. Dardji Noeradi, Department of Geology, Institut Teknologi Bandung (ITB), **Indonesia**
Dr. David McKirdy, School of Earth & Environmental Sciences, The University of Adelaide, **Australia**
Prof. Difan Huang, Key Lab of Petroleum Geochemistry, PetroChina, **China**
Mr. Eddy Purnomo, Exploration Division, Pertamina Upstream Indonesia, **Indonesia**
Dr. Flora Mpanju, ARIPO, Harare, **Zimbabwe**
Dr. Fowzia H. Adullah, Department of Earth & Environmental Sciences, Kuwait University, **Kuwait**
Dr. George I. Unomah, Production Geoscience Special Studies, ExxonMobil Production Company, **U.S.A.**
Mr. George Ngwale, Tanzania Petroleum Development Corporation, **Tanzania**
Prof. Jianping Chen, Key Lab of Petroleum Geochemistry, PetroChina, **China**

Prof. Joseph I. Nwachukwu, Department of Geology, Obafemi Awolowo University, **Nigeria**

Mr. Kinh Vu Van, Vietnam Petroleum Institute, **Vietnam**

Mr. Le Van Hien, Vietnam Petroleum Institute, **Vietnam**

Prof. Linye Zhang, Geology Scientific Research Institute of Shengli Oilfield, **China**

Mr. Lopo Vasconcelos, Geology Department, Eduardo Mondlane University, **Mozambique**

Dr. Mahdi Abu-Ali, Saudi Aramco Exploration, **Saudi Arabia**

Dr. Maowen Li, Geological Survey of Canada, **Canada**

Mr. Mustafa El Jarad, HSEQ Department, General Peoples Committee of Energy, Tripoli, **Libya**

Dr. Neil Sherwood, CSIRO Petroleum, **Australia**

Prof. Ningning Zhong, Basin & Reservoir Research Center, University of Petroleum, **China**

Mrs. N.J. Thomas, ONGC Videsh Limited, **India**

Prof. Noriyuki Suzuki, Earth and Planetary Sciences, Hokkaido University, **Japan**

Dr. Olugbenga A. Ehinola, Department of Geology, University of Ibadan, **Nigeria**

Dr. Oluwadayo O. Sonibare, Chemistry Department, University of Ibadan, **Nigeria**

Dr. Prabhakar Dwivedi, KDMIPE, ONGC Ltd., **India**

Dr R.R. Singh, DGM (Chemistry), KDMIPE, **India**

Mr. Salvator Ntomola, Tanzania Petroleum Development Corp., **Tanzania**

Prof. Samuel Olusegun Akande, Department of Geology, University of Ilorin, **Nigeria**

Prof. Satyapal Bindra, Margeb University & International Energy Foundation, **Libya**

Mr. Shri Shiv Dayal, Petrotech Society, **India**

Prof. Shuichang Zhang, Key Lab of Petroleum Geochemistry, PetroChina, **China**

Mr. Sukumar Pahari, Geochemistry Division, KDMIPE, ONGC, **India**

Ms. Supaporn Pisutha-Arnond, PTT Exploration & Production (Public) Co. Ltd., **Thailand**

Dr. Tae-Jin Cheong, Korea Institute of Geoscience & Mineral Resources (KIGAM), **Korea**

Prof. Tieguan Wang, Basin & Reservoir Research Center, University of Petroleum, **China**

Prof. T.R. Ajayi, Department of Geology, Obafemi Awolowo University, **Nigeria**

Dr. V. Banerjie, Head Geochemistry Division, KDMIPE, **India**

Prof. Vladimir A. Kashirtsev, Institute of Geology of Oil & Gas, Siberian Division of RAS, **Russia**

Prof. Waseem Ahmed, Hydrocarbon Development Institute of Pakistan, **Pakistan**

Prof. Wenhui Liu, Research Institute of Petroleum Exploration & Development, SINOPEC, **China**

Secretary General

Mr. Zhenxi Li, Key Lab of Gas Geochemistry, Institute of Geology & Geophysics, CAS, **China**

1. Petroleum geochemistry and exploration in the Afro-Asian Frontier Basins

Petroleum Geochemistry and Exploration in the Afro-Asian Region –
Digang Liang, Darui Wang & Zhenxi Li (eds)
© 2008 Taylor & Francis Group, London, ISBN 978-0-415-44083-7

Developments in petroleum geochemistry in the last decade

K. Chandra
KDMIPE, ONGC Ltd., Dehradun, India

A. K. Sinha
Geochemistry Division, KDMIPE, ONGC Ltd., Dehradun, India

ABSTRACT: Petroleum Geochemistry came to be known as a branch of science somewhere in 1959 from its beginnings in the application of chemistry to understand origin of petroleum and detect presence of oil and gas in the subsurface accumulations from the anomalous flux of methane through butanes in the surface soils and sediments. The growth in petroleum geochemistry has been directly related to the growth in instrumental methods of analysis. Surface geochemical prospecting, petroleum source rock evaluation, logging, source potential mapping, generative basin mapping, genetic classification of crude oils, natural gases and formation waters, oil to oil and oil to source and gas to gas and gas to source correlations, mapping petroleum quality occurrence trends and estimation of charge of hydrocarbons and geochemical genetic prospect prioritization employing genetic basin modeling are known techniques that are applied today towards risk-reduction and maximization of returns from various portfolios of exploration and development opportunities. A narrative is attempted to describe the developments in petroleum geochemistry in the last decade, 1990 to present day, in the backdrop of (i) first period of modern development and evolution of geochemical concepts of origin and occurrence of petroleum 1960–1980, (ii) the first period of consolidation of knowledge in petroleum geochemistry, 1980–1990 and testing and application of concepts with whatever was possible by various oil companies and academic institutions and (iii) the second period of consolidation of growth of knowledge 1990-todate and prolific application of petroleum geochemistry in as much desirable a fashion as possible by various oil companies and academic institutions witnessing emergence of reservoir geochemistry and highly geoscience integrative migration modeling techniques. Delineation and determining petroleum systems in all the major petroleum provinces of the world has been the most outstanding achievements of the decade. Use of fluid inclusion studies to understand petroleum systems has also been a major development. Introduction of inductively coupled plasma multiple collector mass spectrometry has been a significant development and soon shall bring some revolutionary results on determining petroleum systems based on isotope fingerprinting of organic matter bound trace elements in crude oils, bitumens and kerogens. Petroleum Geochemistry as science has come of age and is one of the imperatives in the trinity of geosciences, geology, geophysics and geochemistry meant to ensure risk reduction and maximization of value creation in petroleum exploration and development efforts

1 INTRODUCTION

Petroleum geochemistry has come to be known as that of branch of science that devotes itself to understanding comprehensively chemical composition and constitution of petroleum and its precursor source organics within the geologic reference frame of sedimentary basins, the ubiquitous natural habitats of petroleum, with a view to achieve better focus on occurrence locales of petroleum and to arrive at better intra and inter regional priorities for exploration and development of petroleum resources.

Petroleum is a fugitive mineral. Petroleum and its precursor source organics and the sedimentary basins themselves continually evolve under the time-temperature controlled spontaneous processes attended with various tectonic and tectono-sedimentary evolutionary histories. Achieving the present stature and capability by petroleum geochemistry to contribute to reduction of risk in exploration and developmental effort has thus been an arduous task that too within a small history of the last 120 years of petroleum geochemistry.

2 THE PERIOD 1960–1980

The nineteen sixties heralded the era of modern petroleum geochemistry with capability to be servicing the industry with the advent of powerful techniques of molecular separations, gas-liquid chromatography, developments in adsorption column chromatography and rapid organic carbon analyzers. These developments reinforced and with time replaced the techniques based on distillation, carbon structural group analysis based on refractive index, density and molecular weight [n-d-m] of homogeneous fractions, fractional crystallization, liquefaction-fractional distillation-combustion-carbon dioxide volumetrics, titration or spectrometry for methane through butanes in surface geochemical prospecting, retorting or dry distillation to estimate the liquids generated from kerogens and coals and oxidative or reductive decomposing and selective defunctionalization, derivative formation and characterization of derivatives by elemental compositions, solubilities, refractive indices, melting and boiling points, viscosities and some confirmatory reactions.

The nineteen seventies and eighties saw the birth of pyrolysis chromatography and precise chromatography principle based robust elemental analyzers suitable to characterize the bacteriostatic, polymeric, lipid solvent insoluble kerogen, the main source-organics, followed by advent of microprocessor based, the magical Rock-Eval-I, the oil show and source rock analyzer. This period also witnessed advent of relatively more robust, solid-state electronics and microprocessor based medium to high pressure high performance liquid chromatographs, ion-chromatographs, flame fluorescence detectors, uv-vis-nir, Raman and laser-Raman and fluorescence and emission spectrophotometers, organic-mass-spectrometers, thermal-ionization and stable isotope ratio-recording mass spectrometers, first generation of GC-MS, FT-IR, FT-NMR/PMR, solid state with magic angle device NMR, High Precision Ratio-recording mass spectrometers, Selective Cathode and ICP atomic absorption spectrophotometers and also host of new generation of laboratory appliances. This period also consisted of raising infrastructure for geochemical work in oil companies, in various universities and schools of technologies.

Substantial quantum of efforts for geochemical data acquisition for testing geochemical concepts on (i) formation of source-rocks, (ii) maturation and maturation processes mimicking the natural time-temperature controlled transformation of sedimentary organic matter through dry pyrolysis and hydrous pyrolysis, (iii) generative basins, (iv) distribution of various homologous series of different families of hydrocarbons especially from C1 to C35, in general and C5 to C10 in particular, (v) biological markers, (vi) primary and secondary migration, (vii) Petroleum systems (viii) in-reservoir alteration and (ix) volumetric-genetic material balance were made in the eighties. Documenting best practices of data acquisition, processing and interpretation and consolidation of knowledge in petroleum geochemistry in the form of text books and human resource development course modules for geoscientist professionals in the industry, were made. Despite some frustration and reactionary situations, petroleum geochemistry carved out a niche for itself. Synergy of geosciences became the watchword for ensuring efficiency and effectiveness in exploration effort.

3 THE PERIOD 1980–1990

The 1980–1990 has been a decade of systematic institutionalization of petroleum geochemistry in the upstream sector in the industry. The period witnessed much improved molecular separation techniques and instrumental analytical techniques capable to take up automated batch analysis. This provided capability to effect a credible timely integration of geochemical inputs with other geosciences to develop exploration and development concepts and to arrive at decisions to either augment or initialize efforts to maximize reserves accretion or reserves replacement opportunities.

During the nineties robust techniques of (i) crude oil characterization, (ii) oil to oil and oil to source and gas to gas and gas to source correlations based on parameters that characterize source depositional environments, source organic matter input and source maturity, (iii) grading of degree of in reservoir bio-degradation, (iv) characterization of other in-reservoir alteration processes like water-washing, gas de-asphalting and formation of reservoir bitumens and (v) characterization of crude oil weathering of oil spills were also developed.

Helium isotopes and carbon isotopes of methane while did show some evidences of endogenic methane and carbon dioxide, yet in conjunction with isotopes of carbon, and dominance of He-4 isotope of helium convincingly supported a sedimentary origin of petroleum.

Among the non-hydrocarbon constituents asphaltene pyroysates, porphyrins and benzothiophenes were found useful in characterizing nature of source environments and source maturity.

The concept of organic matter facies, contrlloing gas vs, oil proclivities, of source rocks was developed and reliable combined techniques of visual microscopic-palynological and coal petrography were developed. Amongst the trace elements nickel and vanadium continued to be the elements of geochemical significance.

The 1980-1990 decade also saw significant developments towards perfecting 3-D genetic Basin Modeling and especially migration modeling, which had been developing since the mid seventies in consonance with the developments in software engineering and fast computing and commercialization of new industry standard user friendly interpretation work stations.

Lots of efforts that were possible in different oil companies and research organization were carried out towards petroleum systems definition in different basins.

Thermodynamic modeling of oil-field waters to predict dissolution or precipitation of solutes with changing pH, temperature and pressure had been a meaningful development. Use of strontium isotopes, and isotopes of carbonate and bicarbonates in formation waters and their cross-plots with total salinity and or coefficient of base exchange were utilized to define stratitigraphic or parastratigraphic spatiation of hydrodynamic systems associated with entrapment and post accumulation alterations of petroleum.

4 THE PERIOD 1990-TO-DATE

The last decade, 1990-2000, and the subsequent current period has witnessed major effort in reconsolidation of the knowledge in petroleum geochemistry in terms of new editions of text books, AAPG memoirs on Basin Evaluation, Petroleum Systems, Source-Rocks in Sequence Stratigraphic Frame Work, Surface Manifestations of Hydrocarbon Migration and Geochemical Prospecting. During this period the numerical models of upward migration provided credible support to the concept of vertical migration through force of buoyancy on hydrocarbon micro-globules. Better appreciation of surface microbiological modification of surface distribution of hydrocarbons was also made.

The decade 1990–2000 also saw availability of improved and more robust and sensitive analytical instruments suitable for petroleum geochemistry. GC-MS-MS, pyrolysis-GC Hydrocarbon Analyzer, Pyrolysis-Thermal-Evaporator-MS Analyzer for integrated adsorbed surface geochemical prospecting, and ICP multicollector mass-spectrometer, multitemperature programming Rock-Eval [Rock-Eval-6] for better estimating the kinetic parameters of kerogen and GC-IRMS both for carbon and hydrogen specific compound fingerprinting of oils and natural gas, syngenetic hydrocarbons in source rocks and hydrocarbons obtained from kerogen pyrolysates, and homologous series of biomarkers. Amongst the biomarkers, diamondoids, C-22-C28 triterpanes, methyl cholestanes and gammaceranes, nor hopanes and distearanes and carbazoles were the new set of molecular correlation tools. ICP-MCMS has been a revolutionary development, it will help in fingerprinting isotopic enrichments of the organically bound trace elements in kerogens, source-rock bitumens and asphaltenes in crude oils providing probably a fairly robust correlation and petroleum system definition.

There have been number of upstream industry oriented major development trends that took place in the last decade. They are summarized hereunder.

The first consisted of revisiting all the major petroleum provinces for a very exhaustive and truly desirable investigations for defining petroleum systems employing best set of geochemical techniques of high-resolution source-rock logging depicting both the total generation and hydrocarbon expulsion potentials, oil to oil and oil to source correlations in practically all the known petroleum provinces of the world. This not only led to identification of new petroleum systems providing optimism and opportunity for further exploration, it also helped in fine-tuning occurrence models and establishing sub-petroleum systems and the extents of their generative basins, delineating crude oil and natural gas compositional occurrence trends and reconstructing migration and mixing and post accumulation alteration histories. Use of aromatic hydrocarbons like the diphenyl methane and dibenzophenyl methane to sub-class one family of oils has been shown. In all the rift basins late synrift and early drift phase lacusterine source-rocks and petroleum systems have been found to be common. The Recent analogues of such source-rocks in the African lakes have been investigated. A few marine source-rock based petroleum system followed by deltaic and coal containing systems in pericratonic and foreland basins have also been common. The South Asian, South East Asian and Far-Eastern basins have provided some important examples of coals and coaly shales acting as effective source rocks of petroleum.

There has been tremendous thrust on the study of fluid inclusions. Such studies have shown promise to establish the timing of migration and filling up of reservoirs, ingress of oil of two systems at two different times leading to identification of presence of additional petroleum systems. Use of fluid inclusions has been successfully demonstrated in geochemical prospecting.

The last decade has also witnessed developments in the field of reservoir and production geochemistry. Crude oils flowing in porous media depending upon the continuity and homogeneity of porosity and permeability are fractionated due to varying degrees of freedom of movement of molecules having different sizes and shapes. Initial success in mapping reservoir continuity and compartmentalization was made on the similarities of star diagrams of six to seven chosen homologous paraffinic hydrocarbons, alkanes and isoalkanes [of same carbon number]. Different workers have used different parameters, and some have been successful with usual parameters that are employed in crude oil correlations. It has been the experience of workers in this field that for each oil and reservoir system based on whole oil chromatography ratios of those set of hydrocarbons should be used which show statistically distinct similarity or variation with known continuity and heterogeneity and such variance in ratios should not be an artifact of precision in replicate analyses and plausible artifact of biodegradation or other in reservoir alteration processes. Such reservoir geochemical characterization has been attempted to help improve water injection schemes for pressure maintenance and enhanced oil recovery.

Some workers proposed to include non-hydrocarbon surface active bio-markers like carbazoles and benzoquinolines and phenathroquinolines, both in reservoir rocks and oils to better describe reservoir flow-units with potential to help in upscaling reservoir models. Use of carbazoles and quinolines has been advocated to decipher spatial extents of migration for well preserved oils of same source maturity.

In the field of reservoir geochemistry attempts have also been made to estimate allocation of daily production of oil from each production string when the compositions of oils from each string are known employing a matrix mathematical numerical modeling. Attempts to estimate kinetic parameters of oil phasing out to gas under reservoir simulated conditions have also been attempted to better understand thermal spontaneous evolution of oils. Stabilities of kerogens and generated oil have been studied under ultra high pressure and temperature conditions in quartz anvil cell coupled with Raman spectrometry.

Other developments of research stage consisted of relating concentrations of alkylphenols and carboxylic acid water ratios in pores and porosity and permeability. Presence of such acids can either lead to decrease or increase of porosity and permeability due to dissolution of carbonate cement or formation of insoluble carboxylates depending on the chemistry of formation waters. The fatty acids in oils and source rocks have also been shown as possible bio-markers suitable for genetic correlations. Use of iodine tracer has been found helpful in understanding hydrodynamics of reservoir and likely quantum of connate and recharged water. Use of carbon isotopes to map secondary dolomitization and porosity development has also been found to show promise.

A yet another important trend of the last decade has been in the area of surface geochemical surveys. A vast industry effort has come to light to very objectively integrate surface geochemical surveys, especially to generate focus areas in the deep and ultra deep waters of the Gulf of Mexico, Gulf of Louisiana, the east coast of Brazil, Venezuela, and the west coast of Africa, the south Caspian sea, the South China sea and many other areas Usually combined gas and bitumen surveys employing fluorescence spectrometry and GC and GC-MS characterization of dispersed hydrocarbons were employed. These surveys not only led to delineation of focus areas for exploration but also could demonstrate the presence of more than one active petroleum system. The work of the last decade in geochemical prospecting has shown that such surveys are worthy of integration at all stages of petroleum exploration.

The significant research area, of the last decade has been to investigate the role of hydrothermal activity in generation of petroleum like constituents in number of offshore areas of the world. The significance of this so-called hydrothermal petroleum in regard to conventional petroleum occurrence is yet to be understood. However, need of filtering effects of hydrothermal activity on the data of sea-bottom core geochemical prospecting has been well identified.

A historic development of the last decade has been perfecting Genetic Basin Modeling which should be a petroleum industry standard like a few reservoir simulation models and which could operate from the basinal and regional scale to a prospect level scale and capable to integrate as much surface geological and structural details as are revealed through 3-D seismics. The model should be capable to be calibrated by the phase behavior reservoir characteristics of the known accumulations to reconstruct parameters associated with dynamics of generation, migration accmulation processes. Such models have come in the market and have been found useful to assess the current level of efficiency of exploration effort in terms of the quantum of resources yet to be found with various risk factors, quick evaluation of exploration acreages, and for assessing geochemical genetic exploration priority for mapped prospects in a given basin and to contribute to maximization of returns from a portfolio of exploration opportunities.

5 CONCLUSIONS

Petroleum geochemistry has come of age. It is now an imperative component of trinity of geosciences, geology, geophysics and geochemistry, to help generate best opportunities and maximizing risk reduction for exploration and development efforts. More work will come to light in future on the role of natural catalysts in petroleum generation processes, on estimation of quantum of oil and gas expelled from source rocks at primary migration stage through micro-fractures caused by hydrocarbon generation over-pressuring and 4-D combined surface geochemical and microbiological prospecting to identify by-passed oil and oil-water front. The coming decades will witness greater role of academia to foster higher education and research in petroleum geochemistry while adopting new developments on the fate of various molecular species in the recent sediments of different depositional environments of different latitudes.

ACKNOWLEDGEMENTS

The authors are thankful to Shri Y.B. Sinha Dir. Exploration ONGCL. Dr. D. Ray, Head KDMIPE ONGCL,

Mrs. Nuzeth Fathima General Manger Geoscience Research, KDMIPE ONGCL and Dr, V. Banerjie, Deputy General Manager KDMIPE ONGCL for support and help provided during preparation of this paper.

REFERENCES

Abrams, M. A., Segall, M., Bartell, S.G., 2001. Best practices for detecting, identifying and characterizing near surface migration of hydrocarbons within marine sediments, SPE offshore Technology Conference (Houston TX, 4/30.2001-5/3.2001).

Alberdi, M., Lopez, L., 2000. Biomarker 18 (Alpha) (H) oleanane: A geochemical tool to assess Venezuelen petroleum systems. Journal of South American Earth Science 13/8, 751–759.

Andrew, S. P., Corvi, P. J., 1995. Simple kinetic models of petroleum formation. Part-I: Oil and gas generation from kerogen. Marine and Petroleum Geology 12/3, 291–319.

Bakr, M. M. Y., Wilkes, H., 2000. The influence of facies and depositional environment on the occurrence and distribution of carbazoles and benzocarbazoles in crude oils: A case study from the Gulf of Suez, Egypt. Organic Geochemistry 33/5, 561–580.

Banerjee, A., Pahari, S., Jha, M., Sinha, A. K., Jain, A. K., Kumar, N., Thomas, N. J., Misra, K. N., Chandra, Kuldeep, 2000, The effective source rocks of Cambay basin, India., AAPG Bull. Vol. 86, No. 3, pp 433–456.

Baskin, D. K., Hwang, R. J. and Purdy, R. K., 1995, Predicting gas and oil and water intervals in Niger Delta reservoirs using gas chromatography, AAPG Bull., Vol. 79, pp 337–350.

Baskin, D. K. and Jones, R. W., 1993, Oil gravity prediction prior to drill stem testing in Monterey Formation reservoirs, offshore California, AAPG Bull., Vol. 77, pp 1479–1487.

Baylis, S. A., 1998, Geochemical comparison of core extracts and oil samples in reservoir, Org. Geochem Vol. 29, pp 463–484.

Behar, R., Vandenbroucke, M., Tang, Y., Marquis, F. and Espitalie, J., 1997, Thermal cracking of kerogen in open and closed systems: determination of kinetic parameters and stoichiometric coefficients for oil and gas generation, Org. Geochem. Vol. 26, pp 321–339.

Bement, W.O., Mc Neil, R. I. and Lippincott, R. G., 1996, Predicting oil quality from side wall cores using PIFD, TFC. And NIR analytical techniques in sandstone reservoirs, Rio Del Rey Basin, Cameroon, Org. Geochem. Vol. 24, pp 1173–1178.

Bennett, B., Larter, S. R., Chen, M., Brincat, D., Gelin, F. J. P., 2000, Fractionation Of Benzocarbazoles Between Source Rocks And Petroleums, Org.Geochem. Vol. 33, No 5, pp 545–559.

Bjoroy, M. and Laberg, R., 1993, Surface Geochemical Studies In The Norwegian Baretnts Sea: Comparison With Drilling Results, Generation Accumulation and Productions III. A. M. Spencer (Ed) Special publication Of The Euoropean Association Of Petroleum Geosciences No. 3, Springer Verlag Heidelberg

Brooks, J. M. and Fleet, A. J., 1987, Marine source rocks, Geological Society Publication, 26, Oxford: Blackwell Scientific publications for the geological Society

Brooks, J. M. and Welte, D. H., 1984, Eds. Advances In Petroleum Geochemistry, London: Academic Press, Vol. 1.

Burnham, A. K., Braun, R. L., Cobum, T. T., Sandvik, E., Curry, D. J., Schmidt, B. J. and Noble, R. A., 1996, An Appropriate model For Well- Preserved Algal Kerogens, Energy and Fuels, Vol. 10, pp 49–59.

Burns, K. A., Brunskill, G. J., Grrnwood, P. F. and Summons, R. E., 2000, Vertical Fluxes Of Hydrocarbons On the Northwest Shelf Of Australia, Org. Geochem. Vol. 32, No 10, pp 1241–1255.

Cai, C., Zhu J., Chu, X., Worden, R. H., Wang, Q. and Xiang, T., 2002, Chemical and Isotopic Evidence For Secondary Alteration Of Natural Gases In Hetianehe Field, Bachu Uplift Of The Tarim Basin, Org. Geochem. Vol. 33, No. 12, pp 1415–1427.

Chandra, K., 2002, Isotopes in Hydrocarbon Research and Exploration, Journal Of Applied Geochemistry, Vol. 4, No. 2B, pp 427–439.

Colletta, B. (Ed.), 2004, Insights Into Petroleum Geochemistry, IFP Publications, Edition Technip, France.

Creaney, S. and Isaksen, G., 2001, Evolution of knowledge and ideas in petroleum geochemistry from "geoalchemy" to petroleum systems analysis and beyond, Annu. AAPG-SEPM Conv. (Denver, Co,6/3-6/2001) Paper Abstr. A 43

Curiale, J. A.; Covington, G. H.; Shamsuddin, A. H. M. and Morelos, J. A., 2002, Origin Of Petroleum In Bangladesh, AAPG Bull. Vol. 86, No. 4, pp 625–652.

Dewaker, Misra, K. N., Chandra, K., Arunachalam, J. and Karunasagar, D., 2000, Isotope Fractionation of Ni60/Ni61 in Kerogen and Bitumen Samples, Current Science Vol. 79, No. 12, pp. 1720–1723.

Dieckmann, V., Fowler, M. and Horsfield, B., 2004, Predicting The Composition Of Natural Gas Generated By The Duvernay Formation (Western Canada Sedimentary Basin) Using A Compositional Kinetic Approach, Org. Geochem. Vol. 35, pp 845–862.

Galimov, E. M., 1988, Sources and mechanisms of formation of gases hydrocarbons in sedimentary rocks. Chemical Geology, Vol. 71, pp 77–95.

Gonçalves, F. T. T., Mora, C. A., Cordoba, F., Kairuz, E. C. and Giaraldo, B. N., 2002, Petroleum Generation And Migration In The Putumayo Basin, Colubia: Insights From An Organic Geochemistry And Basin Modelling Study In The Foot Hills, Mar. Petrol. Geol., Vol. 19, No. 6, pp. 711–125.

Greenwood, P. F. and Summons, R. E., 2003, GC-MS Detection and Significance of Crocetane and Penta methyalicosane In Sediments and Crude Oils, 2003, Org. Geochem., Vol. 34, No. 8, pp 1211–1222.

Gutherie, J. M., Walters, C. C. and Peters, K. E., 1998, Comparison Of macrotechniques used for analyzing oils in side-wall cores to model viscosity, API gravity and sulphur content, In mello, M. R. Yilmaz P. O. (Eds); Intl Conf and Exhibition, 8–11 Nov 1998, Rio de Janeiro, Brazil,

Halpern, H. I., 1995, Development and applications of light hydrocarbon based star diagrams, Bull. AAPG, Vol. 79, pp 801–815.

Herron, S. L.,1991, Insitu evaluation of potential source rocks by wireline logs in R. K. Merill (eds.) Source and Migration Processes and Evaluation Techniques, Tulsa, AAPG, pp 129–134.

Holba, A. G., Dzou, L. I., Wopod, G. D., Ellis, L., Adam P., Schaeffer P., Albrecht P., Greene, T. and Huges W. B., 2003, Application Of Tetracyclic Polyproids As Indicators Of

Input From Fresh Water Environments, Org. Geochem. Vol. 34, No. 3, pp 441–269.
Huang, B., Xiao, X. and Li, X., 2003, Geochemistry and origin Of Natural Gases In The Yinggehai and Qiongdongnan, 2003, Org. Geochem., Vol. 34, No. 7, pp 1009–1925.
Huang, D., 1999, Advances In Hydrocarbon generation Theory (I). Generation And Evolution Model for Immature Oils and Hydrocarbons, Pet. Sci. & Eng. Vol. 22, pp 121–130.
Huang, H., Bowler, B. F. J., Oldenburg, T. B. P., Larter, S. R. and Zhang, Z., 2003, Influence of Biodegradation On Carbazole and Bezocaebazole In Distributions In Oil Column From The Lavhoe Basin N E China, Org. Geochem. Vol. 34, No. 7, 951–969.
Huang, H. and Pearson, M. J., 1999, Dibenzothiophenes In Lacustrine Crude Oils, Bohai Bay Bain, Eastern China, 1999, Org. Geochemistry, Vol. 30, No. 11, pp 1455–1470.
Huc, A. Y., 2003, Petroleum geochemistry at the dawn of the 21st Century, Oil and gas Science and technology, Rev IFP, Vol. 58, No. 2, pp 233–241.
Hunt, J. M., Lewan, M. D. and Hennet, R. J. C., 1991, Modeling oil generation with time-temperature indux graphs based on Arrhenius equations. AAPG Bull., Vol. 75, pp 795–807.
Hunt, J. M., 1995, Petroleum Geochemistry and Geology, 2nd edition, W. H. Freeman and Company, New York.
Hunt, J. M., Philp, R. P. and Kvenvolden, K. A., 2002, Early Developments In Petroleum Geochemistry, Org. Geochem., Vol. 33, pp. 1025–1052.
Jarvie, D. M., Morelos, A. and Zhiwen, Han, 2001, Detection of pay zones and pay quality, Gulf of Mexico: application of geochemical techniques, Gulf Coast Of Geological Societies Transaction LI, pp 151–160.
Jiang, C. and Li, M., 2002a, Diphenylmethane and benzylcyclohexane As Indicator For Oils Derived From The Madison Petroleum System, 2002, Org. Geochem., Vol. 33, No. 7, pp 855–860.
Jiang, C. and Li, M., 2002b, Bakken/ Madison Petroleum Systems In The Canadian Williston Basin, Org. Geochem Vol. 33, No. 7, pp 761–187. [Use of gasoline range light hydrocarbons, saturate aromatic fractions, steranes and hopanes on 122 oils]
Jones, V. T., Mattthews, M. D. and Richer, D. M., 2000, Light Hydrocarbons For Petroleum and Gas Prospecting, Geochemical Remote Sensing of The Subsurface. Edited by M. Hale, Handbook Of Exploration Petroleum geochemistry, Vol. 7 (G. J. S. Govett, Editor), Elsevier Science B.V.
Johnson, C. L., Greene, T. J., Zinniker, D. A., Moldowan, J. M., Hedrix, M. S. and Carroll, A., 2003, Geochemical Characteristics And Correlation Of Oil and Non Marine Source Rock From Mongolia, AAPG Bull., Vol. 87, No. 5, pp 817–846. [Elevated concentration of hexacyclic and heptacyclic polyprenoids]
Karlsen, D. A. and Larter, S., 1990, A rapid correlation method for petroleum population mapping within individual petroleum reservoir: applications to petroleum reservoir description, In Collinson J. D. (Ed) Correlation In Hydrocarbon Exploration, Norweigian Petroleum Society, Graham and Trotman, pp 77–85.
Karlsen, D. A. and Nedkvitne, T., Larter S. R., Bjorlykke K, 1993, Hydrocarbon Composition of authigenic inclusions: applicationto elucidation of petroleum reservoir filling history, Geochimica Cosmochimica Acta Vol. 57, pp 3641–3659.
Kartsev, A. A., 1959, Geochemical methods of Prospecting and Exploration for Petroleum and Natural Gas. (English translation by Witherspon and Romey). University of Calefornia Press.
Katz, B. J., 2003, Hydrocarbon Shows And Source Rocks In Scientific Ocean Drilling, Intl. J. Coal Geol., Vol. 54, No. 1–2, pp 139–154.
Katz, B. J., Robison, C. R. and Chakhmakhchev, A., 2003, Aspects Of Hydrocarbon Charge Of The Petroleum System Of The Yamal Peninsula, West Siberian Basin, Intl J Coal Geol Vol. 54, No. 1–2, pp. 155–164.
Kaufman, R. I., Ahmed, A. S. and Elsinger, R. J., 1990, Gas Chromatography as a development and production tool for fingerprinting oils from individual reservoirs: applications in the Gulf Of Mexico. In GCSSEPB Foundation Ninth Annual Research Conference, October 1, pp 263–282.
Kaufman, R. L. and Robertson, B., 1999, Application of Reservoir Geochemistry In the Iagifu- Hednia Field, Papua New Guinea, APEA J., Vol. 39, pp 421–436.
Klusman, R. W., 1993, Soil Gas And Related methods For Natural Resouece Exploration,, John Wiley & Sons, New York.
Kharaka, Y. K., Carothers, W. W. and Hull, R. W., 1985, Water rocks interactions in sedimentary basins. U.S.G.S. Report, 1985.
Koopmans, M. P., Larter, S. R., Zhang, C., Mei, B., Wu, T. and Chen, X., 2002, Biodegradation and Mixing Of Crude Oils In Eocene ES 3 Reservoirs Of The Liaohe Basin, North Eastern China, AAPG Bull., V. 86, pp 1833–1843.
Kotaraba, M.J. and Lewan, M. D., 2004, Characterizing Thermogenic Coalbed Gas from Polish Coals of Different Ranks by Hydrous Pyrolysis, Org. Geochem., Vol. 35. pp 615–646.
Larter, S., Aplin, A. C., Bowler, B., Lloyad, R., Zwach, C., Hansen, S., Tenaes, N., Sylta, O. and Yardley, G., 2000, A Drain In My Graben: An Integrated Study Of The Hemeidal Area Petroleum System. J. Geochm. Explor., Vol. 69-70, pp 619–622.
Larter, S. R., Head, I., Koopmans, M., Aplin, A., Wilhelms, A., Di Primio, R, Zwach, C., Erdmann, M. and Telenaes, N., 2003, The Controls On the Composition Of Biodegraded Oils In The Deep Subsurface: Part I: Biodegradation Rates In The Petroleum Reservoirs, Org. Geochem. Vol. 34, No. 4, pp 601–613.
Lopatin, N. V., Emets, T. P. and Kataev, O. I., 2002, Lower Jurassic-Paleozoic oil generation and accumulation system in the region of Northern Vartov tectonic terrace. Geol Geofiz Razrabotica Neft Mestorozhd, No. 10, pp 10–27.
Lucach, S. O., Bowler, B. F. J., Larter, S. R. and Frewin, N., 2000, Variation In Alkylphenol Distributions In A Homogeneous Oil Suite From The Dhahban Petroleum System Of Oman, New Insights Into The Petroleum Geochemistry Of NSO Compounds Sumposium, Org. Geochem, Vol. 33, No. 5, pp 581–594. { alkyl phenol as indicator of diatance of migration]
Magoon, L. B. and Dow, W. G., (Eds.) 1994, Petroleum Systems- From Source rock to Trap, AAPG Memoir 60, Tulsa, AAPG.
Mango, F. D., 1997, The Light Hydrocarbon In Petroleum: A Critical review, Org. Geochem, Vol. 26, No.7/8, pp 417–440.

Mann, U. and Zweigel, J., 2003, Prediction of Source Rock Distribution and Quality Variations: The New Of-Mod –3-D Technology, AAPG Intl Conf (Barcelona, Spain, 9/21-24/2003) Proc., A59 Abstract only.

McCaffrey, M. A., Legarre, H. A. and Johnston, S. J., 1996, Using biomarkers to improve heavy oil reservoir management: an example from the Cymic Field Kern County California. AAPG Bull., Vol. 80, pp 898–913.

Mello, M. R., Moldowan, M., Requej, R. and De Miranda, F. P., 2000, Petroleum Geochemistry Applied To teduce Exploration Risk, Brazil Petrol and Gas Inst. Rio, Oil and Gas Conf. (Rio de Janeiro, 10/16-19/2000)

Misra, C. S., Samantha, U., Gupta, A., Thomas, N. J., Misra, K. N., 1996, Hydrous Pyrolysis of a Type III Source: Fractionation Effects During Primary Migration In Natural and Artificially Matured Samples, Org. Geochem. Vol. 25, No. 8, pp 489–505.

Parnell, J., Middleton, D., Honghan, C., Hall, D., 2001, The use of Fluid Inclusion Studies In Constraining Oil Charge, Mar.Petrol. Geol., Vol. 18, No. 5, pp 535–549.

Peters, K. E. and Fowler, M. G., 2002, Application of Petroleum Geochemistry To Exploration and Reservoir Management, Org. Geochem., Vol. 33, pp 5–36.

Peters, K. E., Snedden, J. W., Sulaeman, A., Sarg, J. F. and Enrico, R. J., 2000, A new geochemical stratigraphic model for the Mahakam Delta and Makassa Slope, Kalimantan, Indonesia, AAPG Bull., Vol. 84, pp 12–44.

Petroleum Abstracts, 1990–2003, Tulsa Oklahoma.

Philippi, G. T., 1965, On the depth, time and mechanism of petroleum generation. Geochim. Cosmochim. Acta, Vol. 29, pp 1021–1049.

Price, L. C., 1986, A Critical Over view And Proposed Working Model Of Surface Geochemical Exploration, in Unconvential Methods In Exploration IV, in Davidson (ed.), Southern Methodist University Press, Texas, pp 245–304.

Price, L. C., 1993, Thermal stability of hydrocarbons in nature: limits, evidence, characteristics, and possible controls, Geochimica et Cosmochimica Acta, Vol. 57, pp 3281–3280.

Price, L. C., 1997, Thermal Stability levels and Controlling Parameters of Methane, as determined by C15+ Hydrocarbon Thermal Stabilities, USGS Bulletin 2146-K

Price, L. C., 1998, Origins, Characteristics, Evidence For, and Economic Viabilities of Conventional and Unconventional Gas Resources Bases, USGS Bulletin-1998

Quigley, T. M. and McKenzie, A. N., 1988, The Temperature of oil and gas formation in the sub surface, Nature, Vol. 333, pp 549–552.

Ross, I. M. and Aims, R. I., 1988, Stratification of oils in Columbus basin off Trinidad, Oil and Gas Journal, Vol. 86, 72, 76.

Samanta, U., Chandra, S., Mishra, C. S. and Misra, K. N., 1993, Indian High Wax Oils And The Depositional Environments Of Their Source Rocks, Mar. Petrol. Geology, Vol. 11, No. 6, pp 756–759.

Sassen, R., Ozgul, E., Dfreitas, D. A., Sweet, S. T., Milkov, A. V., Roberts, H. H., Hunt, J. L., Beeunas, M. A., Chanton, J. P., 2003, Gas Venting And Subsurface Charge In The Green Canyon Area, Gulf Of Mexoco, Org. Geochemistry, Vol. 34, No. 10, pp. 1455–1464.

Sheila, J. R. and Nunn, J. A., 1995, Episodic Fluid Expulsion From Geopressured Sediments, Marine Petrol. Geol. Vol. 12, pp 195–204.

Singh, R. R., Gupta, A. K., Tripathy, G. K., Kumar, R. K. and Chandra, Kuldeep, 1985, Stable carbon isotope studies on formation waters of Indian sedimentary basins. In Kumar, R.K,; Dwivedi, P.; Banerjie, V.; Gupta, V. (eds.) Proceedings of Petroleum Geochemistry and exploration in Afro-Asian Region, Balkema, Rotterdam.

Slentz, L. W., 1981, Geochemistry of reservoir fluids as a unique approach to optimum reservoir management, SPE paper 9592, pp 37–50.

Sofer, Z., 1984, Stable carbon isotopic composition of crude oils: Application to source depositional environments and petroleum alteration. AAPG Bull., Vol. 68, pp 31–49.

Stahl, W. J., 1977, Carbon and nitrogen isotopes in hydrocarbon research and exploration., Chemical Geology, Vol. 20, pp 121–149.

Sun, Y; Xu, S., Lu, H., and Cuai, P., 2003, Source Facies Of The Paleozoic Petroleum System Inj The Tabei Uplift, Tarim Basin, NW China: Implications From Aryl Ispoprenoids In Crude Oils, Org Geochem Vol. 34, No. 4, pp 629–634.

Terkin, J. M. J., Frewin, N. L., Indrelid, S. L., 2001, Petroleum Systems Of Oman: Charge Timing and Risksss, AAPG Bull., Vol. 85, No. 10, pp 1017–1045

Tissot, B.; Durand, B.; Espitalie, J.; and Combaz, A., 1974, Influence of nature and digenesis of organic matter in formation of petroleum, Rev. l'inst. francais petrole, Vol. 30, pp 743–777.

Tissot, B. and Welte, D. H., 1984, 2nd edition, Petroleum Formtion And Occurrence, Heidelberg Springer Verlag.

van Krevelen, D. W., 1961, Coal: Typology-Chemistry-Physics-Constitution. Amsterdam: Elsevier Science.

Waples, D. W., 2000, The kinetics of in reservoir oil destruction and gas formation: constraints from experimental and empirical data, and from thermodynamics, Org Geochem Vol. 31, pp 553–575.

Welte, D. H., Horsfield, B. and Baker, D., 1997, Petroleum and basin Evolution: Insights from petroleum geochemistry, geology and Basin Modelling. New York, Springer.

Welte, D. H., Hagemann, H. W., Hollerbach, A., Leythaeuser, D. and Stahl, W., 1975, Correlation between petroleum and source rock. Proc. 9th World Petroleum Congress. Tokyo, Applied Science Publ., London, Vol II, pp 179–191.

Weng, R. F., Huang, W. L., Kuo, C. L., Inan, S, 2003, Characterization of Oil Generation And Expulsion From Coals And Source Rocks, 2003, Org Geochem Vol. 34, No. 6, pp 771–787.

Wenger, L. M. and Isasken, G. H., 2001, Control Of Hydrocarbon Seepage Intensity On Level Of Biodegradation In Sea Bottom Sediments, Org. Geochem., Vol. 33, No. 12, pp 1277–1292.

Xiongqi, P., Li, M., Li, S., Jin, Z., Xu, Z. and Chen, A., 2003, Origin of crude oil in the Jinhu Depression of North Jiangu-South Yellow Sea Basin, Eastern China. Org. Geochem., Vol. 34, No. 4, pp 553–573.

Zhang, S., Liang, D., Song, F., Zhang, D., Wamg, P., Gong, Z., Wu, K. and Li, M., 2003, Geochemistry Of Petroleum Systems In The Eastern Pearl River Mouth Basin, 2003, Org Geochem,Vol. 34, No 7, pp 971–991.

Petroleum Geochemistry and Exploration in the Afro-Asian Region –
Digang Liang, Darui Wang & Zhenxi Li (eds)
© 2008 Taylor & Francis Group, London, ISBN 978-0-415-44083-7

Characteristics and genesis of the western slope thick oil in Songliao Basin, China

Zhaoyun Wang, Caineng Zou, Guanjun Xu & Xiaozhou Wu
Research Institute of Petroleum Exploration & Development, PetroChina, Beijing, China

Xiaodong Sun
Research Institute of Petroleum Exploration & Development, Daqing Petroleum Company, Daqing, China

Jianwei Wang
Research Institute of Petroleum Exploration & Development, Jilin Petroleum Company, Changchun, China

ABSTRACT: The western slope of Songliao Basin spans two oil areas, Daqing and Jilin, and 7 oil and gas fields have been found since 1960s. The western slope oil is characteristic of relatively low density, middle viscosity, richness in non-hydrocarbons and low content of asphalt. Analyses of GC and GC-MS on the sampled western slope oils show three features: the first is high contents of isoprenoids and cycloalkanes, and the special baseline indicates the biodegradation of the oils; the second is a certain amount of n-alkanes, which reflects the feature of multiple fillings by oils; the third is the occurrence of various biomarkers for the western slope oil. Compared with the central depression oil, the content of normal steranes decreases while those of pregnanes and diasteranes increase, normal hopane series obviously decrease and anti-biodegradation biomarkers increase, such as long-chain tricyclic or tetracyclic terpanes and gammacerane. Thus, all the GC and MS data show the biodegradation and the feature of multiple fillings by oils. The carbon isotopic values of the western slope oil are −29 to −32‰, those of kerogen are −22 to −24‰ for the Jurassic coal rock and −27 to −31‰ for the Qingshankou and Nenjiang mudstone. The maturity of source rocks in the western slope is very low, the calculated Ro values of the western slope oil are 0.75–0.95% based on the indices of methyl-phenanthrene and methyl-dibenzothiophene. So, the source rock kitchen of the western slope oils is mainly the Qingshankou lacustrine mudstone of the central depression. The K–Ar chronology data of reservoir authigenic illite in four wells from east to west are 56 Ma, 45 Ma, 35 Ma and 35 Ma, respectively, and they also reflect the migration of the western slope oil from east to west.

1 INTRODUCTION

Songliao Basin is located in northeastern China, including two oil fields, Daqing and Jilin oil fields. The squareness shown in Figure. 1 is our studied area, the western slope of Songliao Basin is bounded by the Taikang uplift in east, the basin border in west, Qiqihaer in north and the southwestern uplift in south. Its prospect area is about $4.3 \times 10^4 km^2$., including 7 oil and gas fields, such as the Fulaerji, Taobao oil fields, Pingyang oil and gas field, Tumuji oil sand, Alaxin, Erzhan and Baiyinnuole gas fields, which have been found early or late since 1960s. All these oil fields are shown in Fig. 1 and the generalized stratigraphic description of the western slope is listed in Table 1. Saertu and Gaotaozi oil beds are main target strata and the oil and gas field type belongs to lithological and structural-lithological ones.

The purpose of the study is to analyze the characteristics and genesis of the western slope thick oil as well as its source rock kitchen, which includes where the oil comes, from the source rock developing in the western slope or in the central depression, and how the oil becomes thick oil or oil sand. The study may play an important role in the resource evaluation and exploration of oil and gas.

2 CHARACTERISTICS OF THE WESTERN SLOPE OIL

2.1 *Physical characteristics*

The physical and group compound characteristics of the western slope oil are listed in Table 2. It has relatively low density (the density of the heavy oil is larger than $0.934 g/cm^3$), middle viscosity, richness in non-hydrocarbons and low content of asphalt (Dun Tiejun, 1995; Hu Jianyi et al., 1994). Compared with the other thick/heavy oils in China, such as the Penglai

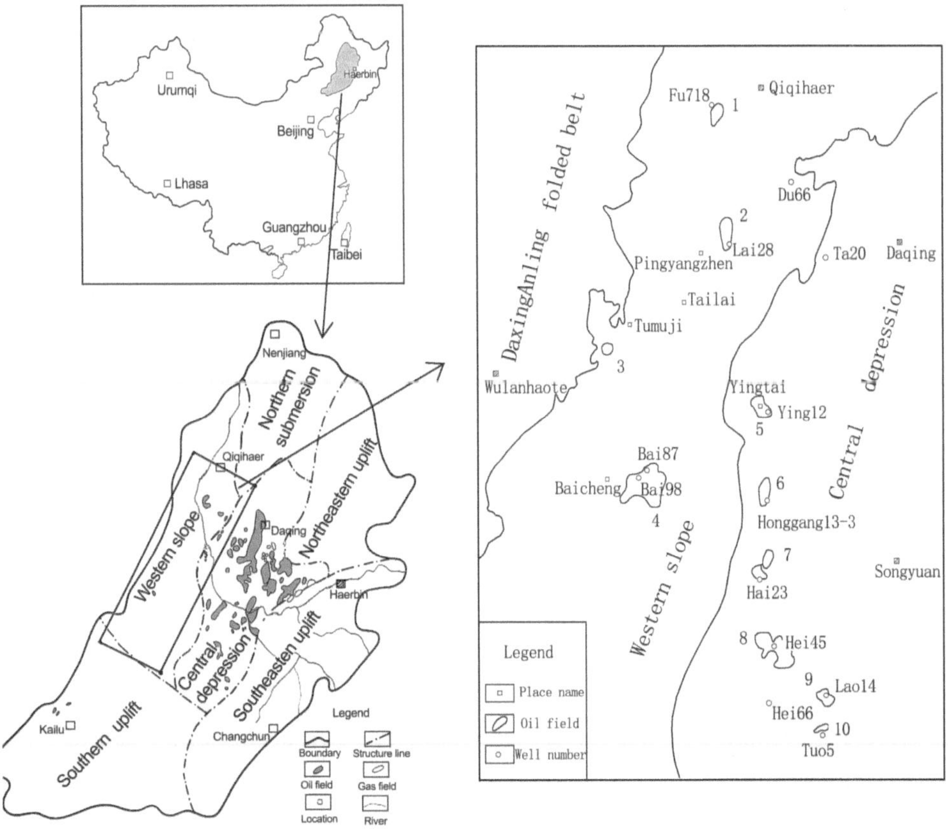

Figure 1. Location of the western slope and distribution of some oil fields in the western slope and central depression of Songliao Basin a- Location of the western slope; b-Distribution of some oil fields in the western slope and central depression 1 – Fulaerji; 2 – Pingyang; 3 - Tumuji; 4 - Taobao; 5 – Yingtai; 6 - Honggang; 7 - Haituozi; 8 - Daqingzijing; 9 – Dalaoyefu; 10 - Shuangtuozi.

19-3 (Guo Taixian et al., 2001), Suizhong 36-1and Chengbei oil field in Bohai Bay (Jiang Peihai et al., 1998), Gaosheng oil field in the Liaohe area (Wang Shuangqing, 2000; Xiong Yongqiang et al., 1998), Heiyoushan and Hongqian well 1in the northwestern edge of Jungaer Basin (Wang Yutao et al., 1998), Da well 3 and Da well 9 in Baiyinchagan of Inner Mongolia (Zhao Mifu et al., 2001), the western slope oil shows the thick oil characteristics of relatively low density and moderate viscosity.

2.2 *Group compound characteristics*

The group compound of the western slope oil is characteristic of low content of saturated hydrocarbons and asphalt, high content of non-hydrocarbons. The group compound triangle of different type oils from the western slope and adjacent central depression is shown in Figure 2. They can be obviously separated into three groups. Tumuji oil sand samples belong to Group I, characterized by contents of saturated hydrocarbons less than 30%, aromatic hydrocarbons less than 5%, and non-hydrocarbons and asphaltum large than 65%. The other thick oil samples from the western slope concentrate in Group II, with contents of saturated hydrocarbons in the range of 30–60%, aromatic hydrocarbons 20–40%, and non-hydrocarbons and asphaltum 20–40%. The oils from the central depression converge in Group III, characterized by contents of saturated hydrocarbons larger than 60%, aromatic hydrocarbons less than 20% and non-hydrocarbons and asphaltum in the range of 10–25%.

2.3 *GC-MS*

The mass chromatograms of TIC (total ion chromatography), M/Z 191 and M/Z 217 for the western slope oil and central depression oil are shown in Figure 3. The analyses of GC and GC–MS show the following three features: the first is high contents of isoalkanes

Table 1. Generalized stratigraphic description of the western slope in Songliao Basin.

Stratigraphy			Lithologic Description	Sedimentary Facies	Relationship
	Q		Brown mudstone, gray siltstone and gravel sandstone		unconformity
T	Taikang		Yellow-brown gravel sandstone and gray-green mudstone		conformity
	Da'an		Mainly gray-green mudstone with sandstone		inconformity
K	Mingshui	II	Dark-purple and gray mudstone interbedded with muddy siltstone	Fluvial	conformity
		I	Gray-green mudstone and dark-purple mudstone interbedded with gray muddy siltstone		conformity
	Sifangtai		Gray-green mudstones and siltstone (partly carbargillite) interbedded with gray muddy siltstone and siltstone		unconformity
	Nenjiang	V	Brown-red and gray-green mudstone and siltstone	Fluvial and	conformity
		IV	Gray-green mudstone and interbedded with siltstone	intergradation	conformity
		III	Mainly mudstone and gravel sandstone		conformity
		II	Gray-black mudstone and gray-black oil shale	Deep lake	conformity
		I	Gray-black mudstone and partly gray-black oil shale, and interbedded with thin muddy siltstone		conformity
	Yaojia	II, III	Mainly dark-red mudstone and interbedded with muddy siltstone	Shallow lake	conformity
		I	Mainly brown and purple mudstone and silty mudstone		conformity
	Qingshankou	II, III	Brown and purple mudstone and silty mudstone, interbedded with gray-brown siltstone	Delta	conformity
		I	Gray and gray-brown siltstone (partly carbargillite) and interbedded with gray-black and black mudstone		conformity
	Quantou	IV	Mainly gray-black mudstone and interbedded with muddy siltstone		conformity

Table 2. Characteristics of the western slope and other thick oils in China.

Physical character Basin	Oil field	Depth (m)	Layer	Density (g/cm^3)	Viscosity (mPa.s)	Content of wax (%)	Content of non-hydrocarbon (%)	Content of asphaltum (%)
The western	Fulaerji	40–460	Ky_{2+3}	0.9347	2634	15	22.7	
slope of	Pingyang		Ky_{2+3}	0.9200	200.7		26.5	0.9
Songliao Basin	Tumuji	5–58	Kn	0.9700	672	17	36	12.4
	Taobao	300–350	Ky_{2+3}	0.9324	243.93	10.1	35.14	
Bohai Bay	Suizong 36-1		E_{3d}	0.953	1195.0	3.47	31.43	8.17
	Chengbei		E_{3d}	0.955	1198.2	6.02	41.02	6.60
	Penglai 19-3			0.915–0.982	9.1–147.8		18.9–29.8	
Liaohe area	Gaosheng	1450	E_2					
Northwest edge	Heiyoushan	<100	T	0.90		1.0		
of Junggar Basin	Well Hongqian 1	>100	T	0.95				
Baiyinchagan	Well Da 3	233.0	K_1	0.9661	1862.7		27.52	7.33
of Inner Mongolia	Da well 9	467.3	K_1				50.75	2.61

and cycloalkanes, especially for cycloalkanes, and the special baseline indicates the biodegradation of the western slope oil, the second is a certain amount of n–alkanes, which reflects the feature of multiple fillings by oils; the third is the occurrence of various biomarkers for the western slope oils and compared with the central depression oil, the content of normal steranes decreases while those of pregnanes and diasteranes increase, normal hopane series obviously decrease and anti-biodegradation biomarkers increase, such as long-chain tricyclic and tetracyclic terpanes, gammacerane[11–20].

The classification rank of the biodegraded crude oil can be referred to Peters and Moldowan (1993)[21], see Table 3. The composition and distribution of the steranes and terpanes will change and some new

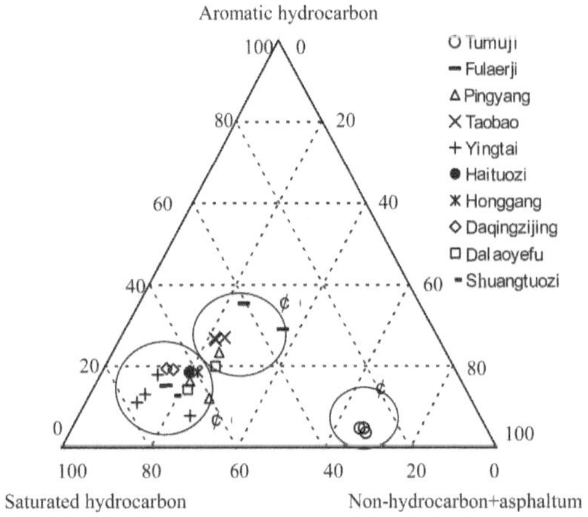

Figure 2. Group compound characteristics of the western slope thick oil and the central depression normal oil.

biomarkers appear with the increase of washing action and biodegradation, indicating the middle metamorphic stage. Most people think that the 25–norhopane series are one of the most important biomarkers and usually used as an indicator of oil biodegradation, but they are absent in biodegraded oils from all oil fields of the western slope, including the Tumuji oil sand. Therefore, the genesis of the 25–norhopane series needs to be studied further.

Studies on 25–norhopane series show that there are three situations to be deserved more attention.[11–12] One is the detection of 25–norhopane series in non-biodegraded oils, such as in the extracts from the non-biodegraded oils of Santanghu Basin, which have whole normal alkane series. The data of geological settings and geochemical analysis reflect non-biodegradation though relatively completed 25–norhopane series have been detected. The second is that 25–norhopane series are abundant in the Upper Permian source rock of Santanghu Basin, which didn't suffer from biodegradation based on analyses of geology settings and geochemistry. The existence and selective distribution of 25–norhopane series in non-biodegraded source rocks reflect that 25–norhopanes series may be not only the product of biodegradation from crude oils, but also related to microbial activities during the early diagenesis. The third is that 25–norhopane series are absent in the oil that has been identified suffering from obvious biodegradation, such as the biodegraded oil from the Gaosheng oil field in Liaohe Basin. Therefore, the above phenomenon suggests that 25–norhopane series may be not the necessary product of biodegraded oils, but also related to the microbial activities during the early diagenesis. The absence of 25–norhopane series for the western slope oil does not mean that the biodegradation and washing action are not very heavy.

In addition, the acid values of the western slope and central depression oils have been measured and listed in Table 4. The total acid value of the Tumuji oil is the highest, namely 7.19 mgKOH/g, which shows an intensity of oxidation. The second is the Well Bai 98 oil from the Taobao oil field, whose total acid value is 4.60 mgKOH/g, reflecting a gradually weak oxidation. The total acid values of the Fulaerji and Pingyang oils are similar, ranging from 0.89 to 1.10 mgKOH/g, and that of the central depression oils is lower and less than 0.11 mgKOH/g, reflecting a successively weak oxidation.

According to the compound characters, distribution shape and the content of biomarkers of the western slope oils, the feature of multiple fillings by oils, and the acid value analysis, the Tumuji oil sand belongs to the very heavy biodegradation ranking, and the other oils in the western slope belong to the moderate one. The biodegradation of the southern oil (the Taobao oil field) is heavier than that of the western oil (the Fulaerji and Pingyang oil fields).

3 SOURCE ROCK KITCHEN OF THE WESTERN SLOPE OIL

3.1 *Carbon isotopic values of oils and source rocks*

The carbon isotopic values of the western slope and central depression oils range, in general, from −29.6 to −31.8‰. The carbon isotopic values for different oils and their group components (including saturated hydrocarbons, aromatics, non-hydrocarbons and asphalt), such as the Tumuji oil sand, Fulaerji, Taobao and Pingyang thick oils in the western slope, Yingtai, Dalaoyefu and Shuangtuozi oils in the central depression, are shown in Figure 4. Different type oils have the same distribution shape and change tendency, reflecting the derivation of the same type of source rocks with the characteristics of lacustrine mudstone.

There are three sets of source rocks developed in the western slope, namely the Jurassic coal rock, Cretaceous Qingshankou and Nenjiang lacustrine mudstones. The distribution range of carbon isotopic values for different formation source rocks and oils developed in the western slope and central depression is shown in Figure 5, which are −22∼−24‰ for the Jurassic coal rock and −27∼−31‰ for the Qingshankou and Nenjiang mudstones while those for the western slope oils are −29∼−32‰. The carbon isotopic values of the Jurassic coal rock are very heavy while those of the Cretaceous lacustrine mudstone are lighter and somewhat identical with that of the western slope oils. Therefore, the source rock for the western slope thick oil is mainly constituted by the Cretaceous lacustrine mudstone.

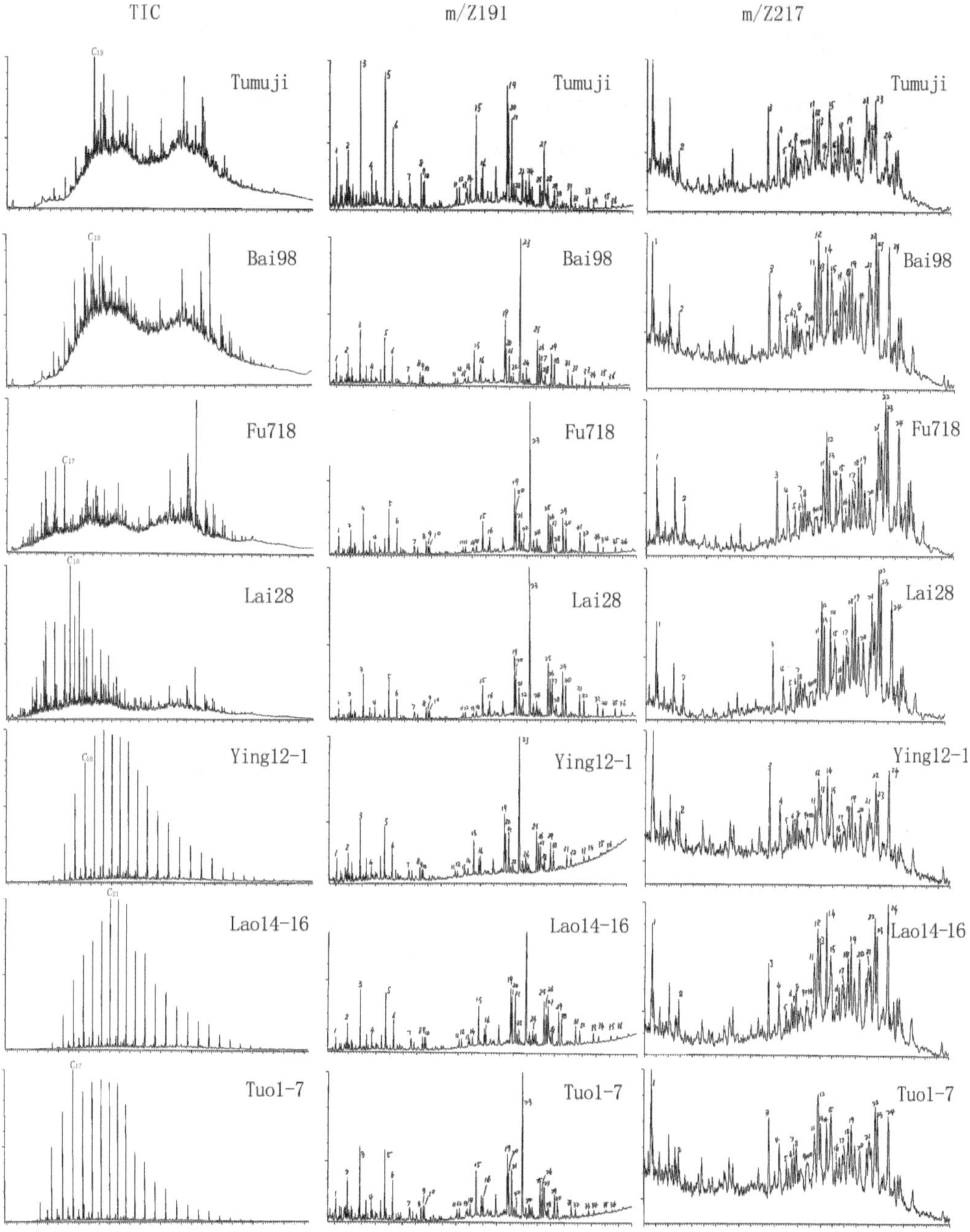

Figure 3. GC-MS chromatograms of the western slope thick oil and the central depression normal oil.

3.2 *Maturity of oils and source rocks*

The calculated Ro value of the western slope oil is 0.75~0.95% based on the index of methyl-phenanthrene. In addition, the calculated Ro value of oils from Fulaerji, Pingyang, Tumuji and Taobao oil field is 0.76, 0.74, 0.77 and 0.81% respectively according to the index of methyl-dibenzothiophene. So, the source rock of the western slope oil is predominated by the Qingshankou lacustrine mudstone in the central depression.

Table 3. Classification rank of biodegraded crude oils (after Peters and Moldowan,1993).

Scale	Composition of biomarkers	Biodegradation ranking
1	Depletion of low carbon numbered normal alkanes	slight
2	General depletion of normal alkanes	slight
3	Trace remains of normal alkanes only	slight
4	No normal alkanes, intactness of acyclic isoprenoids	moderate
5	Disappearance of acyclic isoprenoids	moderate
6(1)	Partial degradation of steranes	heavy
6(2)	Hopane degradation, occurrence of 25–norhopanes	heavy
7	Sterane degradation, intactness of whole diasteranes	heavy
8	Partial degradation of hopanes, no 25–norhopanes	very heavy
9	Absence of hopanes and partial degradation of diasteranes	very heavy
10	Partial degradation of C_{26}–C_{29} aromatic steroids	severe

Table 4. Acid values of the western slope and central depression oils (Unit: mgKOH/g).

Well No.	Tumuji	Bai 98	Fu 718	Lai 65	Lai 28	Ying 12	Ta 9	Tuo 5-14	Lao 14-16
Total acid value	7.19	4.60	0.89	1.10	0.92	0.23	0.09	<0.05	0.11

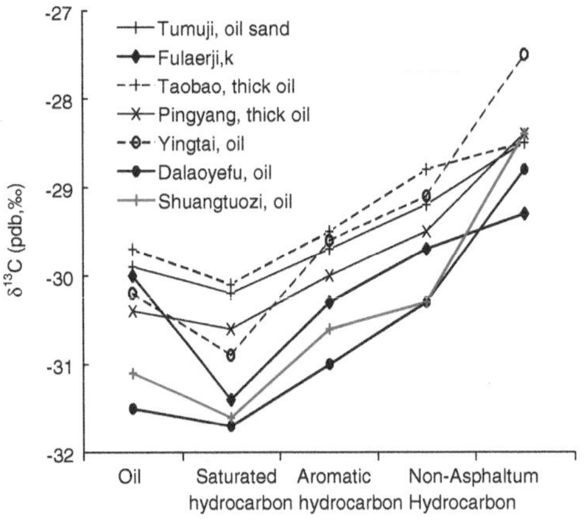

Figure 4. Carbon isotopic values of different type oils and their group components.

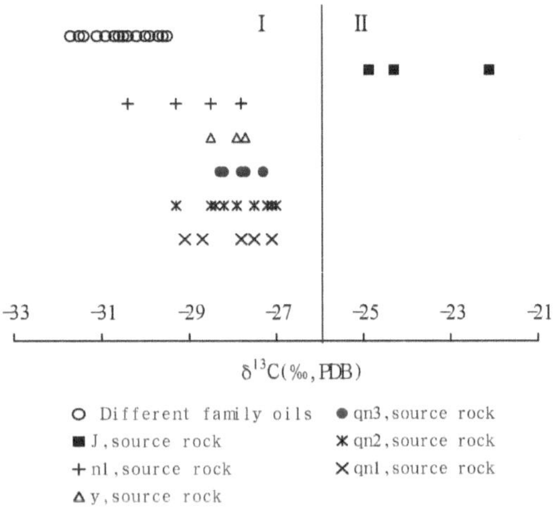

Figure 5. Carbon isotopic values of different family oils and source rocks.

Up to now, there seems no good method to study the maturity degree of oils suffering from biodegradation, washing and oxidation. Since the different biodegradation and the fractionation during oil migration may result in different biomarkers, normal maturity parameter index sometimes can not really reflect the maturity of oils. However, the aromatic structure biomarkers usually have stronger resistance to biodegradation than the saturated ones, the methyl phenanthrene index (MPI) of the western slope oil is in the range of 0.42~0.84, and the corresponding Ro value is 0.75 to 0.95%, so all the western slope oils belong to mature oils. In addition, the application of methyl dibenzothiophene is trusty due to its serious toxicity and strongest resistance to biodegradation. The value of 4–methyl dibenzothiophene/1–methyl dibenzothiophene of the Fulaerji, Pingyang, Tumuji and Taobao oils is 1.82,

Table 5. Division of petroleum systems in the central depression of Songliao Basin.

Petroleum system	Source rock kitchen	Reservoir	Cap rock	Major stage of hydrocarbon generation	Petroliferous layer	Percentage of oil reserves
Qingshankou	1,2,3 sections of Qingshankou	3,4 section of Quantou	1 section of Qingshankou	Nenjiang, Sifangtai Mingshui stage	Fuyang	66.48
		2,3 section of Qingshankou	2,3 sections of Qingshankou and 1 section of Yaojia		Gaotaizi	16.5
		1 section of Yaojia	1,2 sections of Yaojia		Putaohua	
Qingshankou –Nenjiang complex	1,2,3 sections of Qingshankou and 1 section of Nenjiang	2,3 sections of Yaojia and 1 section of Nenjiang	1,2 sections of Nenjiang	Nenjiang, Sifangtai and Mingshui stage	Saertu	Yanjia: 14.4 Nenjiang: 2.55
Nenjiang	1 section of Nenjiang	3,5 sections of Nenjiang	3,5 sections of Nenjiang	Sifangtai and early stage of Mingshui	Heidimiao	

1.39, 2.04 and 3.04, respectively, and the calculated Ro value is 0.76%, 0.74%, 0.77% and 0.81%, respectively.

The maturity of the source rock is very low owing to the uplift of the western slope since early Tertiary. The Ro value of the Jurassic source rock in the depth of 585.80 m from Well Tao 4 is 0.43%, and that of the overlying Cretaceous (Qingshankou and Nenjiang formation) lacustrine mudstones is even lower.

3.3 *Source rock kitchen*

There are three sets of source rocks developed in the western slope itself. The Jurassic source rock is a set of coal rock with carbon isotopic values less than −24‰ for its kerogen. The Cretaceous Qingshankou and Nenjiang source rocks are still immature because of the uplift of the western slope since early Tertiary. The thick oil of the western slope is at a mature stage and its carbon isotopic values show the characteristics of lacustrine mudstone. Therefore, the source rock kitchen for the western slope thick oil is mainly the Cretaceous lacustrine mudstone developed in the central depression.

Songliao Basin has mainly experienced two important stages, namely depression and fault depression. The formation and distribution of source rocks are controlled by these two stages. There were two large scale transgressions during the depression stage. The largest lake occurred at the early stage of the Qingshankou and Nenjiang periods and the Qingshankou and Nenjiang source rocks were deposited during the two processes from transgression to regression. The source rocks are predominated by lacustrine mudstone of the Qingshankou and Nenjiang formations in the Changling depression and three petroleum systems have been formed, respectively, according to the source rock kitchen. One is the lower petroleum system formed by the Qingshankou source rock kitchen and the Yangdachengzi (Y), Fuyu (F), Gaotaizi (G) and Putaohua (P) oil layers. The second is the upper petroleum system formed by the Nenjiang source rock kitchen and the Heidimiao (H) oil layer. The third is the middle petroleum system formed by the Qingshankou and Nenjiang source rock kitchens and mainly the Saertu (S) oil layer (see Table 5).

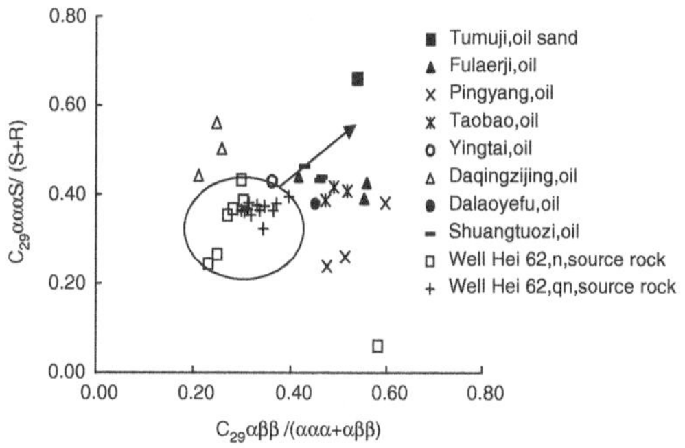

Figure 6. C29$\alpha\beta\beta/(\alpha\alpha\alpha+\alpha\beta\beta)$ vs. C29$\alpha\alpha\alpha$ S/(S+R) of source rocks and crude oils.

The high quality source rock developed in the central depression supplied abundant oil sources. The correlation among parameters, including migration, maturity and difference biodegradation for the Qingshankou and Nenjiang source rocks and the oil in the central depression and the western slope, is shown in Figure 6. For the oils experiencing migration and biodegradation, C29 $\alpha\beta\beta/(\alpha\alpha\alpha+\alpha\beta\beta)$ reflects the characteristics of different migration and biodegradation, and C29 $\alpha\alpha\alpha$S/(S+R) reflects the characteristics of different biodegradation and maturity. Compared with the oils from Daqingzijing, Shuangtuozi and Dalaoyefu in the central depression, the biodegradation of the Tumuji oil is the strongest and the migration characteristics of the Fulaerji, Pingyang and Taobao oils are obvious.

The data of K/Ar isotopic chronological analysis on the authigenic illite in sandstone reservoirs also reflect the migration of the western slope oil from the central depression to the western slope. For instance, the K/Ar isotopic data for the reservoir samples from Well Ta 20, Well Du 66 in the Alaxin gas field, Well Fu 42 in the Fulaerji oil field, and Well Lai 27 in the Pingyang oil field (for well locations, please see Fig. 1) are 56 Ma, 45 Ma and 35 Ma (Wells Fu 42 and Lai 27), respectively, which reflect that there obviously exist three stages of oil migration and accumulation from east to west. The study results coincide with the knowledge that the western oil underwent multiple fillings showed by the biomarker composition characteristics. Faults and muddy crack belts are passages for primary migration of oils, the developed direction and unconformity surface of sedimentary systems are main passages for secondary migration of oils from east to west.

4 CONCLUSIONS

The western slope oil has the characteristics of relatively low density, moderate viscosity, richness in nonhydrocarbons and low content of asphaltum. Analyses of group compounds and GC, GC–MS showed that the oil belongs to mature oil.

The content and distribution of biomarkers are characteristic of biodegradation and multiple fillings by oils, in addition, the acid value of the western slope and central depression oils and the synthetical evaluation showed that the Tumuji oil is severely biodegraded, the Taobao, Fulaerji and Pingyang oils are moderately biodegraded. The biodegradation has an increasing tendency from the east to west and from north to south.

The source rock kitchen study showed that the Qingshankou lacustrine mudstone of the central depression predominates the source of the western slope oil. There are three sets of source rocks developed in the western slope itself, the carbon isotopic value for the western slope oil is $-29\sim-32‰$, and those for the Jurassic coal rock and the Qingshankou and Nenjiang mudstone are -22 to $-24‰$ and -27 to $-31‰$, respectively. The maturity of source rocks in the western slope is very low, the calculated Ro value of the western slope oil are 0.75 to 0.95% based on the index of methyl–phenanthrene and methyl–dibenzothiophene. Therefore, the Qingshankou lacustrine mudstone of the central depression predominates the source rock kitchen of the western slope oil. The K–Ar chronology data of reservoir authigenic illite also reflect the three stages of oil migration from east to west.

ACKNOWLEDGMENTS

This work was supported by the National Natural Gas Project "973"(Grant No.2001CB209100) and National Natural Science Foundation (Grant No. 40672097). Special thanks should go to the leaders of Research Institute of Petroleum Exploration & Development, Daqing Petroleum Company and Jilin Petroleum Company for help in collecting the related information.

REFERENCES

Bao Jianping, Mei Bowen. The abnormal distribution and the Origin of 25-norhopane series. Acta Sedimentology Sinica, 1997, 15(2): 179–183.

Connan J. Biodegradation of crude oil in reservoirs. In:Brooks J, Welte D H, eds.Advances in petroleum Geochemistry 1. London: Academic Press, 198,.299–335.

Dun Tiejun. The resources of thick oil and natural bitumen in Canada. Northwestern Geology, 1995, 16(3): 37–41.

Editorial Committee. Daqing Oil Field Petroleum Geology of China, Jilin Oil Field Petroleum Geology of China. Vol.2, Petroleum Industry Press, Beijing.

Guo Taixian, Liu Chuncheng, Lu Hongzhi. Geological characteristics and hydrocarbon accumulation pattern of Penglai 19-3 Oil field. Petroleum Exploration and Development, 2001, 28(2): 26–30.

Hu Jianyi, Niu Jiayu. Formation and distribution of heavy oil bitumen resources in China. Oil & Gas Geology, 1994, 15(2): 105–112.

Huang Haiping, Ren Fangxiang, R. Larter. The influences of biodegradation to the distribution of benzene carbazole in oil. Chinese Science Bulletin, 2002, 47(6): 1271–1275.

Jiang Peihai, Tang Wenhui. The characteristics of thick oil field in Bohai and exploration importance. Typical Oil and Gas Field, 1998, 5(2): 18–23.

Larter, S. R., Bowler, B. F. J., Li, M. et al., Molecular indicators of secondary Oil migration distances. Nature, 1996, 383: 593–597.

Li M, Larter S R, Stoddart D, et al..Fractionation of Pyrrolic nitrogen compounds in Petroleum during migration: Derivation of migration-related geochemical Parameters. In: Cubitt J. M. and England, W.A. eds. The Geochemistry of Reservoirs. Geol Soc SpecPubl, 1995, 86.103–123.

Liu Luofu, Zhao Jianzhang, Zhang Shuichang, et al..Genetic types and characteristics of the Silurian asphaltic sandstones in Tarim Basin. Acta Petroleum Sinica, 2000, 21(6): 12–18.

Lu S, He W, Huang H. The Geochemical characteristics of heavy Oil and its recovery in the Liaohe Basin. China. Organic Geochemistry, 1990, 16: 437–449.

Peter, K.E., Moldowan, J.M. The Biomarker Guide: Interpreting Molecular Fossils in Petroleum and Ancient Sediments. Prentice Hall, Inc, 1993.

Peters, K.E., Moldowan, J.M., McCaffrey, M.A. et al.. Selective biodegradation of extended hopanes to 25-norhoPanes in Petroleum reservoirs. Insights from molecular mechanics organic Geochemistry. 1996, 24: 765–783.

Wang Shuangqing, Lin Renzi, Mei Bowen. Acidic constituents in heavy oils from the main oil reservoirs of Gaosheng Oil Field, Liaohe Basin. China. Geochimica, 2000, 29(4): 376–382.

Wang Yutao, Jiang Shaobin. An approach to the distribution and origin of heavy oil in the northwest margin of

Junggar Basin. Petroleum Exploration and Development, 1998, 25(5): 18–23.
Xie Xinong, Wang Zengming. Dynamics of basin fluid and its advances. Acta Sedimentology Sinica, 2003, 21(1): 20–24.
Xiong Yongqiang, Geng Ansong.Carbon isotopic composition of individual compounds of asphaltene pyrolysates from biodegraded crude oils in Liaohe Basin. Geochimica, 1998, 27(6): 532–536.
Xu Zhiming, Wang Tingdong, Jiangping, et al.. Probe on the water washing with genesis of high solidifying point crude oil. Geochimica, 2000, 29(6): 556–561.
Yang Yufeng, Wang Zhanguo, Zhang Weiqin. The pattern of total organic carbon and sequences with in mudstone formation, Songliao Basin. Acta Sedimentology Sinica, 2003, 21(2): 340–344.
Zhang Min, Zhang Jun. Effect of water washing on hydrocarbon compositions of petroleum sandstone reservoir. Geochimica, 2000, 29(3): 287–292.
Zhao Mifu, Li Yahui, Xin Quanlin, et al..Forming mechanism of heavy oil in Lukeqin structure belts. Journal of the University of petroleum, China, 2001, 25(1): 51–57.

Petroleum Geochemistry and Exploration in the Afro-Asian Region – Digang Liang, Darui Wang & Zhenxi Li (eds)

ISBN 978-0-415-44083-7

Evaluation of petroleum source rocks of Bengal Basin, India

S. Pahari, I.V.S.V. Prasad, A. Banerjee & Meenu Varshney
KDM Institute of petroleum Exploration, ONGC, Dehradun India.

ABSTRACT: Source potential of Permo-Carboniferous, Mesozoic and Tertiary sedimentary deposits of Bengal Basin has been evaluated to identify the potential source rocks, assess their maturity and amount of hydrocarbon generated/expelled. Nature and maturity of oils are also assessed. Potential source rocks are present in Paleocene and Gondwana sequences in shelf area and also in Miocene, Oligocene and Eocene sections in the hinge A zone and on the east of the platform. In the shelf area, Gondwana and Cretaceous prospects may contain gases generated during 130 to 10 Ma from terrestrial Gondwana source rocks. Although significant hydrocarbon generation has not occurred from the Tertiary sediments in shelf areas due to low maturity, in the hinge zone and on the east of the platform these sediments are mature enough to generate significant hydrocarbon. There is a distinct possibility of better accumulations in Miocene to Paleocene reservoir-trap systems, nearer to areas where Paleogene and younger source rocks are at a maturity of 0.75–0.9% VRo.

1 INTRODUCTION

The Bengal Basin, a pericratonic basin in the eastern part of Indian shield includes four tectonic zones: basin margin, stable shelf, hinge zone and deep basin, separated by NE-SW bound step-faults (Figure 1) and has a total sedimentary area of about 90,000 sq. km comprising both on land and offshore (up to 200 m bathymetry). The geology, tectonics and stratigraphy of the basin are discussed in (Roybarman, 1983). More than fifty years of exploration for hydrocarbons in the Bengal Basin has not yet led to a commercial success. Oil and gas indications in a number of drilled wells and seepages have confirmed hydrocarbon generation in the basin. An attempt has been made in this paper to use the source rock data to evaluate the likely generation/expulsion of hydrocarbon to identify the hydrocarbon charge and accumulation locales. Thus our main objectives are to evaluate the whole sedimentary sequences to identify the potential source rocks, assess their maturity, kerogen type, and hydrocarbon charge potentials.

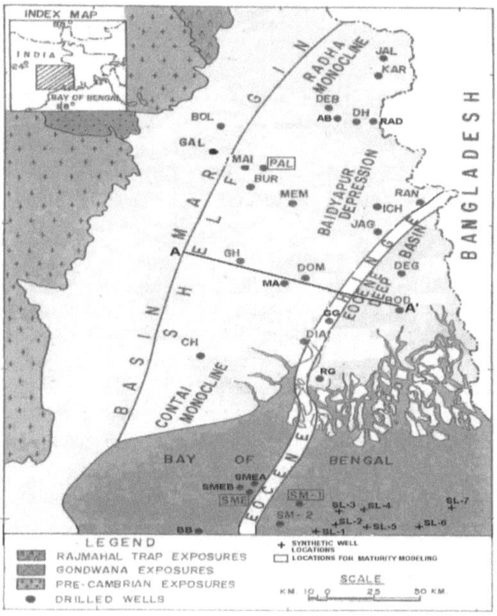

Figure 1. Map of Bengal basin showing location of drilled wells.

2 GEOLOGY AND TECTONICS OF THE BENGAL BASIN

Schematic geological cross section along AA' (Fig.1) of the basin is shown in Figure 2. The Bengal Basin started as a divergent margin basin, initiated by the break up of Gondwanaland at about 126 Ma and separation of the Indian plate. As the Indian plate started drifting northward, Bengal Basin became a marginal sag basin with the central part experiencing maximum subsidence.

The movement and interaction of the Indian, Antarctic, and Australian plates, the Burma platelet, and the Eurasian plates through time are responsible for the pattern of basin formation, sedimentation, and

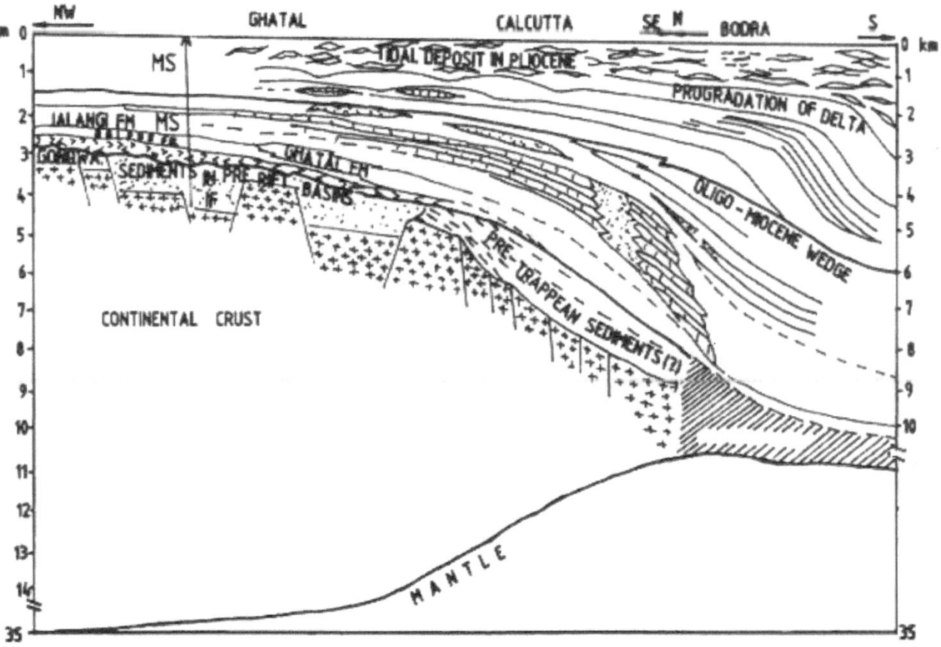

Figure 2. Schematic geological cross section of Bengal Basin along line AA' of figure 1.

structure that define the hydrocarbon potential (Coffin et al. 1998). Towards end of Eocene, Indian plate collided with the Burmese plate giving a southeasterly tilt to the Bengal Basin. Evolution of Bengal Basin broadly consists of four stages. Pre-rift stage (Permo-Carboniferous and Lower Mesozoic time) associated with deposition of continental.

Gondwana sediments in block faulted troughs. Rift-drift stage (late Jurassic-early Cretaceous time) associated with Rajmahal volcanism and break up of Gondwanaland. Post-rift stage (mid Cretaceous-Paleocene time) witnessed clear differentiation of the basin into shelf, slope and rise. After the regressive phase during Oligocene, the basin experienced a late Tertiary deltaic stage of lower Miocene-Pliocene times, showing enormous thickening of clastics towards the basin deep.

Growth faults are common but not extensive in this part. Large-scale shale and salt diaprism are absent. Canyon heads and sub-marine channel levee complexes seem to be present.

3 HYDROCARBON OCCURENCES

Hydrocarbon indications in the drilled locations ICH and GG (Fig.1) from Eocene to Oligocene sediments and hydrocarbon show in CH and BOD drilled locations (Fig.1) have been noticed in Bengal Basin. The gas show in BOD is a dry gas with δ13Cmethane -31 per mil and the oil show in CH is catagenetic (Chandra et al. 2000) in origin.

4 SAMPLES AND METHOD

Source rock evaluation has been done based on the geochemical data obtained from 29 wells (Fig. 1) comprising total organic carbon, Rock-Eval pyrolysis, vitrinite reflectance, GC, GC-MS, and bitumen extraction studies

(Espitalie et al., 1977; Peters, 1986; Peters and Cassa, 1994; Stach et al., 1982; Durand et al., 1986). The study has used computed parameters Source Potential Index (SPI) in metric tons hydrocarbon per square meter (Demaison et al. 1991), hydrocarbon generated (HCG) and hydrocarbon expelled (HCE) to assess source potentials of the sedimentary sequences. The SPI was calculated with the following formula:

$$SPI = h \times (S_1 + S_2) \times d / 1000$$

where, h is the total thickness of the formation in meters, $S_1 + S_2$ is the average sum of S_1 and S_2 values of all samples, and d is the density of rock assumed to be 2.5.

HCG and HCE were calculated using the following formulae (Banerjee et al., 2002):

$$HCG = S_2 \times f \times t \times 2.5/1000 \times (1\text{-}f) \quad (I)$$
$$HCE = HCG- (S_1 \times t \times 2.5/1000) \quad (II)$$

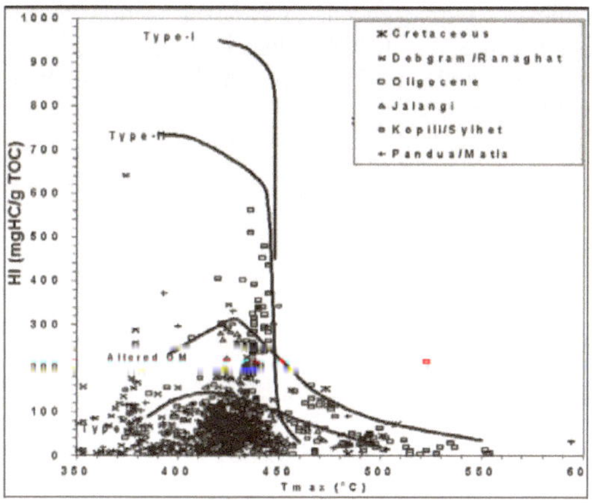

Figure 3. Crossplot of HI vs Tmax for sediments of Bengal basin.

where, HCG and HCE are amounts of hydrocarbon (HC) generated and expelled by a sedimentary column in MT HC/m^2, respectively, t is thickness of sedimentary column in meters and f is fractional conversion at the middle of the column, calculated using Type III vitrinite kinetic parameters (Sweeny et al. 1990). The calculated parameter HCG integrates source potential, thickness, and maturity of a section and is proportional to intensity of hydrocarbon generation in a section. HCE, on the other hand is proportional to amounts of hydrocarbon expelled (if HCE is positive) or hydrocarbons accumulated (if HCE is negative) in a section. HCG and HCE have been used to identify and evaluate hydrocarbon generation and charge potentials of different stratigraphic units and identify kitchen and hydrocarbon accumulation areas. Maturity at various sedimentary sequence boundaries was modeled with the Easy Ro worksheet program (Sweeny et al. 1990) for three locations: one in the shelf area, a second in the hinge zone and a third to the east of hinge area. Measured vitrinite reflectance (VRo) values were compared with modeled values with a uniform temperature gradient 29°C/km, 15°C/km and 15°C/km for the shelf area, the hinge zone and east of the hinge zone, respectively and the surface temperature was assumed to be 25°C.

5 POTENTIAL SOURCE ROCKS OF BENGAL BASIN

The quality of Bengal sediments for various fomations is mostly Type-III kerogen except some Oligocene to Middle Miocene sediments of GG and DOM (Figure 3).

Figure 4 also shows oil prone source rocks in Oligocene, Kopili/Sylhet and Jalangi formations.

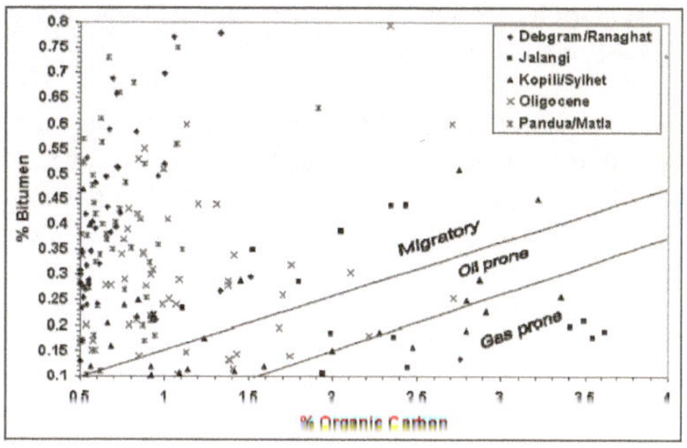

Figure 4. Crossplot of Bitumen (%) vs Organic Carbon (%) for sediments of Bengal basin.

Table 1. Potential source rocks in Bengal Basin.

Well	Depth(m)	TOC (Avg.)	Tmax (Avg.)	S2 (Avg.)	Formation
DOM	4700–4710	1.33	446	1.23	JLNJ
PAL	1925–2035	1.39	435	2.92	JLNJ
PAL	2025–2035	1.31	435	2.41	JLNJ
DOM	3900–3910	1.45	441	1.62	KOP/SYL
DOM	4395–4400	4.6	443	10.91	KOP/SYL
DOM	4625–4635	1.07	443	1.2	KOP/SYL
GG	5222–5230	2.42	437	1.2	KOP/SYL
GG	5548–5555	2.5	444	1.06	KOP/SYL
GG	5674–5680	3.62	435	1	KOP/SYL
MA	3422–3430	1.57	441	1.5	KOP/SYL
DOM	3700–3710	1.22	439	1.03	MMRI/BRDN
GG	4180–4190	1.44	523	3.08	MMRI/BRDN
ICP	4050–4310	1.24	435	1.37	MMRI/BRDN
MA	3248–3255	1.1	435	1.3	MMRI/BRDN
SM2	4576–4580	1.56	434	1.95	MMRI/BRDN
SM2	4849–4860	1.15	445	1.31	MMRI/BRDN
SME	4975–4985	1.46	438	1.17	MMRI/BRDN
SME	5050–5260	2.54	440	1.14	MMRI/BRDN
GG	3760–3770	2.4	484	2.16	PNDU/MTLA
CH	2800–3652	15.6	441	53.88	LOW GON
MAI	3776–4071	18.3	472	13.09	LOW GON
PAL	4459–5500	28.5	486	16.62	LOW GON

Table 1 shows the potential source rock sections in different sedimentary sequences. Fair to good source potential exist within Pandua/Matla, Memari/Burdwan, Kopili/Sylhet, and Jalangi formations (Table 1).

In Gondwana sediments, the organic richness and source potential are excellent, due to presence of coal seams. Tertiary sediments within Jalangi, Kopili, Sylhet, Memari/Burdwan and younger formations, in which some source rock development is observed on the shelf/hinge area are likely to possess better source characteristics in the east of the hinge zone as they are likely to have deposited in a reducing slope environment within the oxygen minimum layer.

6 HYDROCARBON KITCHEN OF BENGAL BASIN

Source potential index (SPI) contour maps for various formations are shown in Figure 5. In all formations from Pliocene to Eocene, SPI values increase towards the hinge and east of the hinge zone. Favourable SPI values in Pandua/Matla, Diamond Harbour/Memari/Burdwan, and Kopili/Sylhet formations indicate better potential of hydrocarbon generation in the hinge and east of the hinge zone.

Hydrocarbon generated (HCG) and hydrocarbon expelled (HCE) contour maps for various formations are shown in Figure 6, Figure 7 in Bengal Basin, only Gondwana sediments of shelf area have high HCG value (HCG: 33) and might be effective hydrocarbon kitchen in shelf areas (Fig. 6). Predominance of coals and carbonaceous shales are the reason for the high HCG values of the Gondwana sediments. In the hinge and east of the hinge zones, Pandua/Matla formations have excellent HCG values (HCG: 8 to16) for oil accumulations. Although average HCG values for Diamond Harbour/Memari/Burdwan, and Kopili/Sylhet formation (HCG: 2 to 4) are not high, these values are very significant for two reasons. First, in Bengal Basin, the drainage style is likely to be dominantly lateral and that requires possibly a lower threshold of SPI/HCG values for oil accumulation. Second, SPI/HCG values are likely to have increased in the deep basin where the sediments are likely to have been deposited in a reducing environment within the oxygen minimum layer. Sediments within Eocene to Miocene sections show significant SPI and HCG values in hinge and east of hinge zone and might be effective hydrocarbon kitchen in those area.

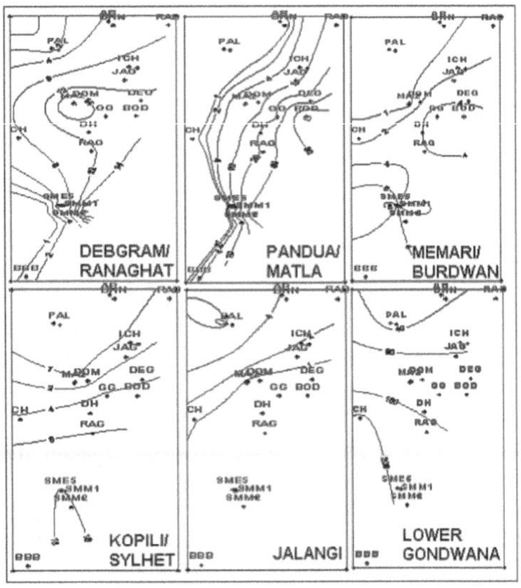

Figure 5. Average SPI contours for Debgram/Ranaghat, Pandua/Matla, Memari/Burdwan, Kopili/Sylhet, Jalangi and Lower Gondwana sediments of Bengal basin.

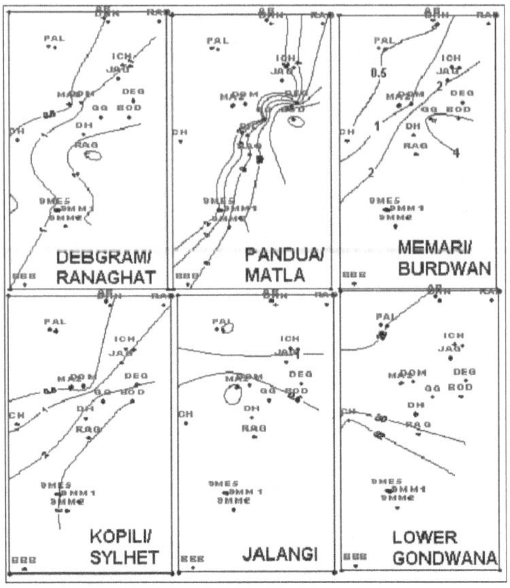

Figure 6. Average HCG contours for Debgram/Ranaghat, Pandua/Matla, Memari/Burdwan, Kopili/Sylhet, Jalangi and Lower Gondwana sediments of Bengal basin.

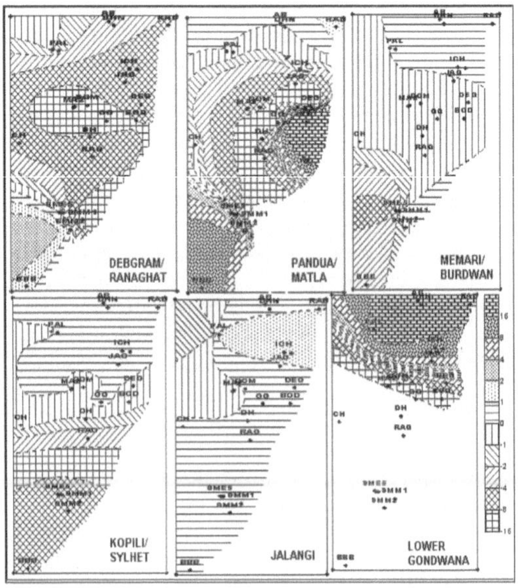

Figure 7. Average HCE fill contours for Debgram/Ranaghat, Pandua/Matla, Memari/Burdwan, Kopili/Sylhet, Jalangi and Lower Gondwana formation sediments of Bengal basin.

7 MATURITY MODELLING OF THE SOURCE ROCKS

To understand maturity progression of the source rocks of different stratigraphic units, maturity at various sedimentary sequence boundaries was modelled with easy Ro worksheet program (Sweeney and Burnham, 1990) at three locations: in stable shelf area (PAL), another in hinge zone (SME) and last in the east of hinge zone (SM1). Use of temperature gradients of 29°C/km at PAL (Fig.8), 15°C/km at SME (Fig.10) and 15°C/km at SM1 and surface temperature of 25°C allowed match between calculated and the observed VRo values at respective locations.

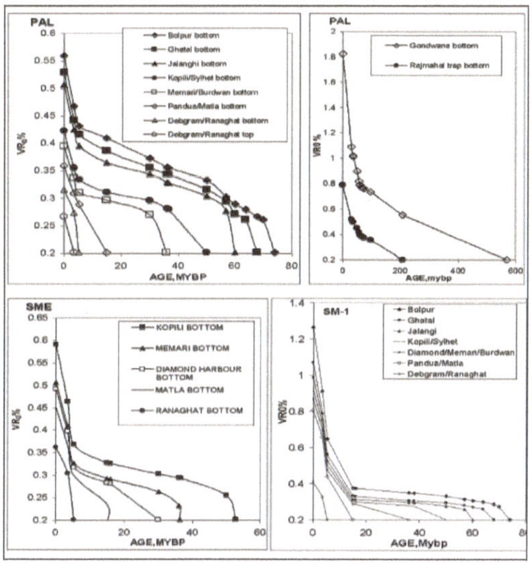

Figure 8. Progression of maturity at location PAL, SME and SM1 of Bengal basin.

Figure 8 shows the progression of VRo values at the formation top levels with age and subsidence diagram respectively at drilled locations PAL, SME and SM1. At PAL, sediments at Gondwana top and bottom attained 0.75% VRo at 10 and 130 Ma, respectively. At SME, none of the drilled sections have attained 0.75% VRo. Sediments at the base of Memari and Sylhet attained 0.5% VRo in the recent past and 3.5 Ma respectively. Post Memari sediments are yet to attain 0.5% VRo, At SM1, Oligocene and basal Miocene sediments are in their peak generation phase. The Eocene, Paleocene sediments are in late mature phase and Cretaceous sediments are in the over mature stage. However, potential source rocks of Paleogene formations could be well within the peak oil generation phase (0.75–0.9% VRo) east of the hinge zone and further down dip.

The maturity modeling shows only source rocks within Paleogene and basal Miocene formations that are likely to have attained peak oil generation maturity east of the Eocene hinge zone at the modeled location. Burdwan/Memari and Pandua/Matla formations are likely to have attained peak maturity basinward from the SME location. In the shelf area, except Gondwana no other section is in peak oil generation maturity.

8 NATURES AND MATURITY OF OIL

Gas chromatography, carbon Isotope and biomarker data for oils of Bengal Basin are shown in Table 2, Table 3.

Pr/nC17 and Ph/nC18 value for Kopili and Sylhet oils from GG and ICH wells suggests the source OM to be peat swamp coal type to coastal swamp type and/or mixed Type II and III (Connan & Cassou, 1980). Pr/Ph value for both oils is about 2.2. One full transgressive and regressive marine cycle in Ghatal sediments followed by deltaic cycle in Oligocene

Table 2. GC and carbon isotopic data of Bengal Basin oils.

Well	Depth (m)	Sat %	Arom %	$Pr + nC_17/$ $Ph + nC_18$	Pr/ Ph	$\delta^{13}C_{Sat}$	$\delta^{13}C_{Arom}$
GG	5770	62.57	16.31	1.39	2.42	−24.75	−24.3
ICH	4346	72.8	23.4	1.57	2.78	−24.62	−24.06

Table 3. Biomarker data for Bengal Basin oils.

Well	Depth (m)	Tm/ Ts	22S/ 22S+R 32H	30H/ 29H	20S/ 20S + R C29 St	$\beta\beta/\alpha\alpha + \beta\beta$ C29 St	Hop/ Sterane
GG	5770	2.26	0.51	1.12	0.53	0.41	9.89

support the presence of such partially oxic environment. Carbon isotopic data (Table 2) suggests that OM rich in marine components sourced these oils. Biomarker maturity parameters (Table 2) indicate the GG oil to be from catagenetic mature stage. These oils are possibly sourced from Paleogene source rocks present in estuarine-marine transitional environment.

9 CONCLUSIONS

Source rock data show presence of potential source rocks in Paleocene and Gondwana sequences in shelf area. In stable shelf, none of the potential source rocks except the Gondwana source rocks have attained peak oil generation maturity and Paleocene source rocks have just entered into oil generation window maturity. Gondwana sediments with HCG value up to 33 and peak oil generation maturity might be the effective hydrocarbon kitchen in shelf areas. In shelf area, Gondwana and Cretaceous prospects may contain gases generated from terrestrial source rocks of Gondwana during 130 to 10 Ma.

Source rock data also show presence of potential source rocks in Miocene, Oligocene and Eocene sediments in hinge and east of hinge zone. Maturity modelling suggests that the better source rocks in Paleogene and younger formations would have reached peak oil window maturity east of hinge zone and may generated hydrocarbon 4 to 6 Ma. In hinge and east of hinge zone, Pandua/Matla formation (HCG: 8 to16) and Diamond Harbour / Memari / Burdwan, and Kopili / Sylhet formation (HCG: 2 to 3) have significant HCG values along with oil generation maturity may likely to have generated hydrocarbon. Potential source rocks in Miocene, Oligocene and Eocene sediments in hinge and east of hinge zone would be effective source rock in hinge and east of hinge zone for hydrocarbon accumulation.

Oils of Bengal Basin are possibly sourced from Paleogene and younger source rocks present in estuarine-marine transitional environment. There is a distinct possibility of better accumulations in Miocene to Paleocene reservoir-trap systems, if present, nearer the hinge zone and east of the hinge zone where Paleogene and younger source rocks are at a maturity of 0.75–0.9% VRo.

ACKNOWLEDGEMENTS

The authors are grateful to Dr. D. Ray, ED- Head, KDMIPE for his constant encouragement. We also acknowledge ONGC for permission to publish this work. The views expressed here are solely those of the authors and not necessarily of the organization where they are working.

REFERENCES

Banerjee, A., Pahari, S., Jha, M., Sinha, A.K., Jain, A.K., Kumar, N., Thomas, N.J., Misra, K.N.& Chandra, K. 2002. The effective source rocks in the Cambay basin, India: American Association of Petroleum Geologists Bulletin 86, 433–456.

Chandra, K., Garg, A.K. & Mishra, C.S. 2000. Forty years of petroleum geochemistry in India: proceedings in the fifth international AAAPG-2000 Conference, p.201–214.

Coffin, M. & Lawyer, L. 1998. Atlas of paleogeographic reconstructions: The Plates Project progress report 215–0798, University of Texas at Austin, 75 p.

Connan, J., & Cassou, A. M. 1980. Properties of gases and petroleum liquids derived from terrestrial kerogen at various maturation levels. *Geochimica et Cosmochimica Acta, 44, 1–23.*

Demaison, G.J. & Huizinga, B. J. 1991. Genetic classification of petroleum systems: AAPG Bulletin, v. 75, p.1626–1643.

Durand, B., Alpern, J.L., Pittjon, B. & Pradier, A. 1986. Reflectance of Vitrinite as a control of thermal history, in J. Burrus, ed. Thermal modeling of sedimentary basins: editions Technip, Paris. Pp. 441–473.

Espitalie,J., J.L. Laporte, M.Madec, F.Marquis, P. Leplat, J. Paulet & A.Boutefeu.1977. Methods rapide decharecterisationdes roches leur degree d'evolution: Revue de I' Institut Francais du Petrole, V.32, .23–42.

Peters, K.E. 1986. Guidelines for evaluating petroleum source rocks using programmed pyrolysis. *American association of petroleum geologists bulletin*, 70,319–329.

Peters, K.E., Cassa, M.R. 1994. Applied source rock geochemistry, in L.B. Magoon and W. G.Dow, eds., The petroleum system- From source to trap:, AAPG Bulletin, Memoir 60, pp. 93–120.

Roybarman, A. 1983.Geology and hydrocarbon prospects of west Bengal, , Petroleum Asia Journal, p.51–56.

Stach, E., Machowsky, M., Teichmuller, M., Taylor, G.H., Chandra, D., & Teichmuller,R. 1982. *Coal petrology*, (3rd ed.) Berlin.

Borntraeger Sweeny, J.J.& Burnham, A.K. 1990. Evaluation of a simple model of vitrinite reflectance based on chemical kinetics: AAPG Bulletin, v. 74, p.1559–1570.

Sweeny, J.J. 1990. BASINMAT fortran program calculates oil and gas generation using a distribution of discrete activation energies: Geobyte, v. 5, p. 37–43.

Petroleum Geochemistry and Exploration in the Afro-Asian Region – Digang Liang, Darui Wang & Zhenxi Li (eds)
© 2008 Taylor & Francis Group, London, ISBN 978-0-415-44083-7

Accumulation history and features of natural gases in the Kuqa foreland basin

Menjun Zhao
Research Institute of Petroleum Exploration and Development, PetroChina

Zhaoming Wang
Tarim Oilfield Company, PetroChina, Korle, Xinjiang

Shuichang Zhang
Research Institute of Petroleum Exploration and Development, PetroChina

Qinhua Wang
Tarim Oilfield Company, PetroChina, Korle, Xinjiang

Yan Song, Shaobo Liu & Shenfei Qin
Research Institute of Petroleum Exploration and Development, PetroChina

ABSTRACT: The components and carbon isotopes for natural gases are mainly controlled by types of organic matter and maturity, but, more and more examples indicate that the accumulation process of natural gases also controls the components and carbon isotopes of natural gases. Although the natural gas in different structural belts of the Kuqa foreland basin has the same original source, namely generated mainly by the Jurassic source rock and subordinately by the Triassic source rock, natural gases in overthrust belts, slope zones and frontier upheaval zones of the Kuqa foreland basin are obviously different in components, contents of dry and wet gases and carbon isotopes. The accumulation history of hydrocarbons in the Kuqa foreland basin shows that the early accumulated hydrocarbons had been destructed and then hydrocarbons generated in late stage accumulated again. Our research indicates that the accumulation process of hydrocarbons obviously controlled the components and carbon isotopes of natural gases in different structural belts of the Kuqa foreland basin.

1 GEOLOGICAL BACKGROUND

The Kuqa foreland basin located in the north of Tarim Basin is an intracontinental basin. Its development has experienced three stages, i.e. late Permian-Triassic foreland basin, Jurassic intracontinental sag basin and Neogene rejuvenated foreland basin. From east to west, the Kuqa foreland basin includes the Yangxia sag, Baicheng sag and Wushi sag. In recent years, there are several great oil and gas discoveries in the Kuqa foreland basin, such as the Kela-2 gas field, Dabei-2 gas field and Yinan-2 gas field in overthrust belts, Dinan-1 and Wucan-1 gas fields in sag and slope zones and Yangtake, Yingmai-7, Yaha and Tiergen gas fields in frontier upheaval zones (Figure 1).

The Kuqa foreland basin is characteristic of richness of natural gases (Zhao Menjun et al., 1999). In recent years, a great progress in researches on the petroleum geology of the Kuqa foreland basin has been made with the development of exploration.

Some geologists studied source rocks of the large gas fields of Kuqa Basin (Li Jian et al., 2001; Zhao Menjun et al., 2002a; 2002b), meanwhile the reservoir and caprock conditions for the large gas fields of Kuqa Basin were also discussed (Zhou Xinxi et al., 2000; Jia Jinhua et al., 2001; Jia Jinhua & Gu Jiayu 2002). Furthermore, some geologists discussed the petroleum geology, the factors controlling large gas fields, the accumulation process of high pressure gas fields and the accumulation features, models, and occurrence regularity of hydrocarbons, respectively (Jia Chengzao et al., 2002; Song Yan et al., 2002; Pi Xuejun et al., 2002; Wang Hongju et al., 2002; Wang Zhenghua, 2001; Tian Zuoji et al., 2001; Zeng Lianbo et al., 2002; Zhao Menjun et al., 2003a).

Although most geologists agree that the hydrocarbon in Kuqa accumulated in multi-stages, there are still some different opinions about the individual times of hydrocarbon accumulations. For example, Zhao Jinzhou considered that the hydrocarbon in Kuqa

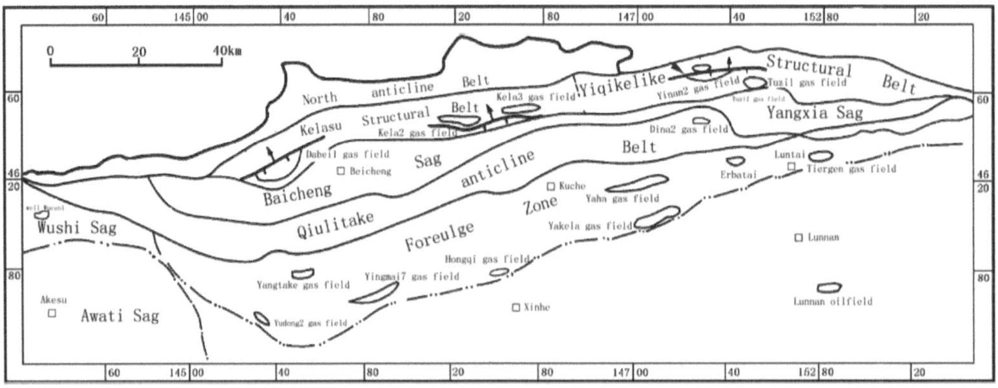

Figure 1. Structural units and major gas fields in the Kuqa foreland basin.

accumulated in three stages, namely early-middle Kangcun of Neogene, late Kangcun-early and middle Kuqa and late Kuqa-Xiyu of Quaternary. Liang Digang maintained that the oil accumulated in early Miocene and gas accumulated after late Pliocene. Moreover, through analyzing the Kela 2 gas field, some scholars proposed that in the early Himalayan epoch oil accumulated and then was destructed, and in the late Himalayan epoch gas accumulated and redistributed (Wang Zhaoming et al., 2002; Zhao Mengjun et al., 2002a, b; 2003a). In this article, based on the above research results, we discussed the accumulation process of natural gases and its control for components and carbon isotopes of natural gases in the Kuqa foreland basin.

2 EXPERIMENTAL

The gas composition and carbon isotopes in fields of different structural belts in the Kuqa foreland basin were measured in the present study. The measurement on the gas composition of 188 gas samples was carried out on HP-5880GC (see Table 1). The $\delta^{13}C$ values of some gas samples were measured on Finngan MAT – 252 mass spectroscope coupled with a GC equipped with a Porapak Q packed column (2 m × 1/8? SS6), the maximum temperature was set at 200°C and the rising rate of temperature was programed at 8°C/min. The analysis errors were less than 0.3‰ as compared with the analytical standard of the GBW04405 referenced gas. The compositional carbon isotopes for 39 gas samples are shown in Table 2. The data of inclusions and reservoir bitumens applied to studying the hydrocarbon accumulation process are cited from Zhao Mengjun et al. (2002a) and Wang Zhaoming et al. (2002).

3 DISCUSSION

3.1 *Geochemical characters and source of natural gases*

3.1.1 *Component characters of hydrocarbons*

The natural gas in Kuqa is predominated by hydrocarbons and non-hydrocarbon components are mainly CO_2 and N_2 (Table 1). Natural gases from the Yinan-2, Kela-2 and Dabei-1 gas pools have the highest hydrocarbon content with the average above 97%. The hydrocarbon content is relatively lower for gases from the Yingmai-7 and Tai-1 gas pools in the frontier upheaval zone. The natural gas in the overthrust belt has a higher methane content, for example, the methane content of the Kela-2 gas field is averagely 96.92%. The methane content of the Yingmai-7 and Tai-2 gas pools in the slope belt is the lowest with the mean below 80% (Zhao Menjun et al., 2002b,c).

(1) The natural gas in the Kalasu structural belt is relatively dry, indicating a similar accumulation process to that of the Kela-2 gas field, namely a two-staged accumulation model with the dominance by the late-stage gas accumulation. Hydrocarbons generated by the Triassic source rock started to accumulate in the early Himalayan epoch while the natural gas generated by the Jurassic-Triassic source rock accumulated mainly in the late Himalayan epoch.
(2) Because the Dabei-1 structure is located in the west of the Kalasu structural belt and the Baicheng sag margin, no matter the hydrocarbon in this structure accumulated in the early period or late period, the maturity is lower than that of the Kela-2 gas field, which is the main reason for the lesser drying coefficient of the Dabei-1 gas.

Table 1. Component contents of natural gases in major gas fields (pools) of the Kuqa foreland basin.

Gas pool	Total hydrocarbon (%)	C_1 (%)	C_2^+ (%)	C_1/C_1^+	CO_2 (%)	N_2 (%)	Total non-hydrocarbon (%)	Number of sample
Yinan2	84.93 – 98.98 / 97.6	82.78 – 94.41 / 89.03	2.15 – 10.02 / 6.26	0.90 – 0.97 / 0.94	0.11 – 2.99 / 1.68	0 – 1.69 / 0.67	0.74 – 4.57 / 2.37	5
Kela2	93.80 – 99.13 / 97.57	93.5 – 98.31 / 96.92	0 – 1.56 / 0.61	0.98 – 1 / 0.994	0 – 2.52 / 1.12	0 – 4.51 / 1.35	0.87 – 6.49 / 2.46	19
Dabei1	94.65 – 99.41 / 98.20	92.16 – 96.18 / 94.87	2.49 – 4.79 / 3.32	0.95 – 0.97 / 0.97	0 – 4.36 / 0.98	0.08 – 2.83 / 0.87	0.59 – 4.44 / 1.83	8
Dawanqi	79.92 – 99.97 / 97.11	65.2 – 96.97 / 88.59	0.09 – 34.16 / 8.52	0.656 – 1.00 / 0.91	0 – 2.61 / 0.3	0 – 20 / 2.55	0.1 – 20.07 / 2.85	77
Yaha	77.85 – 99.13 / 92.58	66.47 – 90.63 / 81.31	1.33 – 24.43 / 11.27	0.73 – 0.98 / 0.88	0.01 – 8.98 / 0.95	0 – 11.53 / 5.13	0.42 – 16.73 / 6.00	69
Yingmai7	63.76 – 97.75 / 92.87	61.90 – 91.53 / 84.75	1.39 – 22.26 / 8.12	0.76 – 0.99 / 0.91	0 – 3.24 / 0.52	2.25 – 33.0 / 6.73	1.67 – 33.36 / 7.24	29
Yangtake	92.08 – 98.87 / 96.39	63.38 – 92.38 / 85.97	2.80 – 27.99 / 10.42	0.71 – 0.97 / 0.89	0 – 0.73 / 0.17	1.1 – 7.86 / 3.22	1.14 – 7.92 / 3.39	17
Tiergeng	91.50 – 98.07 / 95.90	72.30 – 87.47 / 80.20	6.82 – 23.08 / 15.7	0.762 – 0.93 / 0.84	0.05 – 2.63 / 0.57	1.18 – 8.2 / 3.54	2.04 – 8.5 / 4.11	10
Tai2	72.32 – 94.08 / 89.25	65.94 – 78.99 / 72.00	6.38 – 26.18 / 17.25	0.722 – 0.892 / 0.810	0 – 3.14 / 0.852	5.22 – 25.82 / 9.85	6 – 27.68 / 10.7	11
Yudong2	93.91 – 96.17 / 95.04	82.28 – 83.47 / 82.88	11.63 – 12.7 / 12.17	0.868 – 0.876 / 0.872	0.16 – 1.01 / 0.59	3.67 – 5.09 / 4.38	3.83 – 6.1 / 4.97	2
Dongqiu5	85.46 – 98.91 / 95.36	79.85 – 97.24 / 90.24	1.36 – 8.78 / 5.12	0.911 – 0.986 / 0.946	0.07 – 1.69 / 0.64	0.96 – 12.85 / 4.08	1.4 – 14.54 / 4.73	7
Dinan2	96.69 – 99.19 / 98.45	87.0 – 89.38 / 88.6	9.26 – 10.55 / 9.86	0.892 – 0.905 / 0.90	0.33 – 1.81 / 0.52	0.46 – 1.73 / 1.02	0.8 – 3.3 / 1.54	11
Quele1	97.51 – 97.72 / 97.62	80.93 – 81.55 / 81.27	16.17 – 16.58 / 16.35	0.83 – 0.835 / 0.833	0.04 – 0.13 / 0.09	2.17 – 2.44 / 2.30	2.3 – 2.48 / 2.39	3
Wucan1	97.15 – 98.79 / 97.80	82.04 – 83.18 / 82.50	15.11 – 15.61 / 15.30	0.84 – 0.85 / 0.84	0 – 0.56 / 0.27	1.19 – 2.58 / 1.93	1.19 – 2.86 / 2.20	4

Note: $\frac{\text{Minimum} - \text{maximum}}{\text{Mean}}$.

3.1.2 *Characters of carbon isotopes of hydrocarbons*

The $\delta\,^{13}C_1$, $^{13}C_2$ and $^{13}C_3$ values of natural gases in the Kuqa foreland basin mainly are −40‰∼−26.2‰, −26‰∼−16.6‰ and −26‰∼−20‰, respectively. According to the $\delta\,^{13}C_2$, the natural gas in this area is primarily humic. Furthermore, from the frontier upheaval zone to overthrust belt, the $\delta\,^{13}C_1$ increases gradually.

Figure 2 and Table 2 show the characters of carbon isotopes for the Kuqa natural gas. The carbon isotopes increase with the carbon number increasing, basically showing a positive order of carbon isotopes of a biogenetic gas, namely $\delta\,^{13}C_1 < \delta\,^{13}C_2 < \delta\,^{13}C_3 < \delta\,^{13}C_4$. The carbon isotopic inversion of heavy hydrocarbons is common and mainly related with the dissipation and fractionation of natural gases in reservoirs (Qin Shenfei et al., 1999). Natural gases generated from the Kuqa depression but produced in different fields of the Kuqa foreland basin are moderately different. For instance, the carbon isotope of gases in the overthrust belt, especially in the Kelasu structural belt, is obviously heavier than that in the frontier upheaval zone, indicating that gases in the overthrust belt and Kelasu structural belt accumulated in the late high-maturity period (Zhao Menjun et al., 2002c). Because the Dabei1 structure is located in the west of the Kalasu structural belt and the margin of the Baicheng sag, the maturity of hydrocarbons in this area is lower than that in the Kela-2 gas field, this is the main reason for the lesser drying coefficient of the Dabei-1 gas.

3.1.3 *Source of natural gases*

There are two kinds of source rocks in the Kuqa depression, i.e. the middle-upper Triassic and middle-lower Jurassic source rocks characteristic of great thickness and wide distribution. The thickness of the Jurassic and Triassic source rocks is 300–700 m and 200–600 m, respectively. The TOC for the Jurassic source rock ranges from 0.4 to 37.36% with the average of 1.88–4.31% while that for the Triassic source rock is 0.4–10.1% with the mean of 1.07–2.91%. Moreover,

Table 2. Carbon isotopes of natural gases from major gas fields in the Kuqa foreland basin.

Structural belt	Field	Well No.	Strata	Well depth (m)	$\delta^{13}C_1$ (‰)	$\delta^{13}C_2$ (‰)	$\delta^{13}C_3$ (‰)	$\delta^{13}C_4$ (‰)
Thruest belt	Dabei-1	DB1	E	5568.08–5620.00	−29.73	−21.39	−20.80	−21.91
		DB2	K	5658.00–5669.50	−30.83	−21.54	−19.78	
	Kela-2	KL2	E	3499.87–3534.66	−31.10	−16.80	−18.50	−17.80
		KL2	E	3499.87–3534.66	−27.29	−19.41		
		KL201	K2b	3936.00–3938.00	−26.16	−18.09	−19.06	−22.14
		KL201	K2b	3630.00–3640.00	−27.07	−18.48	−19.08	−20.31
		KL201	K1b	4016.00–4021.00	−27.32	−19.00	−19.54	−20.90
		KL202	E	1472.00–1481.00	−28.24	−18.86	−19.25	−20.91
	Kela-3	KL3	E	3544.00–3550.00	−30.83	−17.68	−17.09	
		KL3	E	3104.58–3198.79	−25.14	−18.82		
	Yinan-2	YN2	J	4578.80–4758.00	−34.81	−22.38	−25.87	−21.71
		YN2	J1a	4776.00–4785.00	−32.20	−24.55	−23.07	−22.78
Sag and Slope	Dina-2	DN2	N1j	4598.44–4875.59	−34.27	−20.89	−20.57	−25.78
	Yilake	WC1	K	5917.52–6009.93	−38.00	−28.05	−26.37	
Forebulge	Yudong-2	YD2	E−K	4728.81–4744.8	−37.50	−21.48	−24.47	−23.69
	Yangtake	YT1	E+K	5234.37–5331.92	−38.92	−22.85	−20.91	−22.08
		YT101	E	5329.00–5333.00	−36.17	−23.19	−25.35	−25.27
		YT2	K	5387.00–5390.00	−37.28	−23.03	−26.04	−24.62
		YT5	E	5310.00–5315.00	−37.87	−23.86	−26.52	−27.11
		YT5	E	5294.38–5321.00	−38.73	−24.10	−24.29	−25.18
	Yingmai-7	YM7	E	4583.00–4600.0.	−33.53	−22.05	−23.93	
		YM7	E	4690.00–4700.00	−33.38	−21.89	−20.84	−22.23
		YM701	K	4692.00	−33.52	−22.48	−20.69	
		YM9	K	4945.00–4949.00	−34.14	−25.00	−26.37	−27.42
		YM9	E	4683.00–4801.00	−33.95	−21.50	−20.29	−22.34
	Yaha	YH2	K	5194.00–5196.00	−38.68	−23.47	−22.09	−23.5
		YH3	N1j	4980.00–4983.00	−38.74	−24.69	−21.26	−22.28
		YH4	N1j	4997.00–5001.00	−34.09	−23.50	−21.18	−21.72
		YH401	E	5067–5075	−35.03	−23.51	−22.48	−23.30
		YH5	N1j	5031.23–5103.00	−37.48	−23.68	−21.98	−21.8
		YH5	N1j	5089.50–5110.00	−36.58	−22.77	−24.82	−24.24
		YH6	E	5160.00–5163.00	−36.75	−23.88	−22.58	−21.8
		YH701	E	5203.00–5206.00	−35.12	−22.10	−23.57	−22.76
	Erbatai	TAI2	Nij	5041.00–5045.00	−34.19	−21.44	−18.42	−20.21
		TAI2	E	5097.46–5175.00	−38.76	−22.67	−20.12	−20.64
		TAI2	K	5180.00–5183.00	−39.27	−22.18	−24.43	−22.82
	Tiergen	T1	N1j	4836.50–4847.00	−35.51	−23.74	−21.25	−21.98
		T3	K	5261.00–5264.00	−36.79	−23.12	−25.24	−23.87
		T3	K	5321.00–5320.00	−36.32	−21.81	−24.68	−22.37

the type of organic matter for the Jurassic and Triassic source rocks is type III_1–III_2 and type III_1–II, respectively (Zhao Menjun et al., 2002a). Undoubtedly, the natural gas in the Kuqa depression is mainly generated by these two kinds of source rocks.

According to the research results by Xu Yongchang (1994) and Dai Jinxin (2000), the $\delta^{13}C_1$ of China coal-derived gases is −32∼ −38‰ and the $\delta^{13}C_2$ is −22∼ −28‰. Therefore, the natural gas in the Kuqa foreland basin is humic (Zhao Mengjun & Zhang Shuichang, 2001) and relevant to coal measures source rocks. The carbon isotope of its heavier hydrocarbons indicates that the natural gas in the Kelasu and Yiqikelike structural belts is related with coal and carbonaceous mudstones (Li Jian et al., 2001). It is considered that natural gases of the Kelasu and Yiqikelike structural belts are characteristic of very low content of heavier hydrocarbons and of late-stage gas accumulation, especially for that of the Kelasu structural belt, which is very dry and contains almost no C_{2+} hydrocarbons. Thus, the carbon isotope of heavier hydrocarbons in these gases represents only the contribution of the

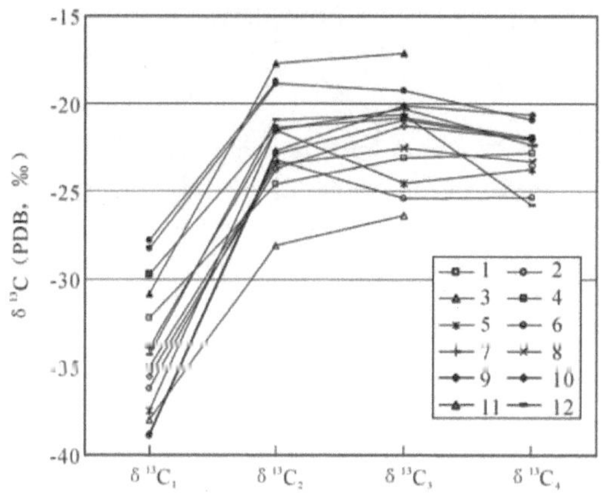

Figure 2. Distribution curve of carbon isotopes of natural gases in the Kuqa foreland basin 1 – Dabei-1 gas field; 2 – Kela-2 gas field; 3 – Kela-3 gas field; 4 – Yinan-2 gas field; 5 – Yudong-2 condensate gas field; 6 – Yangtake condensate gas field; 7 – Yingmai-7 condensate gas field; 8 – Yaha condensate gas field; 9 – Tiergen condensate gas field; 10 – Erbatai; 11 – Yilake condensate gas field; 12 – Dina-2 condensate gas field.

relatively low-maturity Jurassic coal and carbonaceous mudstone. Zhao Menjun et al. (2002b) suggested that the natural gas in the Kuqa depression is mainly generated from the Triassic and Jurassic source rocks and predominated by the Jurassic source rather than the Triassic one.

3.2 *Accumulation process of natural gases in the Kuqa foreland basin*

The geological condition of the Kuqa depression is suitable for two-stage accumulation. In the late Cretaceous-end of Paleocene, structural traps formed in the setting of compression and they mainly captured hydrocarbons generated by the Triassic source rock. After Miocene, the Kuqa rejuvenated foreland basin began to develop with strong tectonic deformation. The traps formed in this period mainly captured the natural gas generating in the late stage. The oil-gas and source rock correlation, inclusions and reservoir bitumen all support the two-stage accumulation model (Wang Zhaoming et al., 2002; Zhao Mengjun et al., 2002a, b, c; 2003a, b). In the Kuqa foreland basin, the hydrocarbon accumulation in the early stage had been destroyed and the later generated hydrocarbon then accumulated and redistributed.

3.2.1 *Accumulation and destruction of hydrocarbons in the early stage*

The early hydrocarbon accumulated mainly by lateral migration. In this stage, the Triassic source rock was at the oil-generative peak and the tectonic stress was less with a few initial structures in the front of the southern Tianshan. So the hydrocarbon generated by the Triassic source rock migrated northward for a long distance and accumulated in the early structure, forming the Kumugeliemu, Misibulake and Heiyingshan oilfields. Meanwhile, the hydrocarbon migrating southward charged the buried-hill reservoirs of the Yaha and Yingmaili and the overlying Cretaceous and Jurassic reservoirs (Figure 3). In this stage, hydrocarbons mainly accumulated in the Cretaceous and Paleocene reservoirs. Due to the early Himalayan movement, structural areas on the edge of the basin uplifted and faults occurred with the permeance of surface water, the reservoirs formed in this stage were, to some extent, degraded and destructed.

3.2.2 *Late accumulation and redistribution of natural gases*

The late hydrocarbon accumulated by both lateral and vertical migration. The hydrocarbon generated by the Triassic and Jurassic source rock with higher maturity migrated in the form of gas phase. In this stage, the depression movement in front mountain was strong. The strong lateral compression uplifted the underlying strata and formed a series of overthrust anticlines, dragging anticlines and decollement folds. Hydrocarbons generated in this stage accumulated mainly in the following two ways (Figure 3).

a. Hydrocarbons migrated southward for a long distance and accumulated in the frontier upheaval zone (Yudong-2, Yangtake, Tiergen and Tai-2). In the same time, hydrocarbons mixed with the early-generated lacustrine oil and formed a series of condensate reservoirs in the Luntai fault-uplift (Yingmai-7, Hongqi and Yaha) and lithologic traps in the slope zone.
b. Hydrocarbons migrated vertically and formed reservoirs in the same strata of source rocks (Yinan 2, J), the near strata (Kela 2, 3, E) or the far strata (Dawanqi, N, Q). The reservoirs in the same strata of source rocks and the near strata are mainly condensate reservoirs and dry-gas reservoirs. The reservoirs with a shallower depth were destructed to become oil seepage or, under the condition of lower pressure and worse caprocks natural gases strongly diffused and the reservoirs were differentiated and formed oilfields (Dawanqi and Yiqikelike).

3.3 *Control of accumulation process on components and carbon isotopes of natural gases*

The accumulation model of hydrocarbons in the Kuqa foreland basin controls the hydrocarbon occurrence in this area (Zhao Mengjun et al., 2003a), furthermore, the accumulation process of hydrocarbons obviously influences components and carbon isotopes of natural gases. Galimov (1988) and Schoell (1988) suggested

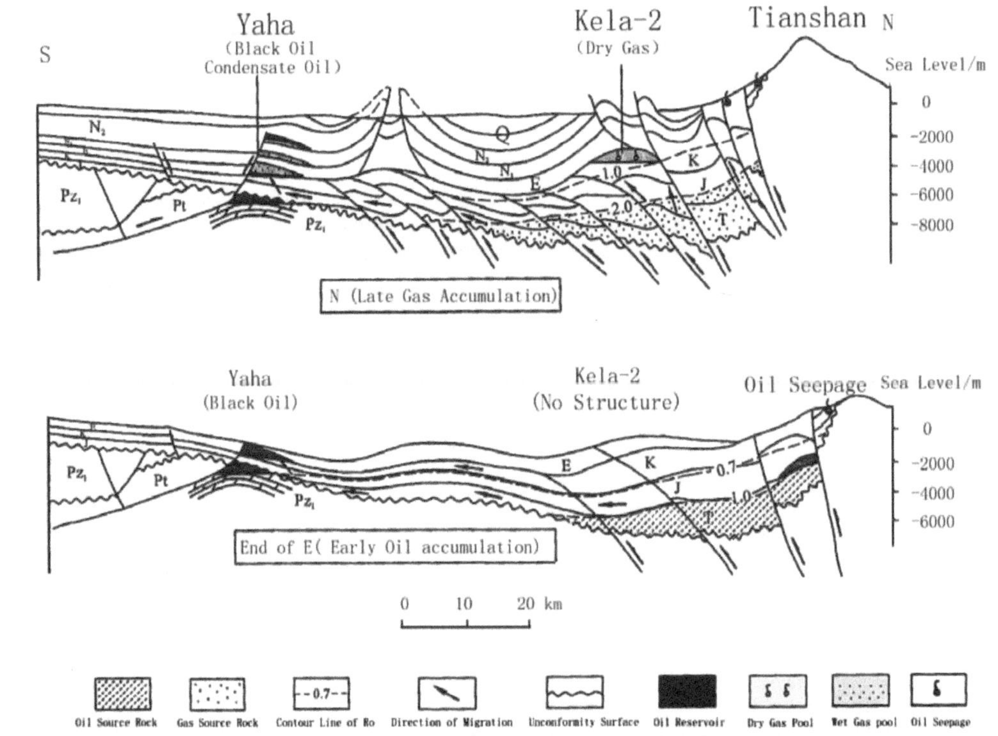

Figure 3. Model for migration and accumulation of hydrocarbons in the Kuqa depression (after Zhao Menjun et al., 2003a).

that the carbon isotope of methane is determined not only by the characters of source matter but also by the accumulation history, namely whether methane is successively trapped in reservoirs or just accumulate in a short period.

Rooney et al. (1995) established the carbon isotope model for successive accumulation and momentary accumulation of methane. Krooss et al. (1995) explained the reasons for the high nitrogen content in some gas fields of north Germany basins using "the mechanism of methane and nitrogen being generated and accumulating in different stages". Obviously, the accumulation process of natural gases greatly influences components and carbon isotopes of natural gas pools. Here we are taking the Kela-2 gas field (the Kelasu structural belt) and Yinan-2 gas field (the Yiqikelike structural belt) in the overthrust belt as well as the Yaha condensate gas field in the frontier upheaval zone as examples to discuss relations between accumulation process and characters of natural gases in the Kuqa foreland basin.

3.3.1 *Kela-2 gas field*

The content of methane in the Kela-2 gas field is 96.9% ~ 98.22% while the content of ethane is very low, ranging from 0.31% to 0.53%, with almost no other heavier hydrocarbons. Therefore, the drying coefficient is almost 1.0. The $\delta\ ^{13}C_1$ and $\delta\ ^{13}C_2$ are heavy, varying in −27.3~−31.1‰ and −16.8~−19.4‰, respectively. According to the $\delta\ ^{13}C_1$ ~ Ro equation by Dai Jinxin, we calculated the maturity of the natural gases. The calculated Ro varies in 2.90%~3.85%, not accordant with the result obtained from the outcrop sample, burial history and thermal history. In the Baicheng sag, the current Ro of the Jurassic and Triassic source rocks is 1.2~2.3% and 0.8~2.6%, respectively (Zhao Menjun, 2002a).

According to the accumulation history of the Kela-2 gas field (Wang Zhaoming et al., 2002; Zhao Mengjun et al., 2002b; 2003a), the hydrocarbon accumulating in the Kela 2 gas field is generated by the Triassic and Jurassic source rocks with Ro above 1.3% and 0.9%, respectively. Because the reservoir for hydrocarbons generated in the early stage used to be destructed, only a part of oils generated by the Triassic source rock accumulates. According to the hydrocarbon-generation model for the type III natural gas, the hydrocarbon generated by the source rock with Ro below 0.9% amounts to 23% of the total generated hydrocarbons while that with Ro above 1.3 accounts for 52% of the total generated hydrocarbons. According to the equation by Dai Jinxin (Dai Jinxin,

2000), $\delta\ ^{13}C_1$ of the natural gas generated by the source rock with Ro of 0.9% and 1.3% is −35.05‰ and −32.79‰, respectively. Therefore, as the result of the dissipation of the early-generated natural gas with lighter carbon isotopes, the carbon isotope of the natural gas accumulating in the late stage could be heavier. Shuai Yanhua (2003), based on the dynamics of methane carbon isotopes, drew the same conclusion.

3.3.2 *Yinan-2 gas field*

The natural gas in the Yana-2 gas field is wet with the methane content of 82.78 ~ 94.41% and C_{2+} content of 2.15 ~ 10.02%. The ratio of C_1/C_{1-5} is 0.92. The $\delta\ ^{13}C_1$ and $\delta\ ^{13}C_2$ are −34.8‰ and −22.4‰, respectively. Compared with the natural gas in the Kela-2 gas field, the natural gas in the Yinan-2 gas field is relatively wet and its carbon isotope is lighter.

Similar to the Kela-2 gas field, the accumulation of hydrocarbons in the Yinan-2 gas field has two stages, namely the destruction of the hydrocarbon generated in the early Himalayan epoch, the secondary accumulation of hydrocarbons in the late Himalayan epoch and the redistribution of the end Himalayan epoch. Because the evolution of source rocks in this area and the Kelasu structural belt is different, the destruction for the early trapped hydrocarbon is relatively weak.

In the end of Paleogene, the Ro value for the Triassic source rock in the Yinan area was 0.8%, however, because in this stage the maturity of the Triassic source rock was higher than that of the Jurassic one, the trapped hydrocarbon was mainly generated by the Triassic source rock. The hydrocarbon accumulating in this stage was then destructed and degraded by the early Himalayan movement. In the middle-end of the Himalayan epoch, with the deposition of overlying strata, the burial depth of the Jurassic source rock increased, the property of the caprock became better and the hydrocarbon effectively accumulated in traps.

In this stage, the accumulated hydrocarbon was mainly generated by the Triassic and Jurassic source rock with Ro above 0.8% and 0.6%, respectively, which is the most important reason why the natural gas of the Yinan-2 gas field is relatively wet and its carbon isotopes are lighter than that of the Kela 2 gas field.

3.3.3 *Yaha Condensate gas field*

The Yaha gas field has mainly trapped condensate gases with the methane content of 66.47 ~ 90.63% and the ethane content of 1.33 ~ 24.43%. The carbon isotope of methane and ethane is −34.09 ~ −38.68% and −22.8 ~ −24.69‰ with the average of −36.6‰ and −23.45‰, respectively.

The accumulation of hydrocarbons in the Yaha gas field has two stages, but because the maturity of source rock in the slope zone was low (Ro below 0.6% and 0.4% for the Triassic and Jurassic source rocks, respectively) in the early Himalayan epoch, the hydrocarbon accumulating in this stage was mainly generated by the deep source rock of the sag. Since the late Himalayan epoch the maturity of the source rock has been moderate to high. The hydrocarbon accumulating in the Yaha structural belt is condensate gas and successively generated by the humic source rock with higher maturity in the slope zone in the late Himalayan epoch. Therefore, the successive accumulation is featured for the hydrocarbon in the Yaha structural belt.

3.3.4 *Features of the hydrocarbon accumulation in the Kuqa foreland basin*

According to Table 3, the accumulation process of hydrocarbons in different structural belts of the Kuqa foreland basin obviously controls the components and carbon isotopes of natural gases. Hydrocarbons in the Kuqa foreland basin accumulate mainly in two stages, namely the early stage (the end of Cretaceous to Paleogene) and the late stage (after Neogene, N_2-Q). Because the maturity of source rocks is lower in the east part of the Kuqa depression, the Yinan-2 gas field mainly traps the natural gas generated by the Jurassic source rock with Ro above 0.6% and the Triassic source rock with Ro above 0.8%. In this area the feature of the late accumulation for hydrocarbons was not obvious. But, the late accumulation of hydrocarbons features largely in the Kela-2 and Dabei-1 gas fields. In the slope and sag zones, the feature of late accumulation is not obvious. Because in the early stage, the source rock in the slope zone was lower in maturity and the evolution of the source rock controlled by the late Himalayan movement mainly occurred in low maturity. The late accumulation of hydrocarbons is almost not found in the frontier upheaval zone and slope zone while the successive accumulation of hydrocarbons from low maturity to medium maturity is the main features in the two areas.

4 CONCLUSIONS

(1) The hydrocarbon content in the Yinan-2, Kela-2 and Dabei-2 gas fields (pools) is higher than those in the Dina-2 and Wucai-1 gas fields of the slope belt and the Yingmai-7, Yudong-2, Yaha and Tiergen gas fields of the frontier upheaval zone. The distribution features of methane content and drying coefficient are similar. Furthermore, the component carbon isotope in the overthrust belt of the Kuqa foreland basin is obviously heavier than those in the slope belt and frontier upheaval zone.

(2) The natural gas in the Kuqa foreland basin is of mainly humic type and generated mainly by both

Table 3. Accumulation features of natural gases in different structural belts of the Kuqa foreland basin.

Structural belt	Major gas field	Component of natural gas	Carbon isotope of natural gas	Features of accumulation
Thrust belt	Kela-2	C_1: 93.5~98.3 C_1/C_{1-5}: 0.99~1.0	$\delta^{13}C_1$: −26.2~−28.2 $\delta^{13}C_2$: −16.6~−19.4	Natural gas generated by T and J source rock respectively with Ro of 1.3 and 0.9, obvious accumulation of natural gas in late stage
	Dabei-1	C_1: 92.2~96.2 C_1/C_{1-5}: 0.95~0.97	$\delta^{13}C_1$: −29.7~−30.8 $\delta^{13}C_2$: −21.4~−21.5	Natural gas generated by T and J source rock respectively with Ro of 1.0 and 0.7, accumulation of natural gas in late stage
	Yinan-2	C_1: 88.0~94.4 C_1/C_{1-5}: 0.90~0.95	$\delta^{13}C_1$: −32.8~−34.8 $\delta^{13}C_2$: −22.4~−24.6	Natural gas generated by T and J source rock respectively with Ro of 0.8 and 0.6, no obvious accumulation of natural gas in late stage
Sag and slope	Dina-2	C_1: 87.0~89.4 C_1/C_{1-5}: 0.89~0.91	$\delta^{13}C_1$: −34.3 $\delta^{13}C_2$: −20.9	Natural gas generated by T and J source rock respectively with Ro of 0.8 and 0.6, no obvious accumulation of natural gas in late stage
	Wucan-1	C_1: 82.0~83.2 C_1/C_{1-5}: 0.84~0.85	$\delta^{13}C_1$: −38 $\delta^{13}C_2$: −28.05	Natural gas generated by T and J J source rock respectively with Ro of 0.6 and 0.4, successive accumulation of natural gas in late stage
Forebulge	Yudong-2	C_1: 82.3~83.5 C_1/C_{1-5}: 0.87~0.88	$\delta^{13}C_1$: −34.8 $\delta^{13}C_2$: −22.4	Natural gas generated by T and J source rock respectively with Ro of 0.6 and 0.4, successive accumulation of natural gas in late stage
	Yangtake	C_1: 75.4~ 92.4 C_1/C_{1-5}: 0.85~0.97	$\delta^{13}C_1$: −36.2~−38.9 $\delta^{13}C_2$: −22.9~−24.1	
	Yingmai-7	C_1: 77.8~91.5 C_1/C_{1-5}: 0.78~0.98	$\delta^{13}C_1$: −33.1~−37.0 $\delta^{13}C_2$: −21.3~−23.2	
	Yaha	C_1: 76.5~90.6 C_1/C_{1-5}: 0.73~0.98	$\delta^{13}C_1$: −34.1~−37.5 $\delta^{13}C_2$: −22.1~−23.9	
	Tiergen	C_1: 72.3~87.5 C_1/C_{1-5}: 0.76~0.93	$\delta^{13}C_1$: −35.2~−39.0 $\delta^{13}C_2$: −21.8~−23.7	
	Yingmai-4,-10	Viscous crude and oil sand		Oil generated by T source rock, no accumulation of natural gas in late stage

the Triassic and the Jurassic source rocks with the predominance of the latter.

(3) The reservoirs in the Kuqa foreland basin used to be destructed in the early stage and then hydrocarbons accumulated again and redistributed in the late stage.

(4) The development of the Kuqa foreland basin in the late stage controls the evolution of source rocks. Therefore, the accumulation process of hydrocarbons obviously influences the components and carbon isotopes of natural gases in different structural belts of the Kuqa foreland basin. The late gas accumulation features obviously the Kelasu structural belt of the overthrust area and moderately the Yiqikelike structural belt as well. The accumulation characters of natural gases in the sag zone and slope zone in the late stage are obviously different. The successive accumulation from low mature gas to mature gas features the frontier upheaval zone.

ACKNOWLEDGEMENTS

This work was supported by the State Key Technologies R & D Program (Grant No. 96-111-01-03), (Grant No. 2001BA605A-06) and Innovation Funds of CNPC for Young Scientists.

REFERENCES

Dai, J.X., Zhong, N.N., Liu, D.H. (2000). The geological fundament and controlling factors for large-medium coal-derived gas field of China.Beijing: Geological Publishing House,11–23.

Galimov, E.M. (1988). Source and mechanisms of formation of gaseous hydrocarbons in sedimentary rocks, In: Schoell M ed. Origins of methane in the Earth. Chemical Geology (special issue), 71:77–95.

Jia, J.H., Gu, J.Y., Guo, Q.Y., Pi, X.J., Wang, Y.H. and Zhou, Y.S. (2001). Sedimentary facies for Cretaceous reservoir in Kela2 gas field of Tarim basin. Journal of Palaeogeography, 3(3): 67–75.

Jia, C.Z., Gu, J.Y. and Zhang, G.Y. (2002). Geological constraints of giant and medium-sized gas fields in Kuqa Depression. Chinese Science Bulletin. Vol. 47 supp. 47–54.

Jia, J.H. and Gu, J.Y. (2002)Control factors and porosity evolution of high-guality sandstone reservoirs of Kela-2 gas field in Kuqa Depression. Vol. 47 supp. 100–106.

Kroos, B.M., Littke, R., Muller, B., Frielingsdorf, J., Schwochau, K. and Idiz, E.F. (1995). Generetion of nitrogen and methane from sedimentary organic matter: implications on the dynamics of natural gas accumulations. Chemical Geology, 126: 291–318.

Li, J., Xie, Z.Y., Li, Z.S., Luo, X., Hu, G.Y. and Gong, S. (2001). Natural gas and source correlation in Kuqa depression of Tarim basin. Petroleum Exploration and Development, 28(5): 29–41.

Pi, X.J., Xie, H.W., Zhang, C. and Tian, Z.J. (2002). Mechanisms of abnormal overpressure generation in Kuqa foreland thrust belt and their impacts on oil and gas reservoir formation. Vol. 47 supp. 85–93.

Qin, S.F. (1999). The genesis for abnormal natural gas in Kuqa Depression of Tarim Basin. Explorationist, 4(3): 21–23.

Qin, S.F., Jia, C.Z. and Tao, S.Z. (2002). The reservoir characters in Kuqa Depression of Tarim Basin. Chinese Geology, 29(1): 104–108.

Rooney, M.A., Cloypool, G.E. and Chung, H.M. (1995). Modeling thermogenic gas generation using carbon isotope ratios of natural gas hydrocarbons. Chemical Geology, 126: 219–232.

Schoell, M. (1988). Mutiple origins of methane in the Earth, In: Schoell M ed. Origins of methane in the Earth. Chemical Geology(special issue), 71:1–10.

Shuai, Y.H., Zhou, Y.R. and Pen, P.A. (2003). The dynamics and accumulation of methane carbon isotope of coal-derived gas in Kuqa Depression of Tarim Basin, Geochemistry, 32(5): 469–475.

Song, Y., Jia, C.Z, Zhao, M.J. and Tian, Z.J. (2002). Controlling factors forlarge gas field formation in thrust belt of Kuqa coal derived hydrocarbon foreland basin. Chinese Science Bulletin. Vol. 47 supp. 55–61.

Tian, Z.J., Zhang, G.Y., Zhou, H.Y., Li, Q.M. and Pi, X.J. (2001). Factor controlling accumulation and accumulation model in the Kuqa petroleum system of Tarim basin. Petroleum Exploration and Development, 28(5): 12–16.

Tian, Z.J., Hu, J.Y., Song, J.G. and Luo, Z.L. (2002). Kuqa foreland basin and its exploration in Tarim basin. Scientia Geologica Sinica, Vol. 37 supp. 105–112.

Wang, Z.H. (2001). The accumulation regularity for the reservoir of Kuqa depression in Tarim basin. Xinjiang Petroleum Geology, 22(3): 189–191.

Wang, H.J. and Hu, J.Y. (2002). The Cretaceous petroleum system and formation of high pressure reservoir in Kuqa depression. Natural Gas Industry, 22(1): 5–8.

Wang, Z.M., Wang, T.D., Xiao, Z.Y., Xu, Z.M., Li, M. and Lin, F. (2002). Migration and accumulation of natural gas in Kela-2 gas field. Chinese Science Bulletin. Vol. 47 supp. 107–112.

Xu, Y.C. (1994). Genesis theory and its application of Natural gas. Beijing: Science Press, 208–211.

Zen, L.B., Zhou, T.W. and Lv, X.X. (2002). The influence of Himalayan movement on the hydrocarbon accumulation in Kuqa depression. Geosciences-Journal of China University of Geosciences, 27(6): 741–744.

Zhao, M.J., Zhou, X.X. and Lu, S.F. (1999). Tarim Basin-a basin for natural gas richness. Natural gas industry, 19(2): 13–18.

Zhao, M.J. and Zhang, S.C. (2001). The genetic groups of natural gas and conditions forming gas reservoir in Tarim Basin, China. In: Jia C Z, Sennikov N V, eds. Symposium on Petroleum Geology of Tarim Basin in China and Siberian Platform in Russia. Beijing: Petroleum Industry Press, 54–62.

Zhao, M.J., Lu, S.F. and Li, J. (2002a). The geochemical characters and source rock of natural gas in Kuqa petroleum system. Petroleum Exploration and Development, 48(6), 4–7.

Zhao, M.J., Lu, S.F., Wang, T.D. and Li, J. (2002b). Geochemical characteristics and formation process of natural gas, Chinese Science Bulletin. Vol. 47 supp. 113–119.

Zhao, M.J. and Zhang, B.M. (2002c). The source rock condition for the formation of large gas field in Kuqa foreland basin. Scientia Geologica Sinica, Vol. 37 supp. 35–44.

Zhao, M.J. and Lu, S.F. (2003a). Hydrocarbon accumulation in two stages and its influence on hydrocarbon occurrence of Kuqa depression. Acta Petrolei Sinica. 24(5): 16–25.

Zhao Mengjun, Song Yan, Liu Shaobo & Qin Shengfei. (2003b). The diffusion influence on gas pool: Dawanqi oilfield as an example. Natural Gas Geoscience, 14(5): 393–397 (in Chinese with English abstract).

Zhou, X.X. (2000). The characters of Tertiary caprock and its control for hydrocarbon accumulation in Kuqa depression. Journal of Palaeogeography, 2(4):51–57.

Petroleum Geochemistry and Exploration in the Afro-Asian Region –
Digang Liang, Darui Wang & Zhenxi Li (eds)
© 2008 Taylor & Francis Group, London, ISBN 978-0-415-44083-7

Biomarker indications as evidence for organofacies variations within the Abu Gabra formation of the Muglad basin, Sudan

O. Ibrahim Mohamed Fadul & A. Mahgoub Mohamed
Greater Nile Petroleum Operating Company (GNPOC), Khartoum, Sudan

ABSTRACT: The Lower Cretaceous Abu Gabra formation is the main source rock of the Muglad basin oils. It has been previously believed that this source rock consists of a single organofacies. This study reveals that at least two different organofacies units exist. A detailed biomarker study of a suite of 57 oil samples was carried out using Gas chromatographic (GC) and Gas Chromatographic/Mass Spectrometric (GCMS) data. Two main oil families ("Family-A" and "Family-B") have been identified based on this. Distinction between these was solely based on variations in Pristane/Phytane (Pr/Ph) ratios, Diahopane indices, Diasterane/Regular sterane ratios, and Regular steranes distributions. "Family-A" being basically restricted to oils from Heglig, Unity, and Bamboo fields, tend to show relatively low (Pr/Ph) ratios in the range of (1.3 to 1.9), low Diahopane indices (4.1–9.6), low diasterane/regular sterane ratio (0.34–0.60), and relatively high C_{28} Regular sterane content (24–32). In contrast, "Family-B" that is typified by the remaining oils, displays a relatively high (Pr/Ph) ratios (1.23 to 2.27), high Diahopane indices (10.8–32.0), high diasterane/regular sterane ratio (0.52–1.30), and relatively low C_{28} Regular sterane content (23–28). The facies unit that generated "Family-A" oils is more algal (high C_{28} Steranes), and was deposited in a more closed, less oxic high-stand system (low Pr/Ph, Diahopane index, and Diasterane/regular sterane) typical of early rift phase lacustrine environments. "Family-B" oil seems to have originated from a more terriginous suite (higher Pr/Ph, low C_{28} Regular Sterane), and was deposited in a more open, oxic-sub-oxic, shallow-stand system (high Diahopane index, diasterane/regular sterane), that is a late rift phase environment.

EXECUTIVE SUMMARY

It's strongly believed that the gradual breakup of the Gondwanaland, and hence the separation of South America from Africa, has resulted in a series of intra-continental rifting systems within the West and Central Africa (Guiroud & Bellion, 1995). These strong and extensive rifts occurred synchroneaously with the opening of the Central and South Atlantic Oceans (Guiroud & Murin, 1992; Binks & Fairhead, 1992).

Sudanese rift basins are considered as part of the Central African rifts system. Muglad basin is perhaps the most important of these, extending up to 800 Km long, 200 Km wide Fig. 1, accommodating sediments that locally attain thickness up to 15 Km in the deepest parts of the basin, and is connected with Anza rift in North Kenya (Bosworth, 1992).

Similar to all Sudanese rift basins, it comprises sediments ranging in age from Early Cretaceous/ Jurassic? to Recent that rest uncomformably on the Pre-Cambrian basement complex.

Muglad basin's known petroleum system elements constitute of: i) the organic rich Early Syn-rift, (Neocomian-Barremian) shale-mudstone succession of the Abu Gabra formation as major effective source

Figure 1. Location map of the study area showing GNPOC concession (1, 2, & 4) plus CNPC's firm (Blk-6) of the Muglad basin, Sudan.

rock [other potential source rocks include Baraka formation (Maastrichtian) and the shales of Nayil (Eocene-Oligocene)], ii) the (Aptian-Cenomonian) Bentiu Sand stone formation plus sands of the Darfur group represent the major reservoirs [others include: Intra-Abu Gabra sands], and iii) the Aradeiba shales (Santonian) as main top seal (Fig. 2).

1 PREVIOUS WORK

Seven oil families were tentatively identified in the Muglad basin based on GC, GCMS, and stable carbon isotopes data (Waples, D.W, 1997). In his report, Waples mentioned that only three true families could actually exist in his study suite. Updating his 1997's report, Waples concluded that there are two oil families in this basin that he referred to as: Family AB and a questionable family from Kaikang-1 well. He further subdivided Family AB into two subfamilies vis; AB1 & AB2 which (according to Moffat, 2000), do not seem to show any geographic or stratigraphic consistency, nor any linkage to the timing of generation or level of biodegradation (Moffat, I.W, 2000).

The Abu Gabra source rock of the Muglad basin has been earlier divided into three subunits (lower, middle, & upper) based on lithofacies patterns and geochemical parameters (Ahmed, Y. M, 1996). According to Ahmed, the lower and upper units represent the same facies, which is characterized by light gray, massive mudstone, common type III organic matter, low TOC (0.3–1.1 weight %), and relatively high oxygen index (avg. 90 mg HC/g organic carbon). While the middle unit consists of dark brown laminated shales, sapropellic amorphous organic matter, high TOC (0.7–6.0), and relatively low oxygen index (avg. 50 mg HC/g organic carbon). He described the depositional environment for the lower and upper facies as shallow, low stand, oxygenated, agitated lake-marginal deltaic environment. Whereas the middle subunit is deposited in high-stand anoxic water/sediment interface, quiet deep-water environment.

A palynofacies and organic geochemical analysis of cuttings samples from eight wells across the Muglad basin concluded at suggesting two organic facies within the Abu Gabra; a lower interval with higher TOC (1.5–2.3 weight %), higher HI (338–546), higher amorphous contents (>80%) and a heavier δ^{13} C_{TOC} value (>−27‰); and an upper, less rich unit (mean TOC 1.4%), mean HI (83), δ^{13} C_{TOC} −28‰. (Mustafa, A.A. & Tyson, R.V., 2002)

2 SCOPE & OBJECTIVES OF THIS WORK

A suite of 57 oil samples from blocks 1, 2, & 4 that are operated by the Greater Nile Petroleum Operating Company (GNPOC) plus 2 additional ones from the block 6 of China National Petroleum Corporation (CNPC) were incorporated in this study (fig. 1). The study is aiming at classifying the oils discovered so far in the Muglad basin through detailed biomarker study and to attempt giving reasonable explanations for any variations that might exist. Gas chromatographic (GC) and Gas Chromatographic Mass Spectrometric (GCMS) data analyzed for both operating companies by the Chinese Research Institute of Petroleum Exploration & Development (RIPED) were used.

3 EXPERIMENTAL PROCEDURES

Gas Chromatography (GC) was performed at RIPED using Hewlett-Packard 5890 instrument (100–325°C°C at 4°C/min, DB5 column, 0.32 mm I.D., 30 m length, & FID detector).

Similarly, Gas chromatography-mass spectrometry (GC/MS) was performed using Fissions Modal MD-800 quadruple system interfaced with a Carlo-Erba Modal 800 gas chromatography. The GC/MS was fitted with a 60 m DB5ms (J & W) fused silica capillary column, 0.25 mm I.D, 0.25 μm film thick, 100°C held 10 minutes, then programmed at 2°C/min to 320°C. Helium carrier gas, head pressure 20 psi, split ratio 30:1, injector 300°C. Transfer line temperature 300°C, EI source 200°C, electron energy 70 V, MID scan mode was cycled at 2 seconds. Identification of the major biomarker families was carried out by comparison with published retention characteristics and by co-elution with reference compounds. Corrected peak areas obtained by baseline subtraction were used to derive biomarker ratios.

4 RESULTS & DISCUSSIONS

4.1 *Biomarker indications*

Biomarkers are defined as "Organic compounds whose carbon structure, or skeleton, is formed by living organisms and is sufficiently stable to be recognized in crude oil or in the organic matter of ancient sediments" (Hunt, J. M., 1996). As such, biomarkers have long been used successfully in the oil industry for hydrocarbons characterization and in correlation studies. The biomarker groups that are most commonly used for this are the Normal Alkanes, Steranes, and Triterpanes groups. Used together, Steranes & Triterpanes can provide most of the information on the "organic facies" system (Waples, D.W. & Machihara, T., 1991).

As previously stated in the above text, this biomarker study is solely based on GC and GCMS data of the saturate fractions of the oils. Both data sets (as will be seen later in this discussion) proved to be highly consistent in characterizing the various organic aspects of the oils under study.

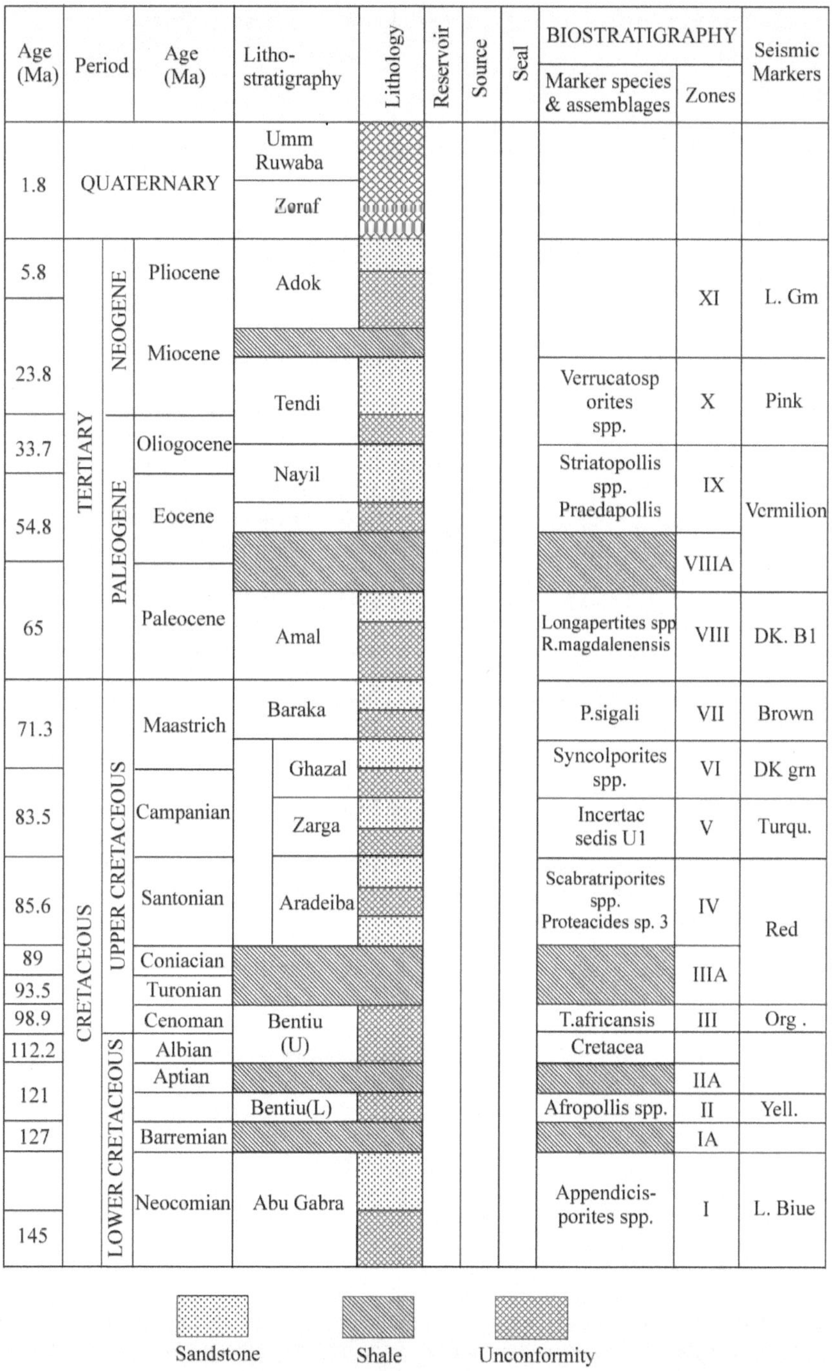

Figure 2. Regional Chrono-stratigraphic chart from Palynological data across Muglad Basin with range of ages for Litho-stratigraphical limits and gap of deposition (after Mohamed Zayed Awad).

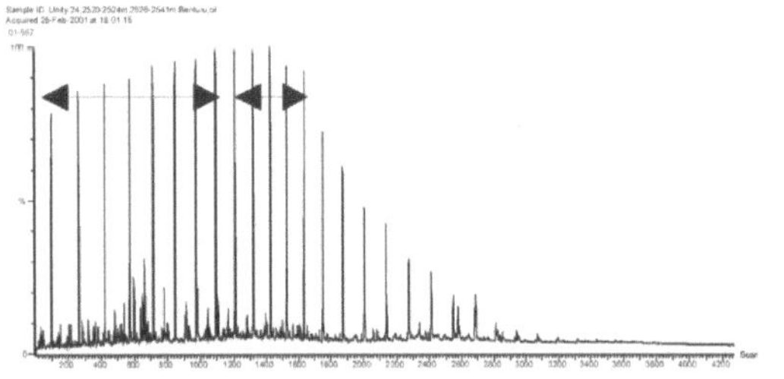

Figure 3. TIC trace of a representative of family-A oils (Unity-24) showing abundant C_{15}–C_{17} n-alkanes compared to C_{19} and higher members.

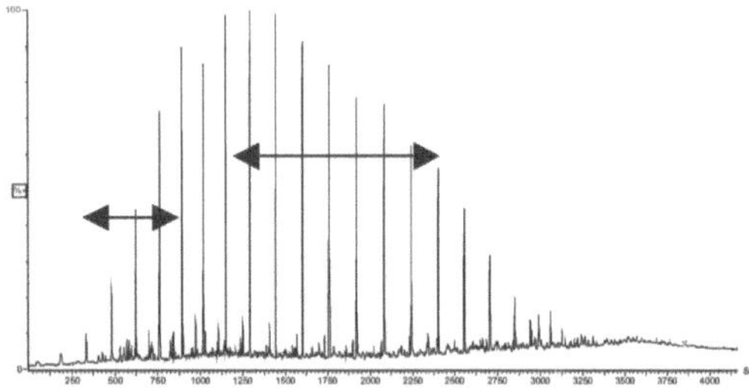

Figure 5. TIC trace of a family-B representative (Shelungo-1) showing abundant (C_{19} & higher) n-alkanes compared to (C_{15}–C_{17}) members.

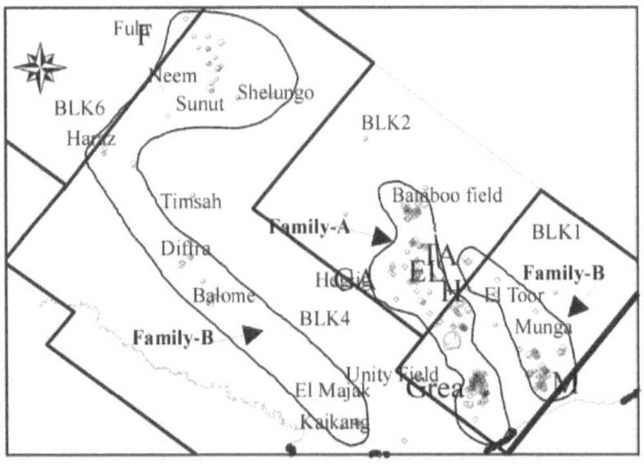

Figure 4. Geographical distribution of family-A and family-B oils in the Muglad basin (Blocks 1, 2, 4, and 6).

4.2 *Gas Chromatography (GC)*

4.2.1 *N-alkanes*

Investigations of the n-alkane distributions of oils from the Muglad basin lead to the recognition of two distinct GC trace patterns. One represented by what is here referred to as Family-A, shows n-alkane distribution with a clear C_{15}–C_{17} predominance (fig. 3).

This kind of distribution, seen mainly in the oils of Unity, Heglig, & Bamboo fields (fig. 4), indicates an origin from bacteria (both photosynthetic and non-photosynthetic) as well as some species of red algae (Hunt, J. M., 1996). This is a clear indication that the depositional environment for this group is a more closed high-stand system.

The n-alkanes distribution in the second group (i.e. Family-B members) characteristically shows a pattern in which C_{19}–C_{31} predominate (fig. 5).

This reflects a stronger influence from higher organisms e.g. Zooplankton lipids and terrestrial organisms on the organic matter of the source rock (Hunt, J. M., 1996). Oils belonging to this family geographically occur along both eastern & western flanks of the basin (i.e. Greater Munga, El Toor, North block 4, Fula (block 6), Diffra, Balome, and Kaikang fields (fig. 4). These organisms flourish in more open and shallow-stand systems.

4.2.2 *Isoprenoids (Pristane/Phyatne ratio)*

Another GC parameter that supports the results driven from the n-alkanes behavior of the Muglad oils is the Pristane/Phyatne ratio (Pr/Ph). This ratio is mainly used as a measure of oxicity of the depositional environment. Family-A oils show a relatively low Pr/Ph values in the range of (1.3–1.9) indicating a less oxic setting for the depositional environment. Such an environment is a favorable habitat for organisms inferred above by the C_{15}–C_{17} skew of the n-alkanes of family-A (i.e. bacteria and algae). Oils of Family-B are however, relatively higher in their Pr/Ph contents (2.23–2.27), which indicate an increase in the oxicity of the depositional environment (Table 1). This reflects the degree of the openness of the system and therefore, its greater chance to receive the more terriginous material shown clearly by the C_{19}–C_{31} n-alkane predominance. Gas Chromatography Mass Spectrometry.

4.3 *Gas chromatography mass spectrometry*

GCMS is another powerful chromatographic tool that is widely used by the oil industry in the hydrocarbon characterization and correlation studies. For application purposes, different GCMS parameters for different biomarker groups that are sensitive to either organic source or maturity are established and frequently used. Among these, the parameters of the most ubiquitous markers; the Steranes, and Triterpanes are well known. Some of these parameters are more source-indicative than maturity and vise versa.

4.3.1 *Steranes*

Steranes are indicators for photosynthetic biota, both terrestrial and aquatic (Waples, D.W & Machihara, T. 1991). Two Steranes groups were investigated in this study; the Regular and the Rearranged Steranes (i.e. Diasteranes). The distribution of the former in the

Table 1. Showing the biomarker parameters used to classify Muglad oils.

NO.	Well Name	Depth/ Interval (m)	Formation	Pr/Ph	C_{27} %	C_{28} %	C_{29} %	DiaHop. in	Gmc. in	NeoHop. in	Ts/ Tm	Dia. Regst	C_{29} Str 20S/ (20S + 20R)	C_{32} Hop 22S/ (22S + 22R)
						Family-A Group								
1	Toma-1	1825–1829	Bentiu	1.33	39	27	34	4.1	16	13	1.9	0.55	0.41	0.58
2	Tayib-1	2565.5	Bentiu	1.33	35	27	38	4.2	15	13	1.5	0.54	0.38	0.60
3	Unity-24	2600–2621	Bentiu 3	1.47	39	24	38	4.9	17	13	1.4	0.35		
4	Unity-31	2311.5–2320	Bentiu 1	1.50	33	28	39	5.1	19	14	1.3	0.36	0.47	0.59
5	Hamra-1	1665.4–1666.6	Bentiu	1.29	38	27	35	5.2	15	13	1.4	0.58	0.38	0.60
6	Toma-1	1635–1639.5	Bentiu	1.40	39	25	36	5.3	15	14	1.5	0.52	0.40	0.60
7	Unity-24	2520–2524	Bentiu 2	1.64	38	28	35	5.4	18	14	1.3	0.34	0.59	0.64
8	KNE-1	2306–2319.5	Aradeiba B	1.64	37	29	34	5.5	19	14	1.3	0.39	0.58	0.60
9	KNE-1	2246–2254	Aradeiba A	1.64	38	29	33	5.5	18	13	1.4	0.42	0.61	0.60
10	KNE-1	2444.5–2449.5	Aradeiba E	1.45	39	29	33	5.6	19	19	1.4	0.39	0.60	0.60
11	Kanga-1	1954–1957	Bentiu	1.40	33	30	38	5.9	17	14	1.9	0.60	0.32	0.60
12	Unity-30	2650–2662.5	Bentiu 1	1.90	36	28	36	6.2	20	16	1.3	0.45	0.57	0.59
13	Garaad-1	1843–1846.5	Aradeiba B	1.41	37	28	35	7.3	18	15	1.5	0.53	0.52	0.61
14	Garaad-1	1811.5–1816.5	Aradeiba A	1.37	38	29	33	7.3	19	15	1.5	0.55	0.58	0.63
15	Khairat-1	2868–2877	Ghazal	1.59	38	28	35	7.7	21	17	1.8	0.46	0.58	0.63
16	Heglig-7	1643–1651	Bentiu	1.33	36	27	37	8.1	16	15	1.4	0.55	0.39	0.60
17	El Harr-1	2728–2745.5	Bentiu 1	1.83	37	28	35	8.9	20	16	1.7	0.52	0.56	0.63
18	Khairat-1	3130–3142	Aradeiba A	1.60	36	27	36	8.9	22	17	1.8	0.53	0.51	0.63
19	Munga N-1	1525.5	FMT	1.75	41	24	35	9.6	16	19	1.5	0.56	0.57	–
20	El Saqr N-1	2479.5	Bentiu	1.67	42	26	33	9.6	18	18	1.6	0.56	0.59	–
21	Koda-1	1859		1.94	41	25	35	10.8	22	17	1.6	0.63	0.68	0.63
22	Bamboo W-2	1294.5	Bentiu	–	29	25	46	11.4	17	16	1.9	0.70	0.42	0.64
23	Bamboo W-1	1448–1459	Bentiu 2	1.50	38	26	37	11.6	22	18	1.9	0.55	0.60	0.68
24	El Mahafir-1	1125–1130	Nayil	–	37	29	34	11.7	23	19	1.8	0.71	0.66	0.63
25	Umm agura-1	1579–1582		1.68	42	23	35	11.8	17	19	1.6	0.55	0.60	–
26	Umm Batutu E-1	1434.5	FMT	1.62	43	26	31	11.8	19	18	1.7	0.60	0.66	–
27	Bamboo W-1	1233–1237	Aradeiba E	1.71	39	25	36	11.9	19	17	1.8	0.78	0.47	0.61
28	Munga-1	1477.5–1486	Bentiu 3	1.78	35	27	38	12.6	13	20	1.5	0.61	0.43	0.61
29	El Bakh-1	1727–1734	Bentiu 1	1.80	38	27	36	13.6	22	19	1.9	1.02	0.64	0.60
30	El Toor-5	1636	Aradeiba E & F	1.85	36	25	39	13.6	15	24	1.7	0.57	0.38	0.58
31	El Full-1	1586–1596	Bentiu 1	1.74	39	25	36	13.8	21	18	1.8	0.81	0.55	0.61
32	Bamboo W-1	1265–1276	Bentiu 1	–	40	24	36	14.0	20	18	1.8	0.92	0.56	0.61

Continued

Table 1. *Continued*

NO.	Well Name	Depth/ Interval (m)	Formation	Pr/Ph	C_{27} %	C_{28} %	C_{29} %	DiaHop. in	Gmc. in	NeoHop. in	Ts/ Tm	Dia/ Regst	C_{29} Str 20S/ (20S + 20R)	C_{32} Hop 22S/ (22S + 22R)
							Family-B Group							
33	Munga-1	1463–1473	Bentiu 3	1.60	46	22	31	14.4	18	23	1.8	0.53	0.43	0.64
34	Hilba-1	2395–2400	Abu Gabra	1.44	42	27	31	15.1	13	22	1.8	1.44	0.73	0.59
35	Neem-1	1223–1227	Ghazal	1.50	38	25	38	18.5	15	23	2.0	1.43	0.43	0.62
36	Fula-1	1191–1202	Aradeiba	–	39	24	37	16.2	15	24	2.6	1.16	0.46	0.60
37	Munga south-1	1596	DST 4	1.67	38	26	36	16.3	15	23	1.6	0.52	0.69	–
38	Shalungo-1	1232.2	Aradeiba	2.27	38	26	36	22.4	16	24	1.6	1.30	0.57	0.58
39	Haraz-1	1395.5	Nayil	1.4	41	25	34	18.1	16	22	2.2	1.31	0.51	–
40	Azraq S-1	731 SWC	Aradeiba	0.58	36	23	41	18.7	17	22	1.8	0.98	0.68	0.58
41	Barki-1	1661.8–1664	Bentiu	1.43	38	25	37	18.3	17	22	2.3	1.24	0.42	0.59
42	Jamouse-1	1065		1.39	36	23	41	20.7	18	26	1.9	0.80	0.62	0.62
43	Sunut-1	2952–2964	Zarqa	1.63	34	26	40	30.5	18	29	3.0	1.86	0.57	0.61
44	Diffra-1	3144–3147		1.83	40	24	35	25.3	18	25	3.1	0.75	0.56	–
45	Nabaq East-1	1409	Bentiu	0.75	38	23	39	20.7	19	28	2.3	0.82	0.68	0.63
46	Wizeen N-1	2050–2058	DST 2	1.78	42	24	33	21.7	19	25	2.2	0.87	0.64	–
47	Assal-1	2215.5	SWC w/shw	1.80	37	27	36	15.9	20	20	1.4	0.77	0.54	0.57
48	Fula-1	1256.3–1286.9	Bentiu	–	38	24	38	15.9	20	23	2.7	0.97	0.51	0.59
49	Shelungo East-1	983	Zarqa shw	1.32	37	26	37	32.0	23	31	1.5	0.86	0.55	0.56
50	El Nar South-1	2320–2323	Bentiu 2	2.00	38	28	34	24.4	23	23	2.2	1.03	0.46	0.61
51	Diffra West-1	2827	Aradeiba	1.62	36	25	38	15.3	23	22	1.6	0.92	0.70	0.59
52	Hamam-1	2982–2990	Bentiu	1.60	41	24	35	18.4	24	23	2.2	0.95	0.73	–
53	Diffra West-1	1611	Aradeiba	1.44	39	25	36	16.3	24	21	1.9	0.90	0.70	0.57
54	Diffra West-1	3002	Bentiu	1.64	36	26	38	18.8	25	22	1.9	1.00	0.71	0.58
55	Balome-1	2295–2297.5	Baraka	–	37	27	36	21.1	28	23	2.8	1.85	0.60	0.61
56	Kaikang-1	1249.3–1261.2	Tendi	1.23	13	40	47	37.9	28	38	4.3	0.95	0.41	0.50
57	Diffra West-1	2893.5	Aradeiba	1.64	38	24	38	21.8	28	24	2.6	0.89	0.78	0.62
58	Kaikang-3	1245–1249	Tendi II	–	31	31	37	32.3	33	35	3.6	2.33	0.64	0.63

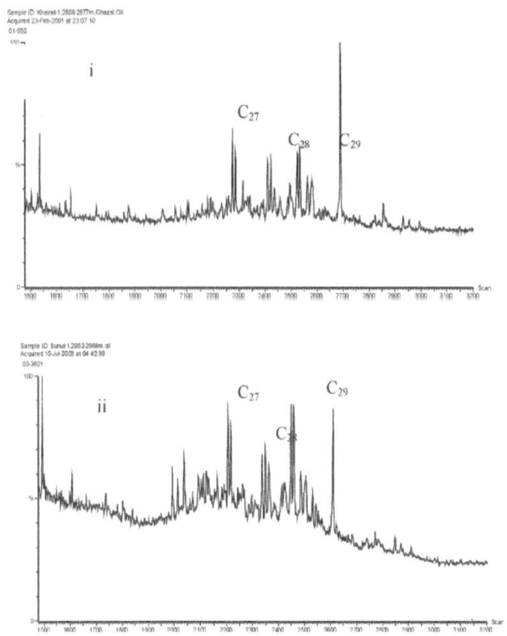

Figure 6. 218 GCMS fragmentograms showing the distribution of Regular Steranes in the two Muglad oil families, i) a family-A member (Khairat-1) & ii) a representative of Family-B (Sunut-1).

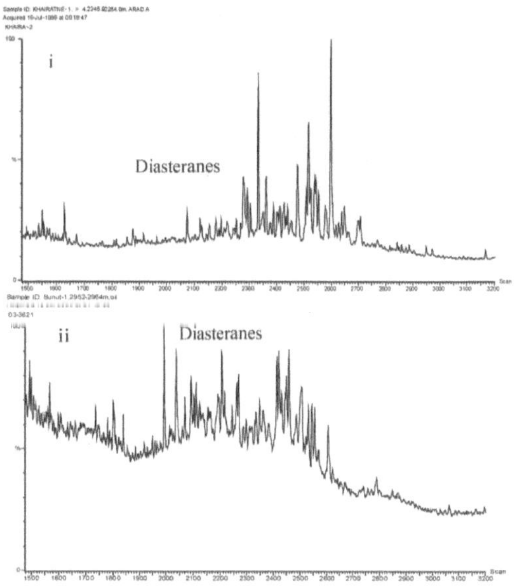

Figure 7. 217 GCMS fragmentograms of the Muglad oils illustrating the distribution of the Diasteranes in the two oil families; i) A representative of Family-A (Khairat NE-1) showing less diasteranes & ii) A candidate from Family-B (Sunut-1) with higher diasterane content.

Muglad oils clearly show two oil families that could be distinguished based on their C_{28} Sterane content (fig. 6 above). Family-A group shows higher C_{28} content (24–32%) compared to (23–28%) for Family-B, which is here interpreted as due to slightly higher lacustrine algae contribution for family-A. This supports the GC results for Family-A as being sourced from a more algal and bacterial source.

(a) Diasterane/regular Sterane ratio:
In general, Diasterane/Sterane ratio is used to distinguish carbonate from clay-rich siliciclastic source rocks. Low ratios in oils indicate anoxic, clay-poor carbonate source rocks, while high ratios are typical of oils derived from clastic source rocks with abundant clay minerals (Peters and Moldowan 1993). The ratio is also found to be maturity dependent, however is not commonly used as maturity indicator in a quantitative sense as it's believed to be influenced by lithology. Oils of Family-A show lower ratios (0.34–0.60) compared to those of Family-B (0.52–1.30) refer to (fig. 7 & Table 1).

This would mean the depositional environment for family-B source is more oxic compared to that of family-A, which is exactly consistent with the conclusion driven above from the Pr/Ph data. Because the source rock for all Muglad basin oils is clastic and clay rich, it is difficult to attribute this variation in the Diasterane/sterane ratio of the two families to any maturity effect since it is affected by this type of lithology.

(b) Sterane maturity:
The 20S/(20S + 20R) Sterane maturity of these oils is highly unreliable as about half of family-A and nearly one third of family-B oils show values that are lower than the equilibrium value of 0.55. Grantham, (1986b) noted the same phenomena in the Far Eastern Tertiary oils that he attributed to some sort of apparent retardation of the isomerization process. This phenomenon may apply to our samples also.

4.3.2 *Triterpanes*
Triterpanes, which are derived mainly from Bacteria, are generally indicators for depositional and diagenetic conditions (Waples & Machihara, 1991). They are mainly classified into Hopanoids (Hopanes, Moretanes, Homo-hopanes etc.) and Non-hopanoids (Gammacerane, Oleanane, Lupane etc.). A group of rearranged hopanes has recently enticed the attention of petroleum geochemists as biological markers that can be applied for geochemical studies of source rocks & oils. Among these, 18^{α}-trisnorneohopane (Ts), 18^{α}-norneohopane (denoted C_{29}Ts) and 17^{α}-diahopane are the most common and rapidly evolving in the literature.

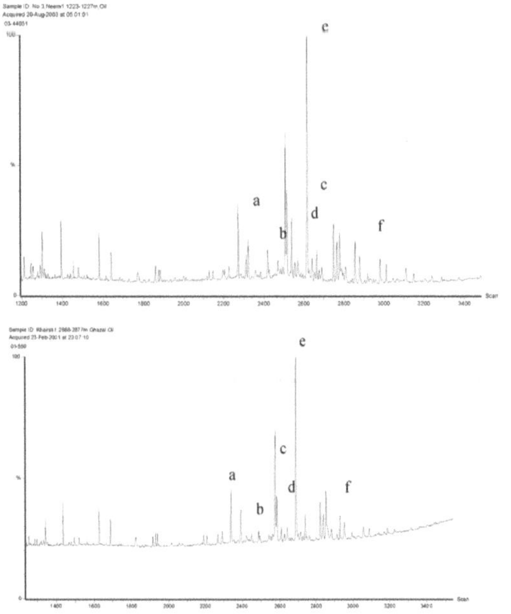

Figure 8. 191 GCMS fragmentograms showing the distribution of the Triterpanes in the two Muglad oil families; an upper Family-B (Khairat NE-1) & a lower Family-A member (Sunut-1). a) Ts, b) Tm, c) C_{29}Ts, d) Diahopane, e) C_{30} Hopane, f) Gammacerane.

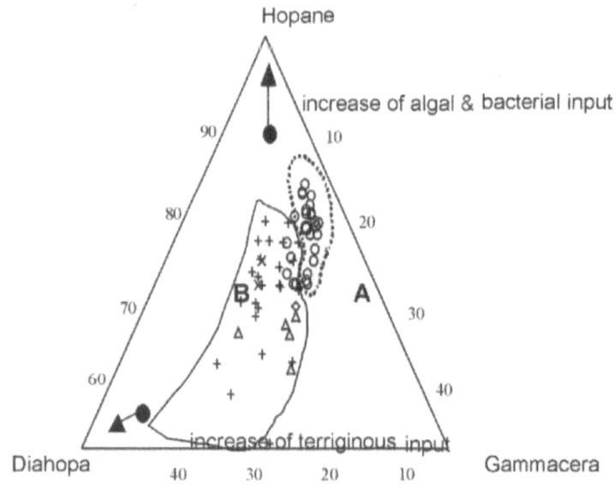

Figure 9. Ternary plot showing Triterpane (hopanes, diahopane, and gammacerane) distribution in the Muglad basin oils of Sudan. Family-A oils are denoted by circles (o), while family-B members are shown in crosses (+). A group of newly discovered family-B oils show noticeably higher gammacerane content (denoted by triangles).

(a)Non-hopanoid Triterpanes:

Gammacerane:

Gammacerane is derived from a tetrahymanol precursor, which is thought to originate in primitive protozoa associated with ciliates that feed on anaerobic bacteria (Damst et al, 1966). Although found in various depositional environments, high concentrations are usually associated with hypersaline lacustrine environments (Moldowan et al, 1985; Hills et al, 1966). Gammacerane indices of the Muglad oils show values that are closely similar in both families (fig. 8 & table 1).

(b)Re-arranged hopanes:

18^{α}-norneohopane (C_{29}Ts) and 17^{α}-diahopane were first identified in a saturate fraction of Brudhoe Bay oil, Alaska using high-resolution two-dimensional magnetic resonance spectroscopy and X-ray crystallography respectively. They are thought to evolve via chemical mechanisms during diagenesis from un-rearranged hopanoid products in rocks with active clays (Moldowan et al. (1991). Killops and Howell (1991) suggested that diahopanes are indicators for bacterial reworking of terriginous higher plant material (Telneas et al. (1992)).

Oils from some newly discovered fields like Diffra, Hamam, and Balome (all belong to family-B), however show relatively higher values (denoted by triangles in (fig. 9).

The geochemical significance of these compounds is not yet obvious, but some workers have used them as maturity indicators (Wenger et al. (1990) and Moldowan et al. (1991)). Wang Chunjiang et al. (2000), concluded that sedimentary organic facies is one of the key controlling factors that govern the relative concentrations of the neohopanes & diahopanes. Some workers argue that they can be used as maturity indicators only on the condition that they represent the same sedimentary organic facies [Wang Chunjiang et al. (2000)].

(i) Diahopane:

Following Moldowan et al, (1991) and Killops & Howell, (1991), this is a diagenetic product of bacterial reworking on terriginous higher plant material with catalysis from active clay particles. From the ternary diagram of (fig.9) above, the two families (A & B) are clearly distinguished from each other based on their 17^{α}-diahopane relative concentrations.

Family-A oils showing lower diahopanes (4.1–9.6), plots far to the upper right corner of the triangle. Oils from family-B in the other hand plot to the lowermost left corner with relatively higher diahopane values (10.8–32.0). This relative diahopane increase in family-B is another support to the results shown earlier by the GC data (n-alkanes & Pr/Ph) that the organic source here is more terriginous. Again, because the source rock for Muglad oils is clastic and clay rich, it is difficult to attribute this variation to any clay content differential of the source. We therefore tend to relate this to the partial oxidation effect due to take place in such sub-oxic environments.

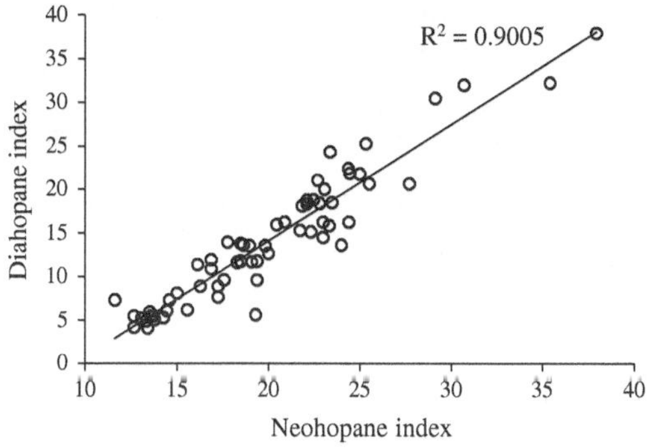

Figure 10. Strong linear relationship between the neohopanes and diahopanes in Muglad basin oils.

(ii) Neohopane:
Like the diahopane, the 18^{α}-norneohopane in the Muglad oils also shows an increase in family-B relative to family-A (fig. 8 & table 1). As mentioned above for the diahopanes, they are most likely products of bacterial reworking of terriginous organic material that are deposited in oxic to sub-oxic environment typical of family-B source. It is evident that these compounds are genetically related judging from their strong linear relationship shown in (fig. 10) above.

(c) Ts/Tm:
Originally proposed by Seifert and Moldowan (1981) as a maturity indicator but Mello et al. (1988) were able to distinguish a lacustrine/saline, marine evaporitic, or marine carbonate depositional environments ($Ts/Tm < 1$) from lacustrine fresh water or marine deltaic ($Ts/Tm > 1$). With the geological background of the basin in mind, the Ts/Tm values of the Muglad oils (>1) therefore confirm a lacustrine fresh water setting for the basin during its Abu Gabra time (Barremian-Neocomian).

(d) Triterpanes maturity:
The 22S/ (22S + 22R) homohopane maturity of these oils is more reliable than that of the steranes. The ratio is calculated from the C32 homohopane. Muglad oils in this study show values that are very close to the equilibrium value of 0.60, which suggests that the 22S/ (22S + 22R) ratio is not affected by the retardation process seen in the steranes.

In fact, the relative concentrations of the neohopanes (Ts & C_{29}Ts), diahopanes, and to a lesser extent, the gammaceranes of the Muglad oils represent the major biological markers upon which this oil classification is based.

5 CONCLUSION

Biomarker indicators proved to be a powerful tool in predicting organo-facies variations within the Muglad basin's main source rock. Based on this study in conjunction with previous studies, the authors have conclusively established that the Abu Gabra source rock of the Muglad basin consists of two organo-facies units. They represent a clear manifestation of the relationship between tectonics and sediment supply in a typical continental rift setting. The first unit responsible for the generation of family-A oils is here interpreted as an early syn-rift, less oxic high-stand sapropellic facies. The second, from which family-B oils originated, is a late phase oxic-sub-oxic low-stand terriginous suite.

ACKNOWLEDGMENTS

We would like to thank top GNPOC management, namely Mr. Kamarshah Harun (GM E & P), Mr. A/Zahir M. A/Zahir (Exploration Manager) and Mr. Khairul Anuar Hussin (Deputy Exploration Manager) for permission to submit the paper. The authors also extend thanks to all who provided advice and help in order to accomplish this work; Dr. Peter Abolins (PCSB), Mr. K.B.Trivedi (ONGC, GNPOC), Mr. Kamil M. Idris (GNPOC) and Mr. H.C.V. Kumar (ONGC, GNPOC).

REFERENCES

Ahmed, Y.M., 1996, Organic matter Characterization and Environmental Control on Organic facies and Lithofacies of Early Cretaceous Lake Assalam, northwest Muglad Basin, Sudan. (A PHD. Thesis).

Binks, R.M., and Fairhead, J.D., 1992, a plate tectonic setting for Mesozoic rifts of west and central Africa, *Tectonophysics*, 213: 141–151.

Bosworth, W., 1992, Mesozoic and Early Tertiary rift tectonics in West Africa, *Tectonophysics*, 213: 115–137.

Chunjiang, W., et al., 2000, Geochemical characteristics and applications of 18α(H)-neohopanes and 17α(H)-diahopanes, *Chinese science bulletin*, v. 45, No. 19, pp. 1742–1748.

Damst J.S.S., et al., 1996, Evidence for Gammacerane as an Indicator of Water Column Stratification, Netherlands Institute for Sea Research and Indiana University.

Grantham, P.J., 1986b, Sterane isomerization and moretane/hopane ratios in crude oils derived from Tertiary source rocks. *Organic Geochemistry*, v. 9, pp. 293–304.

Guiroud, R. and Bellion, Y., 1995, Late Carboniferous to recent, Geodynamic evolution of the West Gondwanian Cratonic, Tethyan margins. In: A.E.M Nairn et al (eds.), *The Tethys Ocean, and Plenum Press, New York*, pp. 101–124.

Guiroud, R. and Maurin, J.C., 1992, Early Cretaceous rifts of Western and Central Africa: an overview, *Techtonophysics*, 213:153–168.

Hills I.R., et al., 1966, An Optically Active Triterpane, Gammacerane, in Green River, CO, Oil Shale Bitumen, *J. Chem. Soc. Chem Commun.*, 752–4.

Hunt, J.M., 1996, Petroleum Geochemistry and Geology, *W.H. Freeman & Company.*

Mello, M.R., et al., 1988a, Geochemical and biological marker assessment of depositional environments using Brazilian offshore oils, *Marine and Petroleum Geology*, 5, pp. 205–223.
Moffat, I.W., 2000, Geochemical and basin modeling study, Central Muglad basin, Sudan, (GNPOC internal report).
Moldowan J.M., Seifert W.K., and Gallegos E.J., 1985, Relationship between Petroleum Composition and Depositional Environment of Petroleum Source Rocks, AAPG Bull., v. 69 No. 8, pp. 1255–1268.
Moldowan, J.M., et al., 1991, Rearranged hopanes in sediments and petroleum, *Geochemica et Cosmochemica Acta*, v. 55, pp. 3333–3353.
Mustafa, A.A., & Tyson, R.V., 2002, Organic facies of Early Cretaceous synrift lacustrine source rocks from Muglad basin, Sudan, Journal of Petroleum Geology, V.25(3), pp.351–366.
Peters, K.E., and J.M., Moldowan, 1993, The biomarker guide: interpreting molecular fossils in petroleum and ancient sediments, *Englewood Cliffs, NJ: Prentice-Hall.*
Seifert, W.K., and Moldowan, J.M., 1981, Paleoreconstrcuction by biological marker, *Geochemica et Cosmochemica Acta*, v.45, pp.783–794.
Telnaes, N., Isaksen, G.H., & Farrimond, P., 1992, unusual triterpane distributions in lacustrine oils, *Organic geochemistry*, v.18, No. 6, pp.785–789.
Waples, D.W., 1997, Oils of the Muglad Basin, Sudan (final GNPOC internal report).
Waples, D.W., and Machihara, T., 1991, Biomarkers for geologists- A Practical Guide to the Application of Steranes and Triterpanes in Petroleum Geology, *AAPG, Methods in exploration*, No. 9.

Petroleum Geochemistry and Exploration in the Afro-Asian Region –
Digang Liang, Darui Wang & Zhenxi Li (eds)
© 2008 Taylor & Francis Group, London, ISBN 978-0-415-44083-7

Accumulation periods of hydrocarbon and its distribution in Dongying Sag, Jiyang Depression, Eastern China

Guangyou Zhu, Shuichang Zhang, Jinxing Dai & Yingbo Liang
Research Institute of Petroleum Exploration and Development, PetroChina, Beijing, China

Qiang Jin
Earth Resources and Information Institute, Petroleum University, Dongying, Shandong, China

Linye Zhang
Geoscience Research Institute, SINOPEC Shengli Petroleum Bureau, Dongying, Shandong, China

ABSTRACT: Within continental rift basins, due to their multi-cycles of deposition and evolution, two or multi-sets of source beds, characterized by various geochemical features and hydrocarbon generating potentials, are generally developed. Resulted from the geological settings of multi-episodes rift movements, the processes of hydrocarbon generation & expulsion often show multi-periods of hydrocarbon charging and accumulation. The formation of oil & gas pools, based on the geochemical features of crude oils and studies on fluid inclusions within reservoirs, can be divided into two different stages. The first stage is the Dongying period of Eogene, then, the upper source rocks of the fourth member (Es_4) of Shahejie Formation distributed in the sag center had entered the oil window and began to expel hydrocarbons. This process lasted about 8 Ma. Later, due to the Dongying episode of the Himalayan movements, the whole area was uplifted and eroded, which led to the hydrocarbon generating process being interrupted and the quantity of generated hydrocarbons being limited, also the maturity being lower. The second stage lasted from the Neogene period to Quaternary. Accompanied by the continuous subsidence and burial, the source rocks of both the lower part of the third member (Es_3) of Shahejie Formation and the upper part of Es_4 further became mature with the enhancement of hydrocarbon generation, which led to the key period for oil & gas migration and accumulation as well as the higher maturity of crude oils. The crude oil distribution of the both stages is of obvious regularity, i.e. the oil from the reservoirs located on the margin of the Dongying sag is of lower maturity and accumulated at the early stage, it is mainly sourced from the upper Es_4source rock; while that from those in the sag center is of relatively higher maturity, primarily derived from the lower Es_3 source rock and accumulated at the second stage. In addition, the source of oil reservoirs located on the transitional belts (slope belts) is rather complex, these reservoirs are, in general, characterized by mixed sources and multi-periods of hydrocarbon charging. The discovery above is a supplement and improvement to the theory on multiple oil accumulation zones.

1 INTRODUCTION

As well-known, a petroliferous basin is not only a giant thermochemical reactor with low temperature, but a natural and complicated reservoir forming system (Anderson et al.,1991 and Tian S, et al., 2001). The former produces the necessary fluids including oil, gas, and water for reservoir formation, while the latter includes the sub-systems constituted by oil sources, carrier systems and traps, etc, with the source rock being the connection between them. Since the hydrocarbon generation is a rather complicated geological and geochemical process, which comes down to both aspects of geological process and kerogen (Wang J, et al., 2003), the multi-cycles of deposition and evolution of a rift basin and the diversity of the elements of kerogen, cannot but lead to the complexity of hydrocarbon accumulation process (He D, et al., 2000; Chen J, et al., 2000; Zhao W, et al., 2000). The embodiment of the complexity includes multi-periods of hydrocarbon charging, the constituent diversity of hydrocarbons and multi-sources of reservoir components. Therefore, starting with the analysis of effective source rocks, to study the generation process and charging periods will not only be helpful to understand the complexity of hydrocarbon generation, but also be important to predict the formation of hydrocarbon reservoirs and their distribution.

As a representative of non-marine dustpan-shaped fault lake basins in eastern China, the Dongying sag, located in the southern part of Jiyang depression, is a third-class negative structure unit of the Mesozoic-Cenozoic rift basin in the Bohai bay. Multi-sets of source-reservoir-cap rock assemblages, with favorable conditions for hydrocarbon accumulation, were developed in Tertiary. In particular, the source beds with huge thickness were developed in Es_3 and the upper Es_4 of the Paleogene formation, which has abundant hydrocarbon resources and heretofore has produced more than 0.4 billion tons of crude oils. Also, the proved reserve of the only Paleogene formation has already accounted for over 45% of the total resources of Jiyang Depression (Jin Q, 2002; Guo Y, et al., 2001). In recent years, along with the change of exploration targets from structural reservoirs to subtle traps, laying an emphasis on studies of basic geology is also becoming more important. In particular, the studies on source rocks should include not only their formation environments and identification approaches, but also their distribution patterns and the consequent influence on hydrocarbon accumulation within traps around (Qiu G, et al. 2001; Jin Q, 2001). Based on the authors' thorough and systematical studies on hydrocarbon generation and accumulation within the Dongying sag, it can be found that the oil fields located on the margin of the sag, with lower crude oil maturity, are mainly sourced by the upper source rocks of Es_4 at the early stage (the Dongying episode of Himalayan movements); while those concentrated in the sag center and with relatively higher crude oil maturity are primarily derived from source rocks of the lower Es_3 and accumulated during the depositional period of Guantao to Minghuazhen formation; whereas, the reservoirs along the slope belt are characterized by mixed sources and multi-periods of hydrocarbon charging. The above discovery is of significance for guiding the petroleum exploration in similar sags.

2 THE DEPOSITIONAL EVOLUTION AND FILLING STAGES DIVISION OF DONGYING SAG

The Dongying sag is a typical Mesozoic-Cenozoic multiple-fault basin in eastern China (Dai J, et al., 1998), whose origin and evolution result from the superposition of two dynamic systems in space and time, i.e. the regional extension system in the direction of NW-SE and another pulling-apart system along the direction of S-N approximately due to the translation of the Tanlu faulting belt (He B, 2001; Yang M, et al., 2002). The above two dynamic systems are comparatively independent and coexist. Therefore, this area is a superimposed basin resulting from the translation, pulling-apart and slips of the Tanlu faulting belt on the basis of regional extension and superposition of Mesozoic-Cenozoic, whose formation and evolution are controlled by the block faulting of the Yanshan period and Himalayan. Along with the evolution of geodynamical settings, the structural style of the basin experienced three development stages, i.e. the Mesozoic inverse block faulting, the Paleogene basin-inclined rifting and the Neogene mass depression, which determine the basin with the geological structure of superimposition.

At the stage of the pre-rifting uplift and the original rifting, the Dongying sag had already shaped basically. The Cenozoic is a stage with intense tectonic movements for the Dongying sag. The basin, controlled by basement faults, experienced three development stages, i.e. the early rift faulting, the medium rift depression, and the late mass depression (Deng H, et al., 1990; Feng Y, et al., 2001). The three rifting cycles occurring in the Paleogene lead to the formation of the most important source and reservoir beds of the Shengli oilfield. According to the larger-scale unconformable surface, sedimentary infilling, volcanic activity and tectonic evolvement, the Paleogene can be called the period of multi-episode rifting and filling. The period can be further divided into three cycles, which include the early rifting episode, the rift extending episode, and the rift converging episode. The lake-basin area experienced three larger-scale processes of expanding → reducing → re-expanding → re-reducing. Meanwhile, the lake water depth also experienced three stages of being deeper → shallower → deeper → shallower. The three cycles are corresponding to the evolutional stages of the sag structure, i.e. the early rifting stage from Kongdian Formation to the lower part of Es_4, the most important rifting stage from the upper Es_4 to the lower Es_2, and the converging stage from the upper Es_2 to Dongying Formation. Among them, the stage from the upper Es_4 to the lower Es_2 is not only the main cycle of the basin evolvement, but the golden stage for development of high-quality source and reservoir beds.

3 DETERMINATION OF EFFECTIVE SOURCE BEDS IN DONGYING SAG AND THEIR CHARACTERISTICS

As is well known, hydrocarbon accumulation can be possible only when source rocks have developed in a sedimentary basin. The previous studies show that the dark argillutite with larger thickness, even located in the major source beds, is not that all can generate and expel hydrocarbons to form oil & gas reservoirs (Liang D, et al., 2000; Zhang S, et al., 2001). Only those that are fine-grained sedimentary rocks with higher abundance of organic matter and have experienced the process of generating and expelling

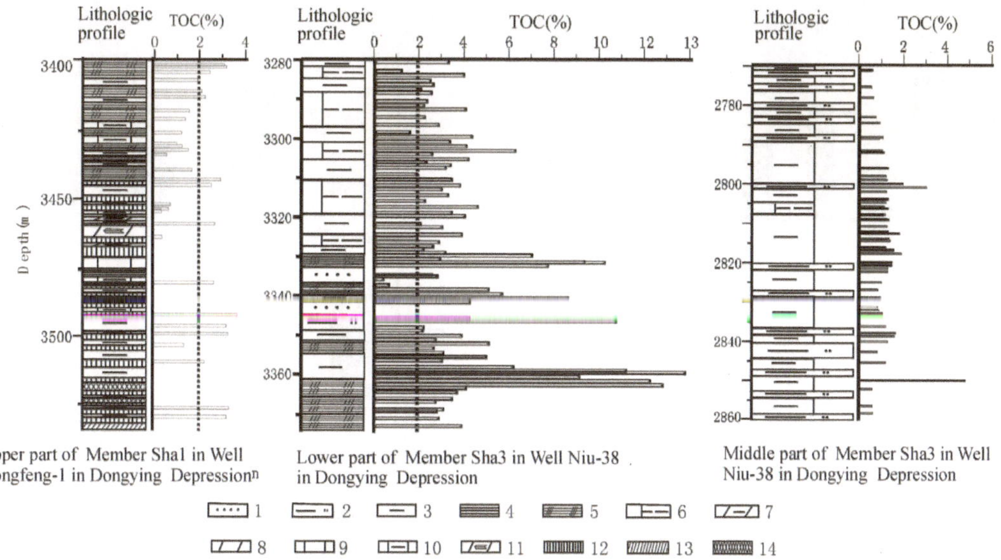

1-sandstone; 2-siltstone; 3-mudstone; 4-shale; 5-oil shale; 6-lime mudstone; 7-dolomite mudstone; 8-dolomite; 9-limestone; 10-lime dolomite; 11-dolomite with gypsum crystals; 12-anhydrite bed; 13-halite; 14-polyhalite bed

Figure 1. Vertical distribution characteristics of lithology and organic matter content for effective source rocks in Dongying Sag.

a great deal of hydrocarbons can contribute to forming oil & gas reservoirs and be regarded as effective source rocks. For a long time, it is generally believed that the dark mudstone of the Paleogene Shahejie Formation (mainly being the upper Es_4, the lower and middle of Es_3), with the thickness varying from 1000 to 1500 m, belongs to effective source rocks. Even in those giant oil-rich basins beyond China, the effective source bed with such a huge thickness has never been found. According to the studies by scholars both at home and abroad on the forming conditions of high-quality and effective source rocks, the distribution of effective source rocks within basins is not with huge thickness, but anisotropic (Bertrand, et al., 1993; Carroll et al., 1999; Derenne, et al., 2000; Zhu G, et al., 2002; Zhang S., 2002; Xia X, et al., 2000). Thereby, it is very necessary to correctly understand the effective source rock within the Dongying sag because it determines the oil & gas exploration arrangement

Based on our detailed core investigation and geochemical analysis of over 4000 m-length mudstone samples selected from more than 20 core holes, and the biomarker analysis of several hundreds source rock and crude oil samples, it can be found that there exist obvious differences in lithology and geochemical features of the three sets of dark mudstones located mainly in the upper Es_4, the lower and middle Es_3. Whereas, it is the source rocks of the lower Es_3 and the upper Es_4 that really contributes to hydrocarbons accumulation and controls the formation of giant oil fields in the Dongying sag.

The upper Es_4 source rock is thought to have been deposited in an intense-reducing depositional environment of saline water (Zhang L, et al.2003; Zhu G., et al., 2003), while the lithology is dominated by dark-gray mudstone and greyish-brown calcareous shale, with thin interbedded dolomite, argillaceous dolomite, gypsiferous mudstone and brown-greyish oil shale.

Also, the deposition sequence of evaporate with thick-layered striation, such as gypsiferous salt and rock salt, is developed in deep parts of the Dongying sag. As a whole, they belong to lake deposits of shallow to semi-deep water, with the organic carbon abundance (TOC) varying between 1.5%~4% (Fig. 1), the kerogen type being dominated by I and II_1 (Fig. 2), the thickness of effective source rocks changing between 150~250 m, and the vitrinite reflectance (Ro) value ranging between 0.7%~0.9% (Table 1). The organic matter within source rocks is mostly derived from hydrobios, with the saturated hydrocarbon showing a unimodal distribution and an obvious phytane predominance (Pr/Ph<0.5). Also, the sequence of long-chain hopanes is characterized by a 'raised tail' and warp gradually, i.e. C35 hopane > C34 hopane > C33 hopane. Whereas, the oleanane is undeveloped while the contents of rearranged sterane and 4-methyl sterane are very low (Fig 2).

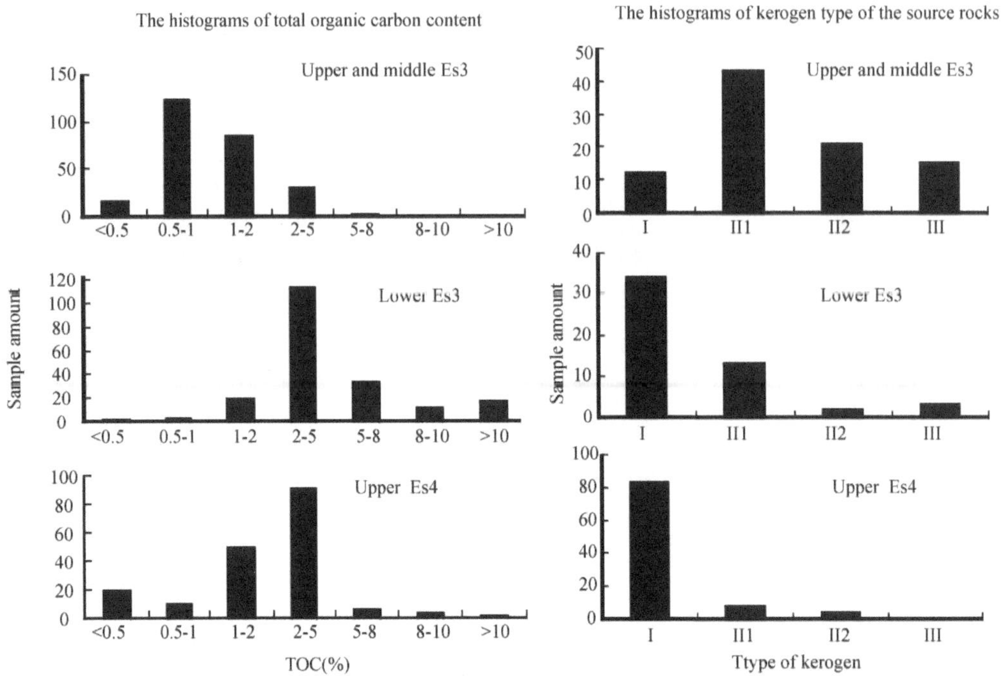

Figure 2. Histograms of the total organic carbon content and kerogen type of the source rocks from the Dongying depression.

Table 1. Characteristic correlation of the upper Es_4, lower and middle Es_3 source rocks in Dongying Sag

Formation	Upper Es_4	Lower Es_3	Middle Es_3
Sedimentary sub-facies	Shallow to half-deep lake	Deep lake	Semi-deep lake, delta
Water Body Environment	Brine-Salt lake	Brackish	Freshwater
TOC(%)	0.6–11.2, mostly > 2	1.5–19, averaging 2.5	1.0±
Chloroform Bitumen "A"(%)	0.3–0.8, mostly > 0.6	0.3–2.3, averaging 0.48	<0.2, averaging 0.08
Kerogen Type	I, II_1	I–II_2, mainly II_1	II_2, III
Organic Origin	Hydrobios	Hydrobios and Terrigenous Higher Plant	Terrigenous Higher Plant
Ro(%)	0.45–0.93	0.36–0.80	0.22–0.57
Pr/Ph	<0.6	0.6–1.6	>1.2
Main Peak Carbon	nC_{16} or nC_{18}	nC_{16} nC_{17} or nC_{27}	nC_{27} or nC_{29}
Gammacerane/C_{31} Hopane	>1.2	1.0–0.41	<0.5
Source Rock Evaluation	effective	effective	Non-effective

The source rocks of the lower Es_3, dominated by dark-grey mudstone, calcareous mudstone, brown-greyish oil shale, are formed in a saline-water depositional environment of deep lake. The TOCs of the lower Es_3 source rocks are relatively higher, mostly ranging from 2%~5%, and while those of oil shales being 5%–13% (Fig.1). In addition, their kerogen types are dominated by II 1 and I, while the Ro values vary between 0.4%~0.8% and the thickness changes from 200~300 m. The normal paraffin hydrocarbons show a bimodal distribution. Compared with the upper Es_4 source rock, the phytane predominance of the lower Es_3 is apparently weak, i.e. the ratios of Pr/Ph vary between 0.6–1.6 (Table 1) and the peak height of the C31~C35 hopanes lowers gradually. The content of gammacerane reduces too, whereas those of rearranged steranes and 4-methyl steranes are very high (Fig. 3).

At the depositional stage of the middle Es_3, since the Dongying delta had been shaped and began to advance

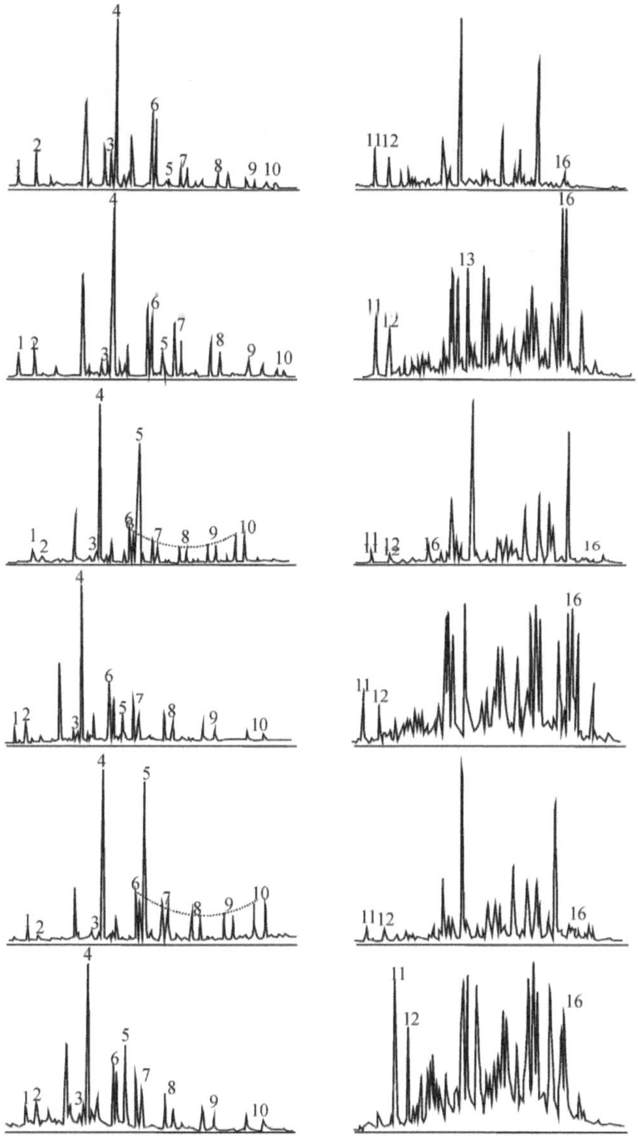

1-Ts; 2-Tm; 3-oleanane; 4-17α(H),21β(H)-hopane; 5-Gammacerane; 6-17α(H),21β(H)-29-homohopanes; 7-17α(H),21β(H)-29-bishomohopanes;8-17α(H),21β(H)-29-trishomohopanes; 9-17α(H),21β(H)-29-tetrahomohopanes;10-17α(H),21β(H)-29-pentahomohopanes; 11,12-diacholestanes; 13-5α,14α,17α-cholestane 20R; 14-5α,14α,17α-ergostane; 15-5α,14α,17α-stigmastane 20R; 16-4- methyl-sterane.

Figure 3. Correlation of hopanes and terpenes of the source rock and crude oil in Shahejie Formation, Dongying Depression.

widely towards the lake basin, the grey and silty mudstone was mainly developed. Mostly, the TOC of the mudstone is less than 1.5%, the kerogen type is mainly III and II 2, while the thickness reaches to 600∼800 m and the Ro values vary between 0.3%∼0.55%. The saturated hydrocarbons are characteristic of a "unimodal" distribution, which indicates a predominant input of terrigenous higher plants. In the middle Es_3, the content of oleanane, which represents the terrigenous input, is higher, while gammacerane lowers obviously and even disappears. Meanwhile, the distribution of high C_{31} hopane is predominant, and the peak distribution of C_{31}~C_{35} hopanes constitutes the shape of 'L', reflecting a weak-oxidizing to weak-reducing environment of delta deposition. Among the sterane sequences, the content of the rearranged sterane is relatively high while that of the 4-methyl sterane is rather low (Fig. 3). In a word, there exist great differences in source rocks of the lower and the middle Es_3.

It is obvious that there exist apparent geochemical differences in the three sets of source rocks of the Dongying sag, which can provide reliable basis for crude oil-source correlation. Based on analysis results of chromatograms and mass chromatograms of a great deal of crude oil samples selected from the Dongying sag, the oil source can be divided into three types (Fig.2), i.e. the upper Es_4 type (crude oil sourced from the upper Es_4 source bed), the lower Es_3 type (crude oil sourced from the lower Es_3 source bed) and the mixed type (crude oil sourced by both the lower Es_3 and the upper Es_4 source beds). Therefore, the crude oil of all the oilfields in the Dongying sag is generated from either source rocks of the lower Es_3 or the upper Es_4, or from both of them. Whereas, there exist great differences between the middle Es_3 source rock and the crude oil produced from each oilfield. Obviously, it is the lower Es_3 and the upper Es_4 source rocks that together control the formation and distribution of the oil & gas reservoirs in the Dongying sag. According to the above analysis, the lower Es_3 and the upper Es_4 dark argillutite, with high organic carbon content and kerogen type of I and II, can be regarded as the effective source rock of the Dongying sag while the oil-source correlation between the lower Es_3 or upper Es_4 source rock and the middle Es_3 source rock is an important approach for determining the effective source rock.

4 ANALYSIS OF HYDROCARBON CHARGING TIME AND PERIOD

Analysis of hydrocarbon accumulation period (HAP) is a hot topic of modern petroleum geology studies. The exact determination of HAP is helpful to correctly understand the formation law of reservoirs. At present, the homogenization temperature approach of hydrocarbon fluid inclusions is relatively mature and mostly applied to studying the HAP (Qiu N et al., 2003). Fluid inclusions are composed of the diagenetic and mineralized fluid trapped in the crystal defect during mineral crystallization. As the original sample and history recordation left behind during the process of fluid generation, evolvement, migration and accumulation, fluid inclusions directly record the conditions and processes of hydrocarbon accumulation within sedimentary basins. Also, as a fossil record, they can be applied to recover the formation and evolvement history of reservoirs. Meanwhile, by means of the integrated analysis on hydrocarbon generation,

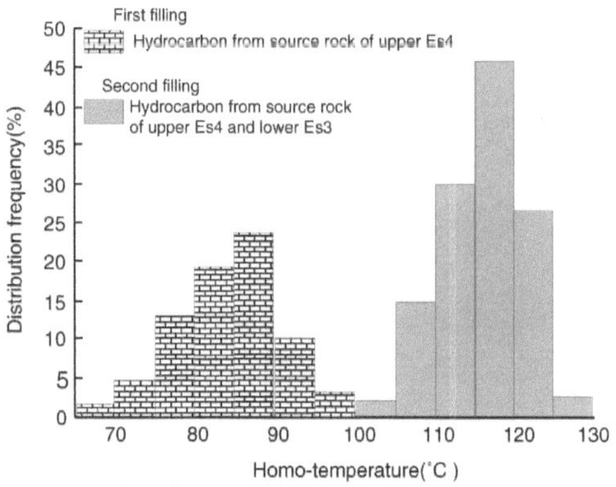

Figure 4. Homothermal histogram of organic inclusions.

expulsion history, and tectonic activities of basins, the periods of oil & gas migration and accumulation can be made clear.

According to the homogenization temperature of fluid inclusions in oil-bearing sandstones selected from many cores in the Dongying sag, the palaeogeothermal values show two main peak spans between 76~92° and 108~124°, respectively (Fig. 4). By means of combining the palaeogeothermal data and taking the surface temperature and the geothermal gradient as 15° and 3.8°/100 m, respectively, the corresponding burial depth of hydrocarbon charging can be calculated to be 1200~1400 m and 2000~2300 m while the charging time 32.8~24.6 Ma and 13.6~2 Ma, respectively (Fig. 4), which correspond to the end deposition of Dongying Formation and the depositional period of Guantao-Minghuazhen Formation, respectively (Fig. 5). These oil-bearing sands are extracted when their inclusions are analysed. By analyzing biomarker components of crude oils, it is found that the maturity of the crude oil charged at the late deposition period of Dongying Formation is relatively low, it has an obviously high content of gammacerane and the predominance of phytane. These characteristics are very similar to those of the biomarkers of the upper Es_4 source rock. Whereas, biomarker components of the crude oil charged at the depositional period of Guantao-Minghuazhen Formation are characterized by diversification, i.e. they are similar either to those of the upper Es_4 source rock or to those of the lower Es_3 source rock, plus some similar to the mixed biomarkers of the lower Es_3 and the upper Es_4 source rocks. The result indicates that the upper Es_4 source rock not only generated and expelled hydrocarbons at the late period of Dongying Formation, but also repeated the process at the depositional period of Guantao-Minghuazhen Formation (Fig. 6).

In addition, according to analysis of the geological evolvement process of the Dongying sag, tectonic

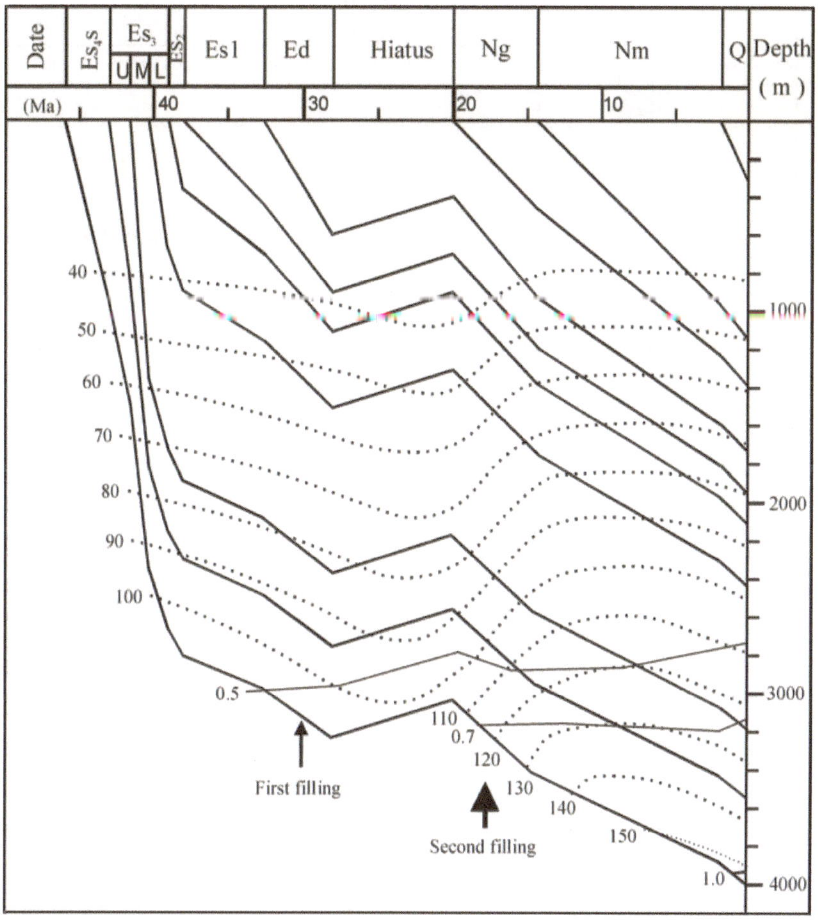

Figure 5. Burial history curves of the Dongying sag.

movement in late Oligocene established its basic structure framework with large-sized structural traps mostly formed during this period, which provide space for large-scale hydrocarbon accumulations. The uplifting and erosion of the total Dongying sag, resulting from the Dongying movement, occurred during the late Paleogene period and led to the interruption of the hydrocarbon generating process of the upper Es_4 source rock.

After undergoing the depositional break of nearly 10 Ma, the basin entered the slow subsiding stage during the Neogene period. Clastic rocks of the Guantao-Minghuazhen Formation, with alluvial-fluvial facies, were deposited around the whole area. Along with the totally depressed subsidence, the deposition of Guantao-Minghuazhen Formation as a whole covered the deposits of Dongying and Shahejie Formation. As the lower Es_3 and the upper Es_4 source rocks were buried further deeply, the upper Es_4 source rock re-entered the oil window while the lower Es_3 source rock went into the oil window for the first time. With the burial depth increasing, most of the lower Es_3 and the upper Es_4 source rocks became mature and reached to the peaks of hydrocarbon generation and expulsion. Meanwhile, weak tectonic activities, occurring during the depression period after the Dongying age, provided necessary conditions for oil & gas migration, which made this period the key charging stage for hydrocarbon accumulations in the Dongying sag (Fig. 5, Fig. 6).

5 DISTRIBUTION CHARACTERISTICS OF OIL & GAS IN DIFFERENT ACCUMULATION PERIODS

The important theory of 'composite oil-gas accumulation zone', developed by China's petroleum geologists on the basis of the long-term exploration practice within continental basins, has fruitfully guided the

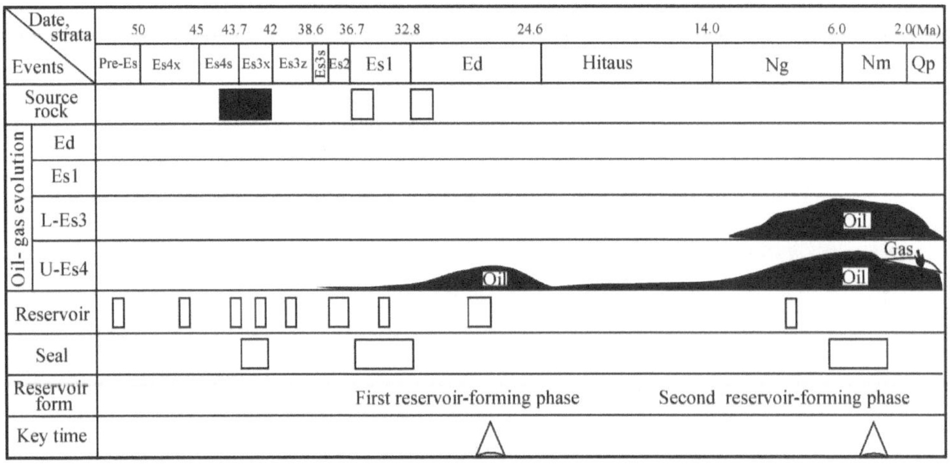

Figure 6. Formation events of oil and gas reservoirs.

hydrocarbon exploration in continental basins, particularly in the faulted basins of eastern China (Hu, J. et al., 1986; Li, D. et al., 2000; Liu, X. et al., 1998; Dai, J. 2001; Li, J. et al., 1998; Li, P. 2001). The core of the above theory emphasizes the 'annular' distribution of various types of reservoirs around the hydrocarbon-generating sag, i.e., lithologic hydrocarbon reservoirs are mainly distributed near the sag center, and buried hill and stratigraphic hydrocarbon reservoirs are located on the margins while fault block hydrocarbon reservoirs concentrate in the slope belts. The theory summarizes the characteristics of the aggregated collection of various hydrocarbon reservoirs distributed in groups and in zones in continental basins with multiple fault depressions, fault blocks and hydrocarbon-bearing strata; whereas it pays little attention to the source of various types of hydrocarbon reservoirs. Based on selected samples and analyses of various types of hydrocarbon reservoirs located in different zones, our study shows that the distribution of various types of crude oils has strong regularity.

On margins of the Dongying sag, the stratigraphic and the buried hill hydrocarbon reservoirs are predominant. The oil-source correlation indicate their crude oils, characteristic of relatively low maturity, the maturity indexes of Ts/Tm, C_{29} sterane 20S/20S + 20R, C_{29} cholestane $\beta\beta/(\alpha\alpha+\beta\beta)$ less than 0.5, 0.35 and 0.40, respectively, and undeveloped pregnane, are mainly supplied by the upper Es_4 source rock (Fig.7). The analysis of fluid inclusions suggests that the crude oils were charged during the late Oligocene period, i.e., the late depositional period of Dongying Formation. Taking the Xin-154 reservoir as an example, its homogenization temperature of oil-sand fluid inclusions varies between 91~99°C, and the late depositional period of Dongying Formation is the period for hydrocarbon charging and accumulating. The biomarkers of the crude oil indicate that the crude oil maturity is relatively lower while the contents of gammacerane and phytane are relatively higher (Fig. 7). The above analyses suggest that the Xin-154 reservoir is charged and sourced from the upper Es_4 source rock at the early stage (Fig. 8).

The lithologic hydrocarbon reservoirs are mainly distributed in the sag center. In general, their crude oil maturity is relatively higher, and the biomarkers of the crude oil are similar to those of the lower Es_3 source rock (Fig.7), the homogenization temperature of fluid inclusions mostly ranges from 108°C to 124 °C, which indicates the hydrocarbon accumulating time is relatively late. These kinds of oil reservoirs, such as the Niuzhuang oil field, were sourced by the lower Es_3 source rock during the depositional period of Guantao-Minghuazhen Formation. Besides higher maturity, the crude oil also has relatively higher contents of pregnane and homopregnane, while the maturity index of C_{29} sterane 20S/20S + 20R exceeds 0.50. For example, the homogenization temperature of fluid inclusions in Well Niu-871 averages about 119°C, while the corresponding period for hydrocarbon accumulation is the depositional period of Guantao Formation.

The source of the hydrocarbon reservoirs located in the slope belts is comparatively complex. There not only exist oil reservoirs sourced by the upper Es_4 source bed, but also those supplied by the lower Es_3 source bed, as well as some charged by the mixed sources of both the upper Es_4 and the lower Es_3. In addition, the distribution range of the homogenization temperature of fluid inclusions is comparatively wider, i.e., some reflecting the early charging while others reflecting the late charging. Certainly, there are also some oil reservoirs that experienced charging twice, such as the Wang-543 reservoir, whose homogenization temperature of oil-sand fluid

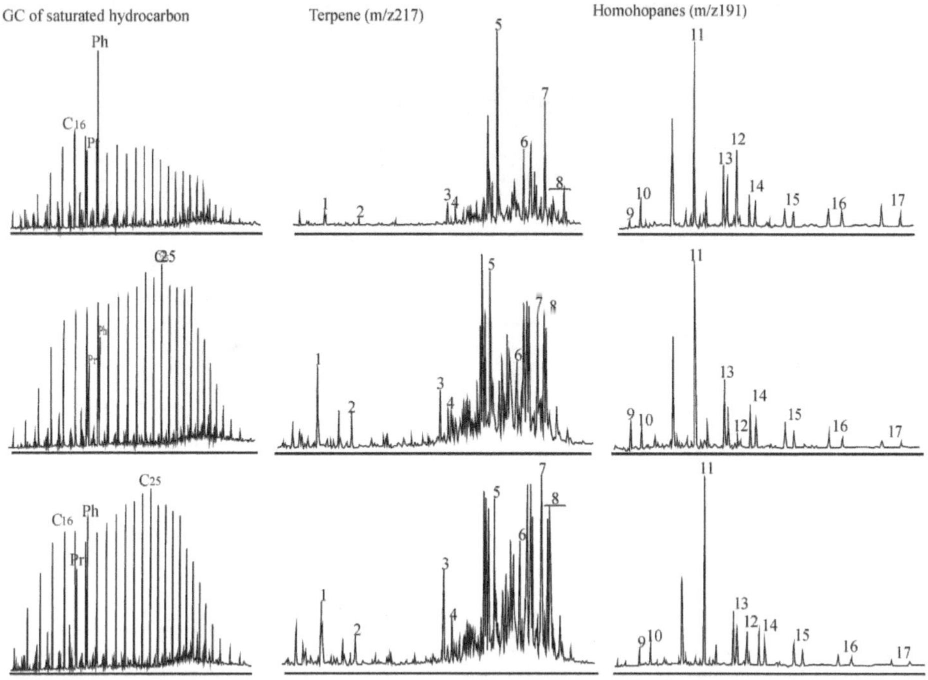

1-pergnane; 2-homopreganane; 3.4-diacholestane; 5-5α,14α,17α-cholestane 20R; 7-5α,14α,17α-stigmastane 20R; 8-4-methyl-sterane; 9-Ts; 10-Tm; 11-17α(H),21β(H)-hopane; 12-gammacerane; 13-17α(H),21β(H)-29-homohopanes; 14-17α(H),21β(H)-29-bishomohopanes; 15-17α(H),21β(H)-29-trishomohopanes; 16-17α(H),21β(H)-29-tetrahomohopanes; 17-17α(H),21β(H)-29-pentahomohopanes.

Upper: early filling in the fringe sag, the oil from the upper Es_4 source rock, low maturity oil (Well Xin154-1, 2956.8-2976.1m);
Middle: late filling in the center sag, oil from lower Es_3 source rock, high maturity oil (Well Niu871, 3270.0-3283.5m);
Lower: dual filling in the transitional area, the early filling oil from the upper Es4 and the late filling oil from the lower Es_3 (Well Wang543, 3173.5-3225.7m).

Figure 7. Biomarker features of crude oils in different hydrocarbon accumulation stages.

inclusions varying between 86~122°C, was charged twice at the end of Dongying period and during the Guantao period, respectively. The evidence of two-periods charging can also be seen clearly from the characteristic pattern of the biomarkers of the crude oil, which show features of both the lower Es_3 and the upper Es_4 source rocks, as well as obvious features of mixed sources, i.e., the normal alkanes showing double-peaks also with higher contents of 4-methyl sterane and rearranged sterane, as well as gammacerane of certain abundance (Fig.7).

The distribution pattern of oil & gas has close relations with the migration style and the lithologic assemblage of the lower Es_3 and the upper Es_4 source rocks. Due to the well-developed beddings and micro-cracks in the upper Es_4 source rock, which is deposited in salt-water settings while the top is sealed by the Es_3 mudstone with huge thickness and the unconformity surface developed in its bottom, its generated hydrocarbons can only migrate laterally. Furthermore, source rocks with salt-lake facies usually became mature at the early stage, which leads to the relatively lower maturity of the crude oil from some oil fields, such as the Bamianhe oil field, etc..

Moreover, beyond that turbidite sandbodies are well-developed in the lower Es_3 source rock deposited under brackish-water conditions, the synsedimentary faults are also well-developed in the Es_3 formation (Zhu, G. et al., 2003). Frequently, these turbidite sandbodies are superposed vertically and continue laterally. They can act as a carrier bed or reservoir and constitute the most favorable assemblage together with source rocks in shapes of alternation and package, which is favorable to hydrocarbons migration, particularly for the vertical migration while the magnitude of oil reservoirs is relatively greater. Within some hydrocarbon reservoirs, due to the faults cutting through the upper Es_4 and the lower Es_3 source rocks, a great number of hydrocarbon accumulation belts are generated around these large-sized and inherited fault zones, such as the Shengtuo oilfield (Fig.8).

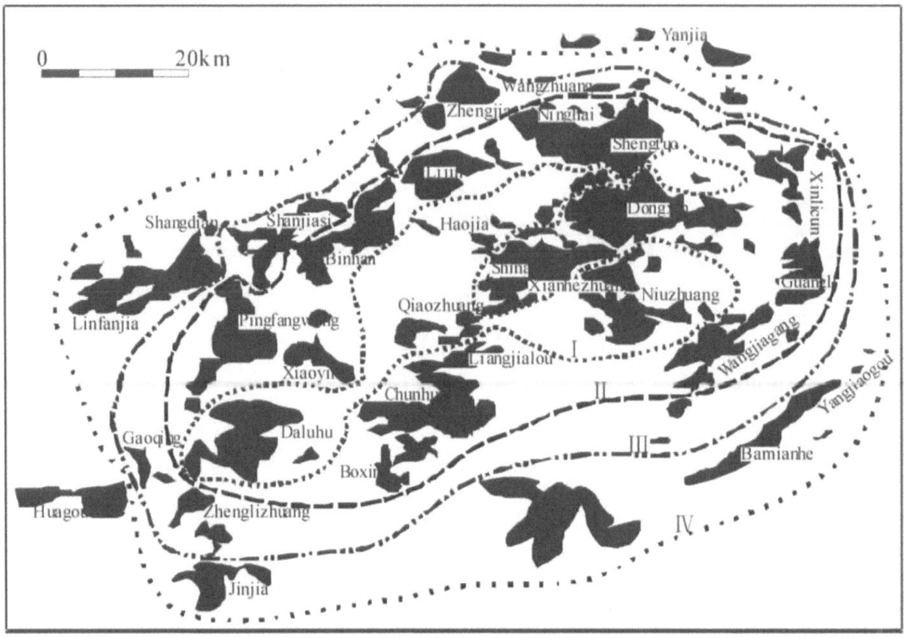

The crude oil in oil pools of Area I is generated from the lower Es_3 source rock. The crude oil in oil pools of Area II is generated from the mixture of the upper Es_4 and the lower Es_3 source rocks, mainly from the lower Es_3 source rock .The crude oil in oil pools of Area III is generated from the mixture of the upper Es_4 and the lower Es_3 source rocks, mainly from the upper Es_4. The crude oil in oil pools of Area Ⅳ is generated from the upper Es_4 source rock. The outside of Area Ⅳ belongs to gas pools.

Figure 8. Oil and gas field distribution features of the Dongying depression.

6 CONCLUSIONS

① Two sets of effective source rocks, i.e., the upper Es_4 source rock of saline-lake facies and the lower Es_3 source rock with brackish deep-lake facies, are developed in the Dongying sag. The oil-source correlation indicates that the middle Es_3 source bed doesn't belong to the effective source rock in the Dongying sag.

② There exist two stages of hydrocarbon charging and accumulation in the Dongying sag. At the first stage, namely the end of the Eogene Dongying period, only the upper Es_4 source rock had entered the oil window and began to expel hydrocarbons with lower maturity. Due to the Dongying episode of Himalayan movements, there exists a long-term hiatus between the Dongying and the Guantao periods, which leads to the interruption of the first hydrocarbon generating process. Resulted from the successive deposition of Guantao and Minghuazhen Formation, the burial depths of the upper Es_4 and the lower Es_3 source rocks increased and then both the source rocks entered the oil window. The corresponding period is the second hydrocarbon generating stage. Also, it is the main hydrocarbon accumulating stage of the Dongying sag.

③ The distribution of crude oils charged at different stages has definite regularity. The oil from the reservoirs located on the margins of the Dongying sag is of lower maturity, mainly sourced from the upper Es_4 source rock, and accumulated at the early stage; while that from those distributed in the sag center, is of relatively higher maturity, primarily derived from the lower Es_3 source rock and accumulated at the second stage. However, the source of the oil reservoirs located on the transitional belts (slope belts) is rather complex, they are generally characterized by mixed sources and multi-periods of hydrocarbon charging and accumulation.

REFERENCES

Anderson R N, He Wei & Hobart M A,1991. Active fluid flow in the Eugene island area, offshore Louisiana. *Geophysic*, 10(4):6~10.

Bertrand P, Lallier V E & Boussafir M,1993. Enhancement of accumulation and anoxic degradation of organic matter controlled by cyclic productivity: a model. *Organic Geochemistry*, 22:5119~520.

Carroll A R & Bohacs K M, 1999. Stratigraphic classification of ancient lakes: balancing tectonic and climatic controls. *Geology*, 27(2):99~102.

Chen Jianyu & Niu Ruiqing, 2000. Organic geochemistry research into multi-phase petroleum migration and accumulation in rift-subsidence basin. *Earth Science*, 25(3):253~259.

Dai Jinxing, 2001. Some problems in the study of petroleum geology. *Advance in earth sciences*,16(5):710~718.

Dai Junsheng, Lu Kezheng, Qi Jiafu, Chen Shuping.1998.The evolution of the Paleogene structural styles in Bohai gulf basin. *Acta Petrolei Sinica*, 19(4):16~20.

Deng Hongwen, Qian Kia.1990. The genetic types and association evolution of deep lacustrine faces mudstones *Acta sedimentologica Sinica*,8(3): 1~20(in Chinese with English abstract).

Derenne S, Largeau C, Brukner~Wein A, 2000. Origin of variations in organic mater abundance and composition in a lithologically hogogeneous maar-type oil shale deposit. *Organic Geochemistry*, 31: 787~798.

Feng Youliang, Li Sitian. 2001. Depositional characteristics of lowstand sand bodies of the third member of Shahejie formation in the Dongying depression and the significance in petroleum geology. *Geological Review*,47(3): 278~286(in Chinese with English abstract).

Guo Yuanling, Zhao Leqing, Shi Hongxia et al. 2001. Characteristic analysis of proven OOIP in Jiyang depression. *Petroleum Exploration and Development*, 28(3): 33~37(in Chinese with English abstract).

He Bin. 2001. On dynamics of the Bohai bay complex faulted basin. *Petroleum Geology & Experiment*. 23(1); 27~31.

He Dengfa, Zhao Wenzhi, Lei Zhenyu, Zhai Hui & Chui Yingliu. 2000. Characteristics of composite petroleum system basins in China. *Earth Science Frontiers*, 7(3):23~37.

Hu Jianyi, Tong Xiaoguang & Xu Shubao,1986. Formation and distribution of complex petroleum accumulation zones in Bohaiwan basin. *Petroleum Exploration and development*,(1):5~9.

Jin Qiang, 2002. Considerations on exploration sustainability in east China enriched petroliferous basin. *Petroleum Geology and Recovery Efficiency*, 2002, 9(1):5~8.

Jin Qiang. Importance and research about effective hydrocarbon source rock. *Petroleum Geology and Recovery Efficiency*, 2001,8(1): 1~4.

Li Desheng, 2000. The progress in the petroleum geology of China towards new century. *Acta Petrolei Sinica*,21(2):1~8.

Li Jinchao, MaYongsheng, Zhang Dajiang, Zhou Kun, Huang Difan, Zhang Shuichang, Cheng Keming, Xu Zhichuan & Li Xiaodi, 1998. Some importants scientific problem.

Li Pilong, 2001. Zonary distribution of hydrocarbon in oil-enriched rifted basins and exploration direction in the Huimin sag. *Petroleum Geology and Experiment*, 23(2):146~148.

Liang Digang, Zhang Shuichang, Zhang Baomin & Wang Feiyu. 2000. Understanding on marine oil generation in China based on Tarim basin. *Earth Science Frontiers*, 7(4):534~547.

Liu Xingcai & Yang Shenbiao,1998. Formation conditions and distribution pattern of the oil fields in the compound oil-gas area, Jiyang. *Journal of Chengdu University of Technology*, 25(2): 276~284.

Qiu Guiqiang, Wang Jufeng, Zhang Xin, et al. 2001. Preliminary study on stratigraphic architecture of middle-Shasan Dongying delta and its significance to hydrocarbon exploration. *Acta Sedimentologica Sinica*, 19(4):569~574 (in Chinese with English abstract).

Qiu N., Cai J., Li Sh & Zeng J., 2003. Thermal evolution and hydrocarbon entrapment in the Weibei sag, Changwei depression. *Chinese joural of Geology*, 38(3):332~341.

Tian S., Chen Y., Zhang X., Cui G & He H, 2001.The fluid dynamics mechanismin migration accumulation dynamics system. *Earth Science Frontiers*, 8(4):329~336.

Wang J., Xiao X., Guo R., Lu H & Gao X. 2003. Study on the generation kinetics of source rocks from the Dongying depression, the BohaiWan basin, 25(4):403~409.

Xia Xinyu & Dai Jinxing, 2000. A critical review on the evaluation of hydrocarbon potential of marine carbonate rocks in China. *Acta Petrolei Sinica*, 21(4):36~41.

Yang Minghui, Xia Wenchen, Zhang Bingshan, Zhang Ning, Peng Xiaoqun. 2000. Parasequence-scale organic facies variations and its significance in nomarine basin. *Acta Sedimentologica Sinica*, 18(2):297~301(in Chinese with English abstract).

Zhang Linye, Kong Xiangxing & Zhang Chunrong, et al.2003. High-quality oil-prone source rocks in Jiyang Depression. *Geochemica*, 32(1):35~42 (in Chinese with English abstract)..

Zhang Shuichang, Zhang Baomin & Wang Feiyu, 2001. The two sets of effective source rock layers in Talimu Basin-I the organic matters characteristics, developing environment and controlling factors [J]. *Progress in Nature Science*, 11(3): 261~268.

Zhang Shuichang, Liang Digang, Zhang Dajiang. 2002. Evalution criteria for Paleozoic effective hydrocarbon source rocks. *Petroleum Exploration and Development*, 29(2): 8~12.

Zhao Wenzhi and He Dengfa. 2002. Characteristics and exploration strategy of composite petroleum systems in China. *Acta Petrolei Sinica*, 23 (6):1~11.

Zhu Guangyou & Jin Qiang. 2002. On source rock heterogeneity—a case study of the well Niu-38, Dongying depression. *Acta Petrolei Sinica*, 23(5): 34~39 (in Chinese with English abstract).

Zhu, G & Jin, Q., 2003c. Geochemical characteristics of the two sets of excellent source rocks in Dongying depression. *Acta Sedimentologica Sinica*. 21(3), 452~460. (in Chinese).

Zhu, G., Jin, Q & Zhou, J., 2003. The Source Rock characters of cycle deep-lake facies in Dongying Depression. *Chinese Journal of Geology*, 38(2), 254~262. (in Chinese).

Petroleum Geochemistry and Exploration in the Afro-Asian Region – Digang Liang, Darui Wang & Zhenxi Li (eds)
© 2008 Taylor & Francis Group, London, ISBN 978-0-415-44083-7

Correlation between contents of TOC and strontium distribution within source rocks of Shahejie Formation in Dongying Sag, Eastern China

Guangyou Zhu, Shuichang Zhang & Jinxing Dai
Research Institute of Petroleum Exploration and Development, PetroChina Company Ltd., Beijing, China

Qiang Jin & Guanmin Wang
Earth Resources and Information Institute, Petroleum University (East China), Dongying, Shandong, China

ABSTRACT: The third member of Shahejie Formation of Eogene (Es_3) is the primary source rock of the Dongying sag, eastern China. The high-quality source rocks of the lower formation of Es_3, usually with laminated structure, are primarily composed of carbonate and clay laminae rich in organic matter. An obvious positive correlation between the content of total organic carbon (TOC) and a microelement, Sr is found in the source rocks. Also, it has been found that Sr is mostly accumulated in the carbonate laminae rather than in the clay laminae. Based on the fact that the carbonate laminae are usually composed of endogenetic thin laminae of carbonates such as calcareous imbricate planks formed by algal blooming, it can be concluded that the source of Sr has a relation with algal blooming. The thin carbonate laminae are resulted from the absorption and assimilation of chemical elements within aquifer by algal blooming. The element of Sr occupies the crystal lattice of Ca, and then accumulates within $CaCO_3$, which results in the synchronous enrichment of TOC and Sr within source rocks.

1 INTRODUCTION

Sr is one of the microelements and has a broad natural distribution (Liu Yingjun, 1984; Zhao Yiyang, 1994). Researches on Sr and its isotopes have become an important approach for studies of the hydrological cycle, the diagenesis of carbonates and the evaluation of palaeoclimate and palaeoenvironments because the variety and differentiation of Sr and its isotopes are closely related to the crystal mineralogy of rocks, salinity and source of solutions, interaction mechanisms between water and rock, and the relationship between biology and environments (Brass, 1976; Hess et al., 1986; Hodell, 1991; Palmer and Elderfield,1985; Brand,1991; Banner,1995; Denison et al., 1998; Blum and Erel, 1995; Depaolo and Ingram, 1985; McArthur et al., 1992; Dix,2001; Richard and McArthur, 1997; Liu Dongsheng, 1985; Chen Jun et al.,1998). The element Sr and other microelements are often used to study the depositional settings of continental rift lake basins. However, along with the in-depth studies, it is observed that there exist some inconsistent phenomena when some indices of microelements are applied to studies of palaeoenvironment (Wang Aihua, 1996; Zhang Zhihuan et al., 1998; Zhu Guangyou et al., 1999). In particular, Sr contents of two series of adjacent samples, although being densely sampled, can show great difference (even orders of magnitude), which raises many questions with respect to the evaluation of palaeoenvironments.

Total organic carbon (TOC) content, controlled by depositional settings, is one of the important parameters for resource evaluation (Tissot & Welte, 1984; Pederson & Calvert, 1990; Bertrand et al., 1993; Jin Qiang & Zha Ming, 2000). The abundance of organic matter within source rocks of lacustrine facies should have a certain relationship with the content of microelements. Recent studies, conducted in the Dongying sag, indicated that the fluctuation of the TOC content within source rocks has surprising comparability with the change of Sr content, especially for the abnormally high content of Sr within high-quality source rocks, which indicates that there may be a relationship between them. However, which factors control the correlation between contents of organic carbon and Sr within lake deposits? Why do abnormally high contents of Sr occur in high-quality source rocks? Which environment conditions do they indicate? The solution of the above questions will provide the theoretical basis for probing the mechanism of transfer and accumulation of Sr within source rocks and the forming conditions of high-quality source rocks. In this discussion, we present a numerical study on the vertical distribution of organic matter abundance and Sr content within the Es_3 deposits of the Dongying rift lake basin as well as their positive correlation. In addition, based on a series of analytical approaches, the enrichment rule of Sr element within high-quality source rocks is clarified, and the positive correlation between TOC and Sr is made clear. Finally, the forming mechanism of high-quality source rocks is further discussed.

2 GEOLOGICAL SETTINGS

Located in the southern Jiyang depression, the Dongying sag is a third-class negative structure unit of a Mesozoic-Cenozoic rift basin in Bohai Bay, eastern China. Also, it is a rift-depression lake basin developed from the palaeotopography of Paleozoic base rocks due to tectonic movements. The Dongying basin experienced four periods of geological evolution that include an early stage of rifting (Ek), a middle stage of rifting (lower-middle Es), a late stage of rifting (upper Es-Ed), and post rifting (late Tertiary). The first-order tectonic sequence of Eogene, bounded by the unconformable surface formed at the end of each rifting stage of Eogene, can be divided into three secondary tectonic sequences (about several Mas), i.e. Kongdian Formation (Ek), the fourth to second members of Shahejie Formation (Es2-4), the upmost member of Shahejie Formation to Dongying Formation (Es1-Ed), which correspond to three sub-rifting cycles. Among those cycles, the Es2-4 formation is the main cyclothem developed within the lake basin (Deng Hongwen et al., 1990). At the early stage of this cycle (Es_4), the strata were mainly composed of shallow-water deposits due to evaporation. With salinity decreasing, the water depth was increasing and the semi-deep lacustrine sediment deposited. The lower part of the Es_3 formation was composed of deep-water lacustrine sediments. During the middle part of the Es_3 formation, the Dongying delta thickened towards the depocenter of the basin untill the late Es_3 when fluvial and deltaic sediments had covered most areas of the whole sag. The dark shale, constituting the main source rock of the Dongying sag, mainly accumulated within the lower to middle parts of the Es_3 formation with the burial depth exceeding 2800 m.

3 SAMPLING AND ANALYSIS

Well Niu-38, located in the southern-central part of the Dongying sag, is a well with the longest continuous core and with the observed thickest shale interval in China so far (Hong Zhihua et al., 1991; Zhu Guangyou & Jin Qiang, 2002), which has provided an uncommon section for this study. Over 70% of the continuously cored interval (2770 m–3376 m, with the total length of 606 m) is the middle Es_3 to lower Es_3 shale. About 25% of the cored interval is sandy shale deposited in prodelta slope by gravity flow (Wang Deping, 1991). A total of 509 shale samples have been collected and analyzed for TOC, and 137 shale samples have been randomly chosen for ICP-MS analysis, some for analysis of GC and GC-MS, together with carbonate content testing, while a few samples are selected for analysis using an electron microprobe and scanning electron microscope.

After the shale samples having been shattered, the contents of several elements such as Sr, Ba, Ga, Zn, were determined using Inductively Coupled Plasma-Mass Spectrometer (ICP-MS) in the Research Institute of Geochemistry, Chinese Academy of Sciences. The GC and GC-MS analyses of saturated hydrocarbons were conducted separately using a Shimadzu GC-9A chromatograph instrument and Finnigan-MAT TSQ-45 GC-MS system in the geochemistry laboratory, Research Institute of Geological Sciences, Shengli Oilfield.

The organic carbon contents were measured on a CS344 carbon-sulphur instrument (Leco Company, USA) in the analysis and testing center, Research Institute of Geological Sciences, Shengli Oilfield. Other analyses were all conducted in the geochemistry laboratory, Research Institute of Geological Sciences, Shengli Oilfield.

4 SOURCE ROCK CHARACTERISTICS OF SHAHEJIE FORMATION

4.1 *Sedimentology features of the source rock*

The dominant lithology of the upper middle Es_3 formation is grayish and gray mudstone with interbeds of grayish silty shale (Fig. 1) and varicolored shale, together with rich phytogenic carbonaceous sand. Within the central part of the middle Es_3 formation, the dark gray shale is dominant with local interbeds of brown-gray calcareous shale, gray silty shale and fossils of ostracoda, sporopollen and alga. As far as the lower part of the middle Es_3 formation is concerned, dark-grey shale and grayish argillaceous siltstone are dominant with the interbeds of grayish fine sandstone while algal fossils increase significantly. The lithology of the lower Es_3 formation includes dark-grey and grey-dark shale, calcareous shale, and brown-gray oil shale (Fig. 1), with local interbeds of turbidite sandstone and extensive algal fossils accumulating in the layers. On the whole, from the lower to the upper formation of Es_3, the lithology changes from fine grain to coarse grain while the biome transforms from aquatic to terrigenous plants, and the depositional environment gradually changes from strong reduction into

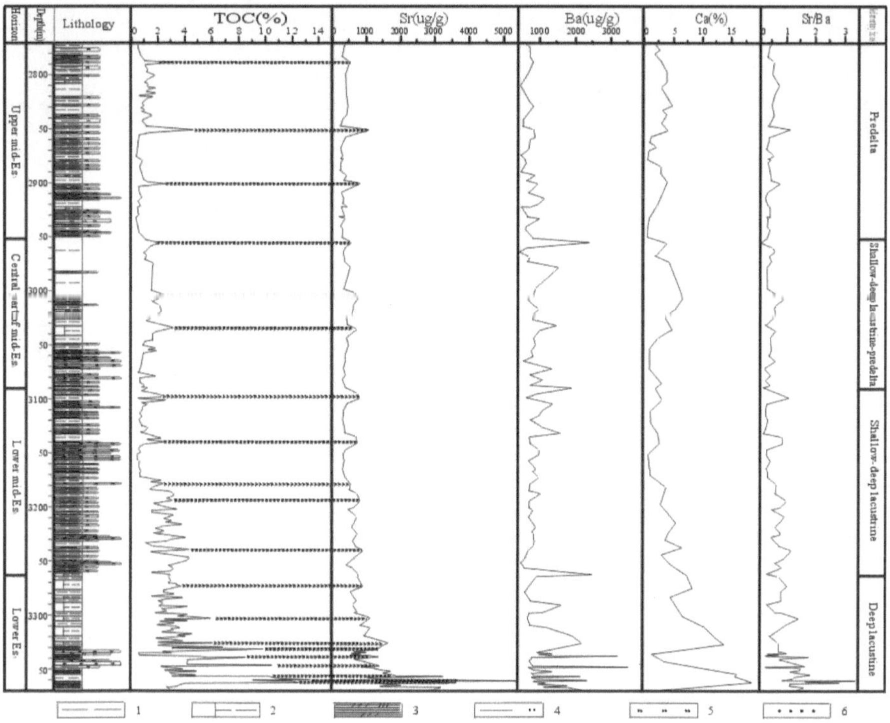

Figure 1. Lithology and geochemical columnar sections of Well Niu-38 in the middle and lower $Es_3$of the Dongying sag.

Table 1. Geochemical characteristics of source rocks in Well Niu-38.

Horizon (m)	Environment	TOC (%)	Organic type	Water salinity	Organic source	Sr (ppm)	Sr/Ba	Ca (%)	Ba (ppm)	Sedimentary facies
Upper mid-Es_3 2770–2952	weak oxidizing-weak reducing	1.01	III, II_2	fresh	higher plant	326.5	0.46	2.26	786.9	prodelta
Middle mid-Es_3 2952–3090	reducing-weak reducing	1.52	II_2, I_1	fresh	higher plant	435.6	0.47	3.46	1016.1	shallow-deep lacustrine-prodelta
Lower mid-Es_3 3090–3263	strong reducing-reducing	2.12	I, II_1	brackish	aquatic organism	533.3	0.59	3.85	1008.8	shallow-deep lacustrine
Lower Es_3 3263–3376	strong reducing	4.38	I	brackish	aquatic organism	1375.7	1.10	9.10	1368.8	deep lacustine

weak reduction, the water depth becomes shallower, and the deep lacustrine facies gradually transforms into shallow lacustrine and delta facies. All the above characteristics reflect the filling process of the lake basin.

4.2 *Geochemistry characteristics of source rocks*

As far as the element composition of the source rock is concerned, the contents of Sr, Ca, Ba, U and Sr/Ba in the lower Es_3 formation are higher than those in the middle Es_3 formation (Table 1), and decline upward, which reflects the salinity decrease of the lake water. The contents of V, Cu, Zn and Ti, however, show the opposite tendency, reflecting the difference between the middle and lower Es_3.

The organic matter abundance in the lower Es_3 (Fig. 1) is rather high with TOC mostly ranging from 2% to 5%. The contents of TOC in oil shale can reach 5%–19%, while those in the dark shale of middle Es_3

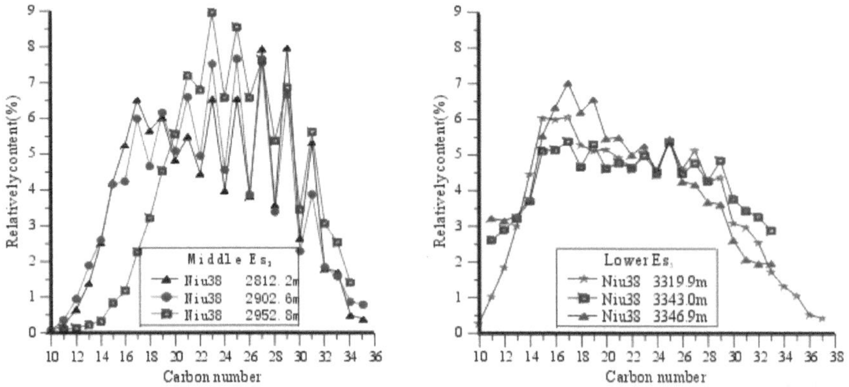

Figure 2. Paraffin distribution correlation of the lower and middle Es2 source rocks of Well Niu 38.

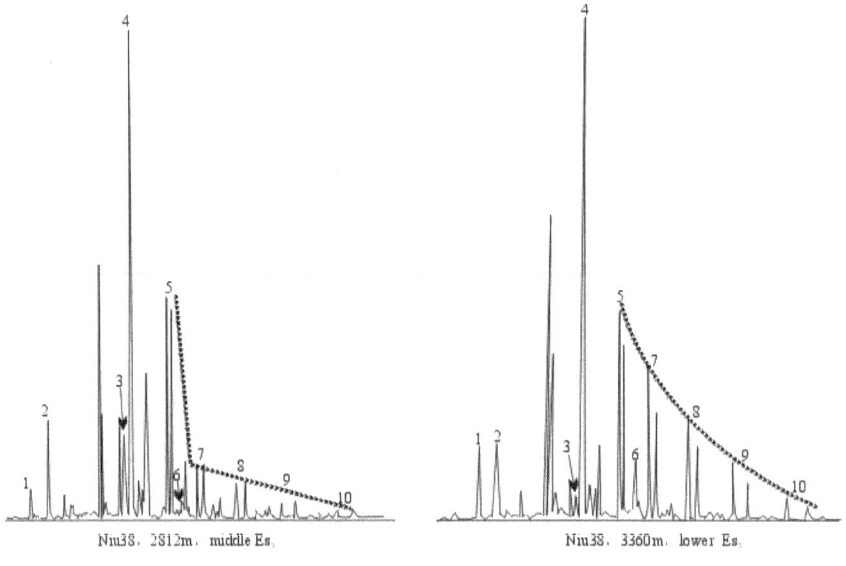

1—Ts; 2—Tm; 3—oleanane; 4—17α(H),21β(H)-hopane; 5—17α(H),21β(H)-29-homohopanes; 6—Gammacerane; 7—17α(H),21β(H)-29-bishomohopanes; 8—17α(H),21β(H)-29-trishomohopanes; 9—17α(H),21β(H)-29-tetrakishomohopanes; 10—17α(H),21β(H)-29-pentakishomohopanes;

Figure 3. Distribution characteristics of hopane series in the source rock from Well Niu-38 of the lower-middle Es3 formation (m/z191).

formation are generally 0.3%–2.1%. The TOC contents in most samples vary between 0.5%–1.2%, and as a whole, they decrease upwards.

The micro-components in source rocks of the lower Es_3 formation are mainly composed of algainite and amorphous matter, and the kerogen is type I and II_1.

The aquatic organism is the main source of organic matter for the source rock.

The normal paraffin hydrocarbon distribution shows a characteristic bimodal pattern, with higher peaks between n-C_{14} and n-C_{23} and mostly at n-C_{16} or n-C_{17}. However, in the middle Es_3 formation, type II_2 and III kerogen is dominant, with an odd carbon-number predominance in n-alkanes. The highest peak of n-alkanes usually is n-C_{25} or n-C_{27} (Fig. 2), which indicates that terrestrial organic matter is the dominant component of organic parent materials.

A moderate concentration of gammacerane is the main feature of the lower Es_3 formation, which disappears in the middle Es_3 formation (Fig. 3), reflecting that the process of the semi-salt water gradually becomes fresh (Seifert W K and Moldowan J M, 1981; Li Renwei et al., 1986). Secondly, the distribution of oleanane is just opposite to that of gammacerane, with a rather low content in the lower Es_3 formation and a gradual increase in the middle Es_3 formation (Fig. 3). Among hopanes, the peak height of C_{31}–C_{35} of the lower Es_3 formation gradually declines in a well-regulated sequence and appears in a lateral arc-like shape. Whereas source rocks of the middle Es_3 formation show predominant C_{31} hopane, and the peak distribution of C_{31}–C_{35} hopanes has a shape of "L", which reflects a delta sedimentary environment of weak oxidization to weak reduction. Such a combined characteristic of biomarker compounds strongly proves the evolution rule of sedimentary environment (Wang Tieguan, 1990; Fu Jiamo et al., 1991), i.e. from the lower to the middle Es_3 formation, the lake water gradually changes from brackish to fresh water while the sedimentary environment changes from strong reduction to weak oxidization or weak reduction, and the infilling process of the lake controls the formation and development of source rocks.

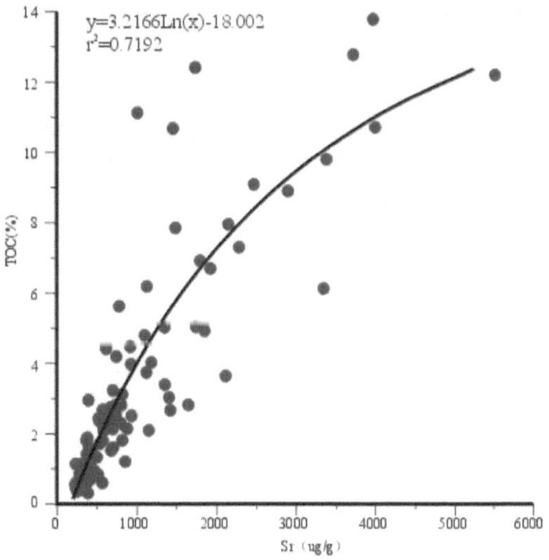

Figure 4. Correlation between contents of Sr and TOC.

5 CORRELATION BETWEEN CONTENTS OF TOC AND SR WITHIN SOURCE ROCKS

5.1 *Vertical distribution of contents of TOC and Sr within source rocks*

According to the analytical results of a number of source rock samples, the vertical distribution of TOC ranges from 0.32% to 12.78%, which indicates strong anisotropy of the Es_3 source rock (Derenne S et al., 2000; Zhu Guangyou et al., 2002). The general tendency, however, is that the TOC content gradually increases from the middle to the lower Es_3 formation and reaches the maximum in the lower Es_3 formation (mostly exceeding 10%). The distribution pattern of Sr content, varying between 223.5 μg/g and 5512.7 μg/g, is similar to that of TOC while the maximum value equals to 24-times the minimum value, also the content of Sr in the lower Es_3formation is apparently higher than that in the upper Es_3 (Fig. 1).

It is worth notice that the fluctuation of TOC content was synchronous with that of Sr, which indicates that there exists a significant positive correlation between them (Fig. 4).

5.2 *Geochemical characteristics and source of Sr*

The abundance of Sr in the continental sediment of China is about 200 μg/g and nearly 375 μg/g in the earth's crust (Zhao Yiyang, 1994), whereas the abundance of Sr within source rocks of the Es_3 formation in the Dongying sag is significantly higher than the above values, especially within high-quality source rocks with higher TOC content in the lower Es_3. So where does this additional Sr come from?

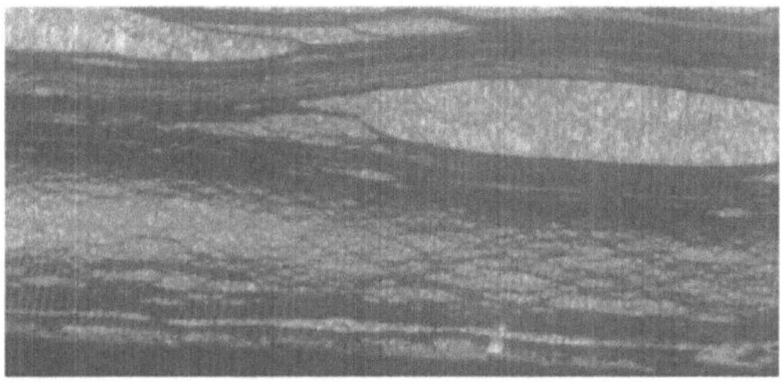

Figure 5. The interbedding of black lamina of organic matter and grey carbonatite (18×48 mm^2).

It is known that the chemical composition of sediments is codetermined by source rock components, sedimentary processes and environments. As far as the lower and middle Es_3 in Dongying rift-lake basin is concerned, the depositional process is stable with same source for the clastic fragments, whereas the only difference is the environment of deposition. In detail, the lower Es_3 Formation was formed in semi-salt environment (Li Chengfeng and Xiao Jifeng, 1988), while the upper being formed in fresh water. As far as the particle size of sediment is concerned, the shale particles in the lower Es_3 Formation, with higher calcium contents and more laminations including thin carbonate laminae, are significantly thinner than those of the upper which almost contain no calcic composition and carbonate lamina (Fig 5). A positive correlation between the contents of Sr and carbonate is obvious ($r2 = 0.71$, Fig 6). At the same time, the samples with higher Sr contents usually have more bedding composed of a mixture of thin carbonates and clay and organic matter laminae (Fig 5). Electron microprobe scanning confirms that the Sr accumulates preferentially within the carbonate laminae and thus shows a strong correlation with Ca (Fig 7). Both theoretical and experimental study has shown that Sr2+ accumulates by replacing Ca2+ in the carbonate lattice (Banner J L, et al., 1994) since Sr2+ and Ca2+ have similar ion radius and electrovalence. The endogenetic deposition of calcium carbonate within the lake is the main reason for abnormally high Sr content.

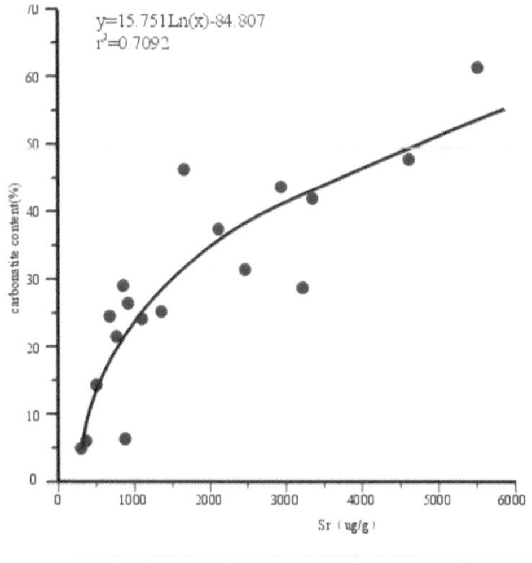

Figure 6. Correlation between contents of Sr and carbonatite.

5.3 *Synchronous accumulation relationship between organic matter and Sr*

Trace elements have a very close relationship with the life activities of organism. Some organisms need specialized circumstances such as rich silicon in diatom and more calcium carbonate in particle alga(Ye Lianjun, et al., 1995). The effect of aquatic organisms on the chemical composition of lake water, by catabolism, accumulation, conversion, storage and release, can directly influence the con- centrations of many substances dissolved in the water. The photosynthesis and respiration of planktonic organism in lake water can also exchange substances and energy with water, therefore changing the composition of water body. The influence of aquatic organism on basic components of water body also includes the chemical element absorptive action and assimilation of organism such as alga (Wu Fengchang, et al., 1987). It is generally thought that the assimilation of CO_2 during photosynthesis of planktonic organisms is the important factor for authigenetic calcium carbonate deposition within lakes (Chen Jingan, et al., 2002). With the temperature being higher and the photosynthesis of planktonic organism being stronger, a great deal of CO_2 absorbed by lake water causes the reaction balance of $Ca2++2HCO_3- = CaCO_3+H_2O+CO_2$ to move toward the right, which generates deposits of calcium carbonate by attaching around the outside of alga cell (Robbins L L and Blackwelder P L,1992). At the same time, high temperature favors the intense

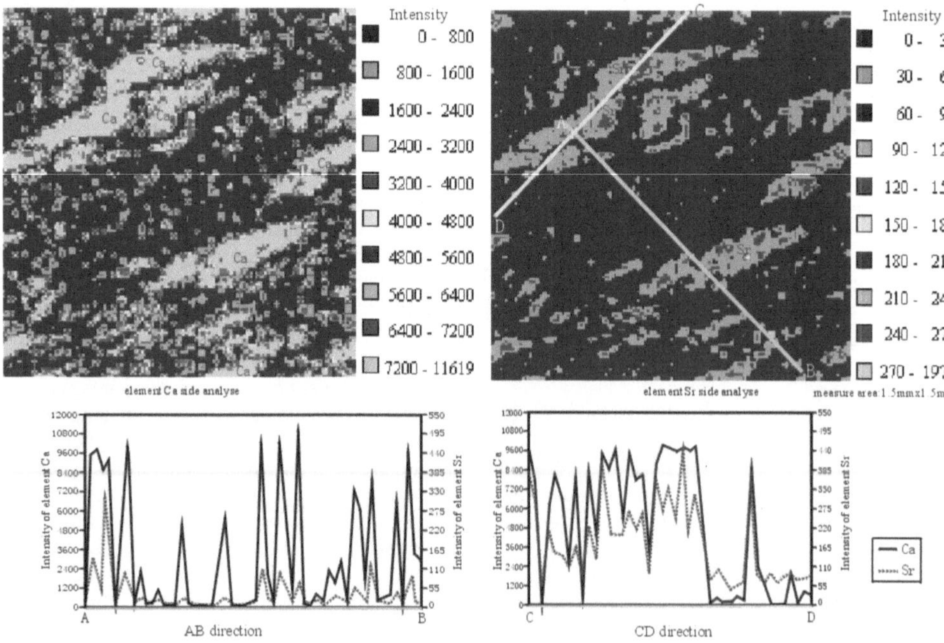

Figure 7. Plane and line analysis of Ca and Sr using electron probe.

evaporation of water resulting in the deposition of calcium carbonate. For a lake with a stable supply of terrigenous clastic sediments, the authigenetic precipitation of calcium carbonate within lake water will directly influence the content of carbonate in sediments. In particular, in a semi-salt confined lake as the lower Es_3 Dongying sag, the deposits of authigenetic calcium carbonate can account for a rather higher proportion of the overall sediment. The high quality source rocks in lower Es_3 Formation usually have laminated structure while the phenomena of algal blooms are rather common (Liu Chuanlian, et al., 2001). On a microcosmic view, the source rocks usually have a three-layer structure wherein the primary composition of the bottom is coccolith (Zhang Linye, et al., 2003) or calcite, and those of the middle-layer are organic matter and clay, with grey pyrite particle in thinner upper-layer (Fig 8). Usually, with increasing temperature, the planktonic organism such as alga bloom gradually, which strengthen the photosynthesis and result the deposit of carbonate. In summer and autumn, with the organism and alga growing up and blooming, the lamina with higher content of organic matter from the remains of alga and animal (waxy yellow in microscope) and those with higher content of Ca can be generated. Whereas in winter, since the lower productivity, the deposits of grey-black grain assemblages were mainly composed of pyrite particle (Fig 8), which recorded the whole geological process of alga blooming, death and burial. Carbonate lamina is commonly formed through the absorption and assimilation of chemical elements within water body due to alga blooming (Kelts K and Hsu K J, 1978), with the grey laminae of pure calcium carbonate viewed via the scanning electron microscope are mainly composed of archetypal crystalline grains of calcsparite. Since Sr mainly accumulates in carbonate laminae, therefore, the source of Sr has a relation with alga blooming. The blooming of organism (such as alga), on one hand, increased the deposit amount of calcium carbonate within lake water and resulted a great deal of carbonate deposition (Dean W E, 1999) by absorbing more CO_2, while the Sr easily occupied the crystallographic site of Ca and built up in $CaCO_3$. On the other hand, the death of alga not only increased the supply of organic matter into lake water, but also the preservation amount of organic matter, which caused the generation of high quality source rock. Therefore, it appears the phenomenon of synchronous accumulation of abnormally high contents of TOC and Sr.

6 SIGNIFICANCE OF ENVIRONMENT INDICATING

For a stable lake with terrestrial supply, the content of calcium carbonate can effectively indicate climate change, i.e., the rising temperature can strengthen the photosynthesis and result in lake water oversaturated with carbonate, finally generate the deposits of carbonate (Chen Jingan, et al.,1999).

The reason for annual laminations in lake sediments is that the changes of temperature and rainfall lead to the difference of chemistry of lake water and its original productivity (Kelts K and Hsu K J, 1978). The grey interbeddings, mainly composed by crystals of calcite formed due to biochemical effect, belong to the substances deposited in spring and summer of moist climate zone. Whereas, the black interbeddings, with rich organic matter and composed of the remains of alga and planktonic organism, mostly belong to sediments of autumn. The changing of some factors, such as seasonal climate and biologic activities etc, lead to the corresponding change of the composition and amount of sediments and form the annual lamination, which records the information of seasonal environment changing. Therefore, the annual lamination has high time discernibility, which has unique value for study of environmental diversification. Evaporation due to temperature rising has relations with seasonal deposition of $CaCO_3$ (Kilham P, 1990). The change of climate (such as temperature and rainfall)

A. Color lamina formed by different components under microscope (tint lamina are mainly made up of coccolithophore plate scale or cryotocrystal calcspar, yellow lamina are composed of organic matter and clay, black lamina are mostly made up of iron pyrite grains)

B. iron pyrite grains

C. algal blooming

D. carbonate laminas and clay laminas with rich organic matter viewed via the scanning electron microscope

Figure 8. Microcosmic structure features of calcic mudstone lamina (A, B and C are under microscope, D is under scanning electron microscope).

can change the concentration of water body by adjusting the amount of lake water (Wu Fengchang, et al., 1987).

The broadly distributed annual laminations in source rock of the lower Es_3 Formation of Dongying sag indicate that the water body has obvious seasonal or long-term phenomena of delamination, and the surface of sediment has no disturbance sign since weak water recycling and biologic disturbance. In addition, the long-time anoxic environment of bottom water is favorable to effective preservation of sedimentary organic matter. The perfect ichthyolite preserved in good condition during coring observation indicates a stable reducing environment According to the distribution of carbonate content in Es_3 Formation, the erosion occurred during the depositional period of Es_3 was weak while the sediment of $CaCO_3$ resulted from evaporation strengthen the deposition of endogenesis calcium carbonate in lake, which made carbonate lamination blooming in source rock of the lower Es_3 Formation. Meanwhile, the weaker erosion within the drainage area and the stronger deposition of endogenetic calcium carbonate indicate the environmental conditions of lacking flowing water on the surface and the lake water being evaporated and concentrated. Also, it indicated the climate at that time just being warm-dry period with features of drought and high temperature. The strong erosion within drainage area, occurred during the depositional period of middle Es_3 formation, and weak deposition of endogenetic calcium carbonate reflect the conditions of the flowing water on the surface within the drainage area of lake basin being blooming, and the lake water being desalted, the lake surface being expanded, and the sedimentary process quickened, which indicate the regional climate being cold-wet period with rainy and low temperature. Therefore, the content of Ca in deposit of lake is controlled by the diversification of climate. Since the chemical characteristic of Sr is similar to that of Ca and both belong to alkaline-earth metals, the deposits of $CaCO_3$ were frequently concomitant in lake water. Therefore, Sr can be used as the index of climate and environment. The forming of endogenetic calcium carbonate lamination in lake reflects the delaminated status of lake water, which is just the key condition for generation of high-quality source rock.

ACKNOWLEDGEMENTS

We are gratitude to Xu Jinli and Chen Zhilin, Senior Engineers of the Research Institute of Geological Sciences, Shengli Oilfield (SINOPEC) for their support, to Institute of Geochemistry, Chinese Academic of Sciences and Geochemistry Lab of Research Institute of Geological Sciences in Shengli Oilfield for a great deal of analyses, to Wan Congli, a doctoral student of the University of Petroleum, China for assistance in coring observation and sampling, and to Prof. Ma Zaiping for assistance in observation by scanning electron microscope.

REFERENCES

Banner J L, Musgrove M & Capo R. Tracing groundwater evolution in a limestone aquifer using Sr isotopes. Geology, 1994, 22:687–690.

Banner J L. Application of the trace element and isotope geochemistry of strontium to studies of carbonate diagenesis. Sedimentary, 1995, 42: 805–824.

Bertrand P, Lallier-Verges E & Boussafir M. Enhancement of accumulation and anoxic degradation of organic matter controlled by cyclic productivity: a model. Organic Geochemistry, 1993, 22:511–520.

Blum J D & Erel Y. A silicate weathering mechanism linking increasing in marine87Sr/86Sr with global glaciation. Nature, 1995, 373: 415–418.

Brand U. Strontium isotope diagenesis of biogenic aragonite and low-Mg calcite. Geochim Cosmochim Acta, 1991, 55:505–513.

Brass G W. The variation of the marine 87Sr/86Sr ratio during Phanerozoic time. Geochim Cosmochim Acta, 1976, 40:721–730.

Chen Jingan, Wan Guojaang, Wang Fushun, Zhang D D, Huang Guirong, Zhang Feng & Schmidt R. Environmental records of carbon in recent lake sediments. Science in China(Series D), 2002, 45(10):875–884.

Chen Jingan, Wan Guojiang, Chen Zhenlou & Huang Ronggui. Chemical elements in sediments of lake Erhai and palaeoclimate evolution. Geochimica, 1999, 28(6): 562–570.

Chen Jun, Ji Junfeng, Qiu Gang,et al. Geochemical studies on the intensity of chemical weathering in Luochuang loess sequence, China. Science in China(Series D), 1998, 41(3):531–536.

Dean W E. The carbon cycle and biogeochemical dynamics in lake sediments. Journal of palaeoclimates. Journal of Palaeolimnology, 1999, 21(4):375–393.

Deng Hongwen & Qian Kia 1990. The genetic types and association evolution of deep lacustrine faces mudstones. Acta sedimentologica Sinica, 8(3):1 ~ 20 (in Chinese with English abstract).

Denison R E, Kirkland D W & Evans R. Using strontium isotopes to determine the age and origin of gypsum and anhydrite beds. Journal of Geology, 1998, 106:1–17.

Depaolo D J & Ingram B L. High-resolution stratigraphy with strontium isotopes. Science, 1985, 227:938–941.

Derenne S, Largeau C & Brukner-Wein A. Origin of variations in organic mater abundance and composition in a lithologically hogogeneous maar-type oil shale deposit. Organic Geochemistry, 2000, 31:787–798.

Dix G R. Origin of Sr-rich magnesian calcite mud in a Holocene pond basin. Journal of Sedimentary Geology, 2001, 71(1):167–175.

Fu Jiamo, Sheng Guoying, Xu Jiayou, et al 1991. Application of biomarker compounds in assessment of paleoenvironments of Chinese terrestrial sediments. Geochemica, (1):1 ~ 12 (in Chinese with English abstract).

He Xiubin. Recent advances in the trace element and isotopic geochemistry of strontium and application perspectives. Advance Earth Sciences, 1997, 12(1):15–19.
Hess J, Bender M L & Schilling J G. Evolution of ratio of strontium-87 to strontium-86 in seawater from Cretaceous to present. Science, 1986, 231:979–984.
Hodell D A. Variations in the strontium isotopic composition of seawater during the Neogene. Geology, 1991, 19:24–27.
Hong Zhihua, Chen Zhilin, Yin Mo & Zhang Chunrong. Distribution characters of organic matter in middle-lower part of Sha-3 member in Niu38 well, Dongying depression. 1991, 12(3):292–299.
Jin Qiang & Zha Ming 2000. Co-sedimentation of tertiary evaporates and oil source rocks in the western Qaidam basin. Chinese Journal of Geology, 35(4):465 ~ 473 (in Chinese with English abstract).
Kelts K & Hsu K J. Freshwater Carbonate Sedimentation. In: Lerman A. ed. Lakes: Chemistry, Geology, Physics. Berlin: Spinger-Verlag, 1978, 295–320.
Kilham P. Mechanisms controlling the chemical composition of lakes and rivers: data from Africa, limnol. Oceanogr.1990, 35(1):80–85.
Li Chengfeng & Xiao Jifeng.1988. The application of trace element to the study on paleosalinities in Shahejie formation of Dongying basin Shengli oil field. Acta Sedimentologica Sinica, 6(4):100 ~ 107 (in Chinese with English abstract).
Li Renwei, Lin Daxing & Wang Zhizhen, et al.1986. The new criteria used by judging high-salinity environment J. Chinese Science Bull. (8):604 ~ 607 (in Chinese).
Liu Chuanlian, Xu Jinli & Wang Pinxian 2001. Algal blooms: the primary mechanism in the formation of lacustrine petroleum source rocks. Geological Review,47(2):207~210(in Chinese with English abstract).
Liu Dongsheng. Loess and the Environment. Beijing: Science Press, 1985:336–348.
Liu Yingjun, Cao Liming & Li Zhaolin,et al. Element Geochemistry: Beijing, Science press, 1984, 360–365.
Mc A J M, Burnett J & Hancock J M. Strontium isotopes at K/ T boundary discussion. Nature, 1992, 355(6355):28.
Palmer M R & Elderfield H. Sr isotope composition of seawater over the past 75 Ma. Nature, 1985, 314:526–528.
Pederson T F & Calvert S E. Anoxia versus productivity: what controls the formation of organic-carbon-rich sediments and sedimentary rock? AAPG Bulletin, 1990, 74:454–466.
Richard J H, Mc Arthur J M. Statistics for strontium isotope stratigraphy: a robust LOW ESS fit to marine Sr-isotope curve for 0 to 206Ma, with look-up table for derivation of numeric age. Journal of Geology, 1997, 105:441–56.
Robbins L L, Blackwelder P L. Biochemical and ultra structural evidence for the origin of whitings: A biologically induced calcium carbonate precipitation mechanism. Geology, 1992, 20:464–468.
Seifert W K, Moldowan J M. Paleoreconstrucation by biological markers. Geochimica et coschimica acta, 1981, 45:783–794.
Tissot B P, Welte D H. Petroleum Formation and Occurrence. Berlin Spinger-Verlag, 1984:339–416.
Wang Aihua. Discriminant effect of sedimentary environment by the Sr/Ba ratio of different exising forms. Acta sedimentologica Sinica, 1996, 14(4):168–173.
Wang Deping.1991. The sedimentation and formation mechanism of lacustrine endogenic debris flow. Acta Geologica Snicia, 65(4):299 ~ 316 (in Chinese with English abstract).
Wang Tieguan. 1990. A contribution to some sedimentary environmental biomarkers in crude oils and source rocks in China. Geochimica, 19(3):256 ~ 262 (in Chinese with English abstract).
WuFengchang, WanGuojiang, CaiYurong, WanXi. Geochemical mechanisms controlling basic chemical compositions of lake water mass. Advances in Water Science, 1997, 8(2):169–175.
Ye Lianjun, Fan Pu, Ma Baolin, Huang Difan, Cheng Keming. Biogeochemical circulation, paleoenvironment and biomineralization of the meso-cenozoic non-marine sedimentary basins in China. Advance in Earth Sciences, 1995, 10(5):427–431.
Zhang Linye, Kong Xiangxing, Zhang Chunrong, et al.2003. High-quality oil-prone source rocks in Jiyang Depression. Geochemica, 32(1):35 ~ 42 (in Chinese with English abstract).
Zhang Zhihuan, Yang Fan, Li Dongming, et al. Biomarker assemblage characteristics of source rocks and associated crude oil in saline lake facies of Cenozoic in China[J]. Acta sedimentologica sinica, 1998, 16(3):119 ~ 123.
Zhao Yiyang, Yan Mingcai. Geochemistry in sediments of the China shallow sea:Beijing, Science press, 1994, 43–46.
Zhu Guangyou, Jin Qiang.2002. On source rock heterogeneity – a case study of the well Niu-38, Dongying depression. Acta Petrolei Sinica, 23(5):34 ~ 39(in Chinese with English abstract).
Zhu Guangyou, Zhong Jianhua & Zhou Yaoqi, et al. The dinosaur eggshell fossil of the late Cretaceous period from Xixia basin, Henan proince, China: Supper high content of strontium and its significance. Acta sedimentologica Sinica,1999,17(4):659–662.

Petroleum Geochemistry and Exploration in the Afro-Asian Region –
Digang Liang, Darui Wang & Zhenxi Li (eds)
© 2008 Taylor & Francis Group, London, ISBN 978-0-415-44083-7

Comparison of Proterozoic basins of India with similar basins of the world: Implications to hydrocarbon resource prospects

C. Vishnu Vardhan, Smitha K. Panicker & B. Kumar
National Geophysical Research Institute, Hyderabad, India

ABSTRACT: Hydrocarbon discoveries in Proterozoic basins from different parts of the world viz. Lena Tunguska, Siberia; Amadeus and Mc Arthur basins, Australia; Yangtze Craton, China and Huqf basin, Oman etc. have demonstrated that Proterozoic marine biomass had potential for generation, entrapment and accumulation of hydrocarbons. The wells drilled in Proterozoic Vindhyan and Bikaner-Nagaur basins of India have yielded gas shows and commercial find of hydrocarbons. Some of the Proterozoic basins of India are geologically and structurally similar to the other producing basins of the world and contain all the prerequisites for petroleum generation i.e. sufficient sediment thickness, evidence of rich stromatolitic and fossil assemblages and presence of organic matter etc. There are reports of presence of fair to good source rocks, reservoir rocks and cap rocks from some of the Proterozoic basins of India. These basins are characterized as Frontier basins (category III and IV) with indeterminate hydrocarbon prospects. Exploration efforts in these basins comprising geological, geophysical and geochemical studies are in progress as part of India Hydrocarbon Vision-2025. The hydrocarbon prospects of selected Proterozoic basins of India, viz. Vindhyan and Cuddapah in analogy with selected producing basins of the world are compared and discussed.

1 INTRODUCTION

The occurrence of commercial oil and gas accumulations in Proterozoic formations of Siberia, Australia, Oman, china etc. have proved that Proterozoic biomass was capable of generating hydrocarbons and older strata should not be omitted in the everlasting search for petroleum. The six Proterozoic basins of India viz. Vindhyan, Cuddapah, Chattisgarh, Bastar, and Bhima -Kaladgi are grouped under category III and IV i.e. prospective to potentially prospective basins. These basins offer favorable structural geometry for hydrocarbon generation and accumulation viz. large open anticlines, buried domal structures, inversion structures, cross folds which are typical of giant oil fields (Jokhanram, 2003). The Proterozoic Basins of India are almost undeformed and unmetamorphosed with aggregate sediment thickness varying from 0.5 to 10 Km. Presence of rocks which can act as good reservoirs and seals are reported from these basins. Hydrocarbon occurrences in Proterozoic basins have been reported by presence of gas shows from the test wells drilled in Vindhyan basin of stratified, unmetamorphosed sandstones, shales and limestones. The sedimentary sequence has been divided into two lithostratigraphic units, the Lower Vindhyan (Semri Group) and the Upper Vindhyan (Kaimur, Rewa and Bhander groups), separated by a (Srivastava et. al 1983) and presence of bitumen and oil seeps from Cuddapah basin (Kuldeep Chandra et al., 1999), respectively. This shows that basic elements necessary for hydrocarbon generation might have existed during the Precambrian. The paper reports geological, geophysical and geochemical aspects favorable for hydrocarbon generation from two Proterozoic basins of India viz. Vindhyan and Cuddapah basins and compares with other Proterozoic basins of the world.

2 GEOLOGY AND TECTONICS OF VINDHYAN AND CUDDAPAH BASINS

2.1 *Vindhyan basin*

The Vindhyan basin is an intracratonic basin, having an areal extent of 165×10^3 km^2 of which about 40% lies under the Deccan Trap volcanic flows (Fig. 1). Sediments of this super group attain a maximum thickness of about 5250 m (Srivastava et al., 1983) deposited in a shallow marine domain and comprise well-marked erosional unconformity. Stratigraphy of the Vindhyan Super Group is given in Table 1. The Vindhyan sediments are least metamorphosed and referred as the best preserved Proterozoic sedimentary succession of India.

The basin has a characteristic rhombohedral configuration bounded by major faults and is divided by

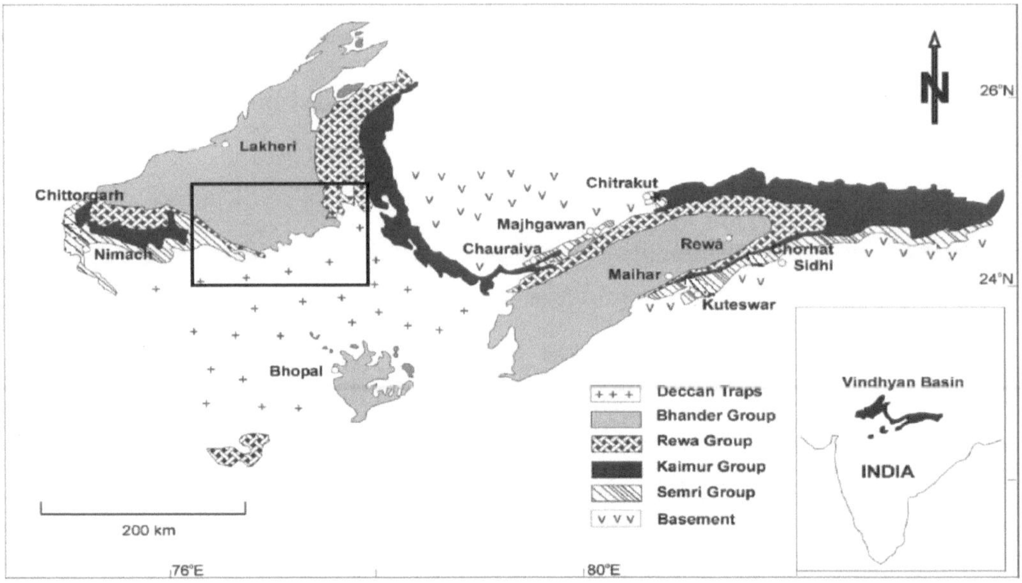

Figure 1. Geological map of Vindhyan Basin (modified after Ray et al., 2003). The demarcated area indicate the anomalous zone from adsorbed soil gas surveys.

Table 1. Generalised Stratigraphy of Vindhyan basin (modified after Sastry and Moitra, 1984; Bhattacharyya, 1996).

Group	Subgroup	Formation
UPPER VINDHYAN		
Bhander		Bhavpura
		Balwan Lst.
		Shikaoda Sst.
		Sirbu Shale
		Bundi Hill Sst.
		Lakheri Lst.
		Ganurgarh Shale
Rewa		Govindgarh Sst.
		Jhiri Shale
		Asan Sst.
		Panna Shale
Kaimur		Dhandraul Sst.
		Mangesar
		Bijaigarh Shale
		Ghaghar Sst.
		Susnai Breccia
		Sasaram
....	Unconformity	....
LOWER NDHYAN		
	Rohtas Subgroup	Bhagwar shale
		Rhotas Limestone
Semri	Kheinjua Subgroup	Rampur Fm.
		Salkhan Limestone
		Koldaha Shale
	Mirzapur Subgroup	Deonar Fm.
		Kajarahal Lst.
		Arangi Fm.
		Deoland Fm.

Bundelkhand Massif into two parts, the Chambal Valley in the west and the Son Valley in the east. Based on Gravity-Magnetic data and the seismic reflection data, the basin dips southward, the deepest part being close to southern basin margin. The Lower Vindhyan (Semri) is comparatively more structurally deformed than the Upper Vindhyan. Normal faults arranged in an en-echelon pattern, along the son Narmada lineament; series of step faults along the southeastern boundary of Bundelkhand massif and NW-SE trending tight folds around Chittorgarh are reported. In addition to intra-basinal faults there are many exposed folds associated with major fault systems, which are aligned in the direction of the adjoining faults.

2.2 *Cuddapah basin*

The Cuddapah (Fig. 2) is an epicratonic basin covering an area of 44.5×10^3 km^2. (Vivek S. Kale, 1991). The sediments have an aggregate thickness of about 12 km and the thickness increases from west to east signifying the deepening of basin towards east. The sedimentation in Cuddapah basin took place in a shallow marine environment and the sequences are dominantly argillaceous and arenaceous with subordinate calcareous sediments. The Cuddapah Super Group is divided into Papaghni (2100 m), Chitravati (6000 m) Nallamalai (3500 m) and Kurnool (520 m) groups separated by an unconformity (Table 2). Each group starts with quartzite and ends with a shale unit representing cyclic repetition of quartzite and shale sequence reflective of transgression and regression in

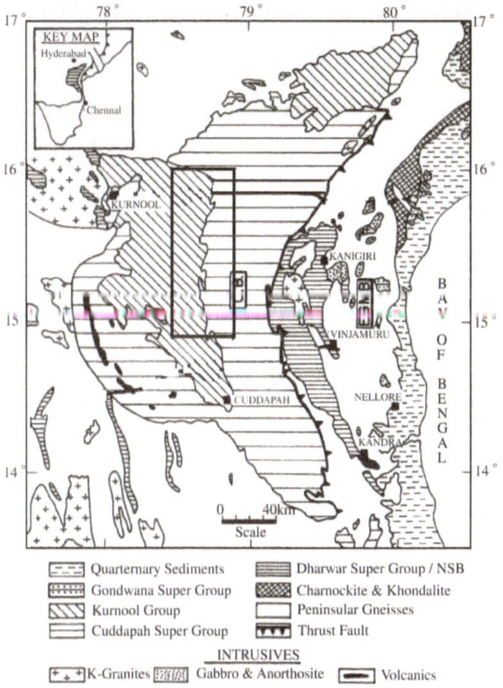

Figure 2. Geological map of Cuddapah basin (modified after Singh and Mishra, 2002). The demarcated area indicate the anomalous zone from adsorbed soil gas surveys.

an episodically sinking basin. The sediments in the basin are mostly mature (Naqvi and Rogers, 1987) and unmetamorphosed except on the eastern part due to the thrusting of Eastern Ghat Mobile Belt.

The western half of the Cuddapah basin is less deformed with sub horizontal or gentle quaquaversally dipping beds (Nagarajarao, et al., 1987). The beds are generally dipping towards east, and affected by a series of faults. The Papagni and Chitravati Groups are mostly undisturbed but the Nallamalai Group is highly disturbed with isoclinal and recumbent folds, complex faulting and also thrusting with the eastern ghat mobile belt. Culminations and depressions are common which have developed either due to superposed folding or due to uneven basement. Iswarkuppam, a N-S elongated dome is an important structural feature in the north-central part of the basin (Ramam et al., 1997). Faults in the basin are mostly normal and reverse types, whose periodic reactivation may have played a major role in the evolution of Cuddapah basin. The eastern margin of the basin is intensely faulted and affected by thrusts, which has been confirmed by deep seismic studies (Ramam et al., 1997).

Table 2. Stratigraphy of Cuddapah basin (modified after Ramam et al., 1997).

Group	Formation	Lithology
CUDDAPAH		
Kurnool	Nandyal Shale	Shale
	Koilkuntla Lst.	Limestone
	Paniam Qrtzite	Quartzite
	Owk shale	Shale (ocherous)
	Narji Lst.	Limestone
	Banganapalli Qrtzite	Conglomeratic Quartzite
	Unconformity...	
	Srisailam Qrtzite	Quartzite
	Unconformity...	
	Cumbum (Pullampet)	Phyllite,shale,
Nallamalai		qtzite, dolomite
	Bairenkonda (Nagari Quartzite)	Bairenkonda qtzite, shale,Nagari congl., qtzite & shale with intrusives
SUPERGROUP		
	Angular Unconformity	
	Gandikota Qtzite	Qtzite, shales, sills of olivine dolerite
Chitravati	Tadpatri	Shale, tuffs, qtzite, dolomite with Intrusives (flows, sills, ignimbrites, tuff)
Papaghni	Pulivendla Qtzite	Congl. & qtzite
	Disconformity..	
	Vempalle	Stromatolitic dolomite, mudstone, chert, Breccia, qzite with basic flows and intrusives.
	Gulcheru Quartzite	Congl. arkose, qtzite and shale

3 PALEOLIFE

3.1 *Vindhyan basin*

Microbial life in Lower and Upper Vindhyans is represented by the well-preserved *algal stromatolites, primitive brachiopods* etc. Stromatolites are invariably present in the chemogenic sediments of all groups of Vindhyan Supergroup. Three different groups of *stromatolites* have been recognized in early Proterozoic, Riphean and Vendian sediments of Vindhyan Basin. The Semri Group yielded *Kussiella-Colonella* and *Conophyton-Colonella* assemblages and Bhander Group contain *Baicalia-Tungassica* assemblage. *Kusiella-Colonella* assemblages are characteristic of

Table 3. Comparison of Vindhyan and Cuddapah basins of India with Lena- Tunguska and Sichuan of USSR and China.

	Vindhyan	Cuddapah	Lena-Tunguska	Sichuan
Basement	Precambrian	Precambrian	Precambrian	Precambrian
Basin Type	Intra-cratonic	Epicratonic	Intra-cratonic	Intra-cratonic
Tectonics	Structurally Controlled	Structurally controlled	Structurally controlled	Structurally controlled
Sediment Thickness	~ 5 Km	~ 10 Km	~ 4 Km.	6–10 Km.
Depositional Environment	Marine	Marine	Marine	Marine
Lithology (Major)	Limestone,shale, sandstone and dolomite	Quartzite, limestone, shale & sandstone	Dolomite, Limestone, Shale, Sandstone, Siltstone and Conglomerate	Dolomite, Limestone, Shale and Sandstone
Source Rocks	Probable source Rocks are Limestones and Shales	Probable source Rocks are Limestones, Shales and Dolomites	Sandstone	Carbonate and black shale
Volcanism	Present in the Southern part	Absent	Present	Present
Status	Yet to be Explored	Yet to be Explored	Producing	Producing
Toc Content	Upto a maximum of 8.08%	....	5–10%	0.4–0.5%
Nature of Hydrocarbons discovered	Reported presence of gas shows	Reported Presence of Bitumen & oil seeps	Gas & oil	Gas

Lower Riphean of USSR and *Conophyton-Colonella* assemblages are comparable to that from Middle Riphean of USSR. *Annelid* and *Arthropod* remains are also reported from Semri and Bhander groups. Both planktonic and benthic microfossil assemblages are present in the basin and are well documented in Semri Group (Srivasthava et. al. 1983). *Chuaria sp., acritarchs, algal* and *fungal* remains, macrofossils and vascular plants have been reported from Vindhyan sediments (Srivastava et al., 1983; Venkatachala et al., 1996). About 50 records of *metaphytes* and *metazoa* are also recorded from different groups of Vindhyan Supergroup.

3.2 *Cuddapah basin*

The sediments of the Cuddapah Basin have yielded rich *stromatolitic assemblages, blue green algae* etc. Lower Cuddapah (Papaghni & Chitravati Groups) carbonates consist of stratigraphically controlled columnar *stromatolites*. The Vempalle formation of Papaghni Group is characterized by the *Collenia sp., Cryptozoon sp., Gymnosolen, Jurusania, Jacutophyton, Conophyton, Kussiella and Inzeria Oncoithic sp.*, and *Stratifera sp.* etc demonstrating the extent of proliferation of microbial life during Cuddapah sedimentation (Schopf and Prasad, 1978; Raha et. al 1982; Gururaja and Chandra, 1987). *Ediacaran* fossils from Paniam quartzite of Kurnool group; *Sabellidites* a worm tube and *chuaria* a mega fossil in Owk shale of Kurnool group have been reported (Gururaja et al., 2000). These indicate the prolific extent of life during Proterozoic times in the basin. In the southwest part of the Cuddapah basin (Thummalapalle-Gandankipalle), the shale contains thin stringers of brown colored organic matter (bitumen ?) and humic acid produced by decay of algal colonies (Kuldeep chandra et al., 1999).

4 COMPARISON OF PROTEROZOIC BASINS OF INDIA WITH PRODUCING BASINS OF THE WORLD

A comparison of Vindhyan and Cuddapah basins of India with Lena-Tunguska and Sichuan basins of USSR and China are given in Table 3. It is evident that both Vindhyan and Cuddapah basins are stratigraphically, structurally and tectonically similar to these producing basins. Other factors that favor the generation and entrapment of hydrocarbons in these basins are discussed below.

4.1 *Vindhyan basin*

The Vindhyan sediments have considerable thickness, organic content, suitable structures, good reservoir and cap rocks, which are favorable for the generation and entrapment of hydrocarbons. Normal faults (anticlinal structures) arranged in an en-echelon pattern along the Son Narmada lineament, and series of step faults along the southeastern boundary of Bundelkhand massif may act as favorable locales for hydrocarbon accumulation. Krishnan (1997) recorded that the lower Vindhyan shales have geochemical parameters with possible source-rock potential. The limestones and sandstones have developed fracture porosity and secondary porosity to form good reservoir rocks. The cap rock conditions are better in Upper Vindhyans due to the presence of alternating shale, sandstone, and limestone sequences. Vindhyan sediments extending in the northern part below the alluvium have hydrocarbon generation potential. Considerable TOC content (upto 8.08% in Kajarahat Limestone, 6.43% in Bijaigarh Shale, 1.09% in Carbonaceous shales) and the fair amount of organic matter indicate the development of potential source rocks in Vindhyans. The limestone and shale sequences have been found to possess marginal to good source rock potential on the basis of organic carbon distribution (Corg. Upto 1.76%). Carbonate rocks of the Nagod Formation have been characterised by marked positive $\delta^{13}C$ signatures (4.1±0.9‰) suggesting higher organic productivity / burial which is one of the prerequisites for the existence of hydrocarbon source rocks (Kumar et. al, 2000). Adsorbed Soil Gas Surveys carried out by National Geophysical Research Institute in the western part of Vindhyan basin (Chambal Valley) have shown that the region in and around Baran-Jhalawar-Bhanpura-Garot has anomalous concentration of C_1–C_4 (Kumar et al., 2002) in shallow soils. Tectonically, Vindhyan basin is comparable to Proterozoic basins of USSR (Lena Tunguska) and Australia (Amadeus), which are producing.

4.2 *Cuddapah basin*

Cuddapah is the thickest of all the Proterozoic Basins of India. The Vempalle limestone Formation of Papaghni Group and Tadpatri shales of Chitravati Group contain abundant stromatolites and can be potential source rocks in Cuddapah basin. Further the Vempalle limestone/dolomite may act as good reservoir rocks with its vuggy porosity and fracture porosity. The Nagari and Pulivendla quartz-sandstones (orthoquartzites) of Chitravati Group also can form potential reservoirs. The structures like synclines, anticlines, fault closures etc. can play an important role in hydrocarbon generation and entrapment. Kuldeep Chandra, et al., (1999) have reported that it would be ideal to look for the traps below the Nallamalai group of sediments, which are highly disturbed. The Nandyal shale of the Upper Kurnool Group is an ideal cap rock and may be a likely target for petroleum exploration. The thickness of the sediments where the Nadyal shale is exposed is in the range of 9000–6000 m. (Kuldeepchandra et al., 1999). Adsorbed gas surveys carried out by NGRI have indicated that the northwestern region around Nandyal in the Cuddapah Basin may form a prospective zone (Kumar et al., 2005).

5 SUMMARY

- Vindhyan and Cuddapah basins of India are stratigraphically and structurally similar to Lena-Tunguska and Sichuan basins, which have yielded commercial hydrocarbons.
- Both the basins have basic pre-requisites for the generation of hydrocarbons such as presence of good sediment thickness, fair amount of organic content, suitable structures, reservoir rocks, cap rocks etc.
- Carbonate rocks of the Nagod Formation (Limestone Formation) of Bhander Group have been characterised by marked positive $\delta^{13}C$ signatures suggesting higher organic productivity / burial which is one of the prerequisites for the existence of hydrocarbon source rocks in Vindhyan Basin.
- Adsorbed Soil Gas Surveys in Vindhyan Basin have shown that the region in and around Baran-Jhalawar-Bhanpura-Garot has anomalous C_1–C_4 concentrations.
- Adsorbed Soil Gas Surveys in Cuddapah Basin have shown that the northwestern region around Nandyal may form a prospective zone.

Structures present along the southeastern part of Bundelkhand Massif and along the southern margin, bounded by Narmada Son Lineament in Vindhyan Basin, and favorable structures in Papagni, Chitravathi and Kurnool Groups of Cuddapah Basin could be ideal sites for hydrocarbon entrapment.

ACKNOWLEDGEMENTS

The authors are thankful to Director NGRI for granting permission to publish this paper. Shri Kuldeep Chandra, Ex. Executive Director, KDMIPE, ONGC is gratefully acknowledged for useful discussions. One of the authors (Smitha K. Panicker) is thankful to NGRI-CSIR for providing Senior Research Fellowship.

REFERENCES

Bhattacharyya, A., 1996. Recent Advances in Vindhyan Geology. 331p.

Gururaja, M.N., Ashok Kumar, P., Ch. Venugopal Rao and Chavhan, R. H. 2000. Precambrian-Cambrian boundary strata in Cuddapah basin, Andhra Pradesh, Geol. Surv. Ind. Spl. Publ. 55: 155–162.

Gururaja, M.N. and Chandra, A., 1987. Stromatolites from Vempalle and Tadpatri Formations of Cuddapah Proterozoic Supergroup, A.P. In: Purana Basins of India, Geol. Soc. India Memoir 6: 399–427.

Jokhan Ram 2003. Frontier Basins: A major thrust area in Indian energy scenario, "New Challenges in Hydrocarbon Research and Exploration" CSIR Diamond Jubilee workshop III, NGRI, Hyderabad.

Jyotiranjan S., Ray, Veizer, J. and Davis, W.J., 2003 C, O, Sr and Pb isotope systematics of carbonate sequences of the Vindhyan Supergroup, India: age, diagenesis, correlations and implications for global events, Precambrian Research, 121:103–140.

Krishnan, P.V., 1997. Hydrocarbon exploration in Madhya Pradesh. GSI Misc. Publ. 63:77–89.

Kuldeep Chandra, Punati S., Rao, B. N., Prabhu and N. J., Thomas 1999. Hydrocarbon Potential of the Cuddapah basin of India in the light of global Proterozoic hydrocarbon habitats, Proc.Third Int. Pet. Conf. & Exbn. Petrotech – 99, New Delhi. 131–148.

Kumar, B., Das Sharma, S., Sreenivas,B., Rao, M.N. and Chawla, B.R. 2000. Isotope event stratigraphy of the Proterozoic carbonates of Vindhyan basin, India: Relation to organic carbon productivity/burial and hydrocarbon resource prospects. In: Garg, A.K., Banerjie, V., Swamy, S.N., and Dwivedi, P. (Eds.) Proceedings of the fifth international conference and exhibition on petroleum geochemistry and exploration in the Afro-Asian region, organized by ONGC, 25–27 Nov., New Delhi, 561–566.

Kumar, B., Das Sharma, S., Patil, D.J., Sreenivas, B., Raju, S.V., Kalpana, G., 2002. "Adsorbed Soil Gas Surveys for Hydrocarbon Research and Exploration in Chambal Valley, Vindhyan basin" Unpublished Technical Report No. NGRI-2002-Exp-361.

Kumar, B, Dayal, A.M., Patil, D.J., Kalpana, G., Vishnu Vardhan, C., Smitha K. Panicker, Devleena Mani, Aparna, B., 2005 Adsorbed Soil Gas Surveys for Hydrocarbon Research and Exploration in Cuddapah Basin. Unpublished Technical Report No. NGRI-2005-Exp-499.

Nagaraja Rao, B.N., Rajurkar, S.T., Ramalingaswamy, G. and Ravindra Babu, B. 1987. Stratigraphy, structure and evolution of the Cuddapah basin, Purana Basins of Peninsular India, GSI. Mem. 6: 33–86.

Naqvi, S.M, and John, J.W. Rogers, 1987. Precambrian Geology of India, Oxford Monographs on Geology and Geophysics. 6.

Raha, P.K., and Sastry, M.V.A., 1982 Stromatolites and Precambrian stratigraphy in India. Precam. Res., 18: 293–318.

Ramam, P.K and Murthy, V.N. 1997. Geology of Andhra Pradesh, Geological Society of India, Bangalore.

Sastry, M.V.A., and Moitra, A.K., 1984. Vindhyan Stratigraphy-A Review. Memoir. Geol. Surv. Ind. 116 (2):109–148.

Singh, A.P., and Mishra, D.C. 2002 Tectonosedimentary evolution of Cuddapah basin and Eastern Ghats mobile belt (India) as Proterozoic collision: gravity, seismic and geodynamic constraints. Journal of Geodynamics. 33:249–267.

Srivasthava, B.N, Rana, M.S and Narendra, K. Verma 1983. Geology and Hydrocarbon prospects of the Vindhyan Basin, in Petroliferous Basins of India, Petroleum Asia Journal.

Venkatachala, B.S., Sharma, M. and Shukla, M. 1996. Age and life of the Vindhyans- facts and conjectures, In: Recent advances in Vindhyan Geology. (ed) A. Bhattacharya, Mem. Geol. Soc. India. 36: 137–166.

Vivek S. Kale 1991. Constraints on the evolution of the Purana basins of Peninsular India. Jour. Geol. Soc. Ind. 38: 231–252.

William Schopf, J, and Prasad K.N., 1978. Microfossils in Collenia-like stromatolites from the Proterozoic Vempalle Formation of the Cuddapah basin, India. Precam. Res., 6: 347–366.

Petroleum Geochemistry and Exploration in the Afro-Asian Region – Digang Liang, Darui Wang & Zhenxi Li (eds)
© 2008 Taylor & Francis Group, London, ISBN 978-0-415-44083-7

Hydrocarbon-related diagenetic zones and its implication for the Yingnan 2 gas pool, Eastern depression of Tarim Basin

Yanxia Li
School of Petroleum Resources, Petroleum University of Xi'an, Shaanxi, China

Ningning Zhong & Zhihuan Zhang
Faculty of Natural Resources and Information Tech., Petroleum University of China, Beijing, China

ABSTRACT: The Yingnan 2 gas pool, located in the Yingjisu sag in the eastern depression of Tarim Basin, NW China, is sourced from the lower Jurassic sandstone. The gas hydrocarbons are composed predominately of CH_4(75.1%), C_2^+ (9.27%), and relatively higher N_2 (13.87%~16%). Direct caprock of the gas pool is abnormally tight sandstone that is strongly cemented with carbonate. Calcite of the tight sandstone shows two luminescence phases under the cathodoluminescence microscope. The former phase of calcite shows nacarat while the latter shows saffron, and the former is replaced by the latter. Stable carbon and oxygen analysis by laser micro-pyrolysis method reveals that, on the one hand, the carbonate-cement has a strongly negative $\delta^{13}C$ value (−7.17 to −26.15‰), which has not only an inorganic origin but also a distinct characteristic of biogenic origin. On the other hand, the $\delta^{18}O$ value of the calcite cement varies from −12.44 to −21.85‰, which reveals the presence of sedimentary seepage of meteoric water. Therefore, all the evidence demonstrates that the tight sandstone is a product of the hydrocarbon-related diagenetic zones (HRDZs). The data on chemical composition, content ratios and stable carbon isotopes of fluid hydrocarbons reveal that the formation of the HRDZs in the Yingnan 2 gas pool is simultaneous with the process of hydrocarbon seepage and biodegradation.

In the Yingnan 2 gas pool, an obvious hydrocarbon leakage takes place at the early stage and hydrocarbon accumulation occurs at the late stage. The compactness of tight sandstone as cap-rock develops gradually along with the gaseous hydrocarbon accumulation. Finally, a secondary condensate gas pool such as the Yingnan 2 reservoir is developed after the formation of the HRDZs in this region.

1 INTRODUCTION

The tight sandstone is the product of Hydrocarbon-Related Diagenetic Zones (HRDZs) (O'Brien et al., 1995, 1999). The formation of HRDZs is a process in which hydrocarbon slowly biodegrades. The carbon isotopic compositions of the carbonate cements show negative value related with biological origin, and the forming process has a lot to do with the oxidation of methane-oxidising bacteria from migrating hydrocarbon sweepage (Peckmann et al., 1999, 2001; Sassen et al., 1987, 1991; Stakes et al.,1999).

2 GEOLOGICAL SETTINGS

2.1 *Geologic structure and sediments*

The Yingnan 2 gas pool is located in the northeastern part of Tarim Basin (Fig. 1a). The study region includes five structural units (Kongque River Slope, Manjiaer Depression, Yingjisu Sag, Tadong Salient and Tanan Uplift) (Fig. 1b). The Yingjisu sag located in the eastern part is a Mesozoic-Cenozoic rift depression, which was generated during the late Yanshanian orogenies and the Hercynian orogenies on the background of the long-term denudated Palaeozoic terrane. The Yingnan 2 gas well is the only discovered pool in this region.

The Paleozoic sedimentary rocks disclosed in the Yingnan 2 well include the Ordovician and Silurian strata, and the Carboniferous, Permian and Triassic strata are missing in this region according to the fossil evidence. The Jurassic sediments were deposited in the rift stage while the Cretaceous-Tertiary strata were formed in the depression stage. More than 900 m sediments were mainly developed in the Lower Jurassic, which contains Ahe Formation and Yangxia Formation. The Jurassic sediments are mainly land fluvial-lacustrine deposits in lithology, including green gray or gray siltstone, sandstone and conglomeratic sandstone embedded with brown siltstone, mudstone and coal-bearing sequence. The Lower Cretaceous sediments

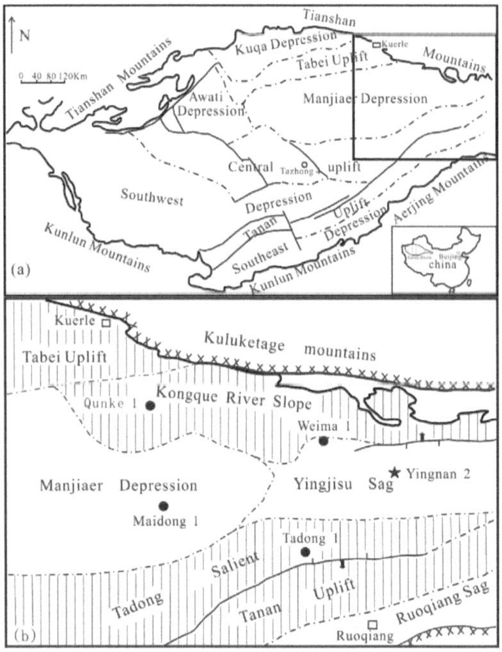

Figure 1. Map showing geological tectonic units of Tarim Basin (a) and the location of the Yingnan 2 gas pool (b).

are mainly composed of fluvial-deltaic deposits, including brown-gray conglomeratic sandstone, fine sandstone, varicolored gravel embedded with brown mudstone in lithology.

2.2 *Reservoir fluids*

The Yingnan 2 well has penetrated 57 oil and gas pay zones in the Jurassic and Silurian strata, ranging from 3292 to 4737 m in depth. The total pay thickness is around 328 m. Condensates are recorvered mainly from the lower part of the Lower Jurassic to the top of the Silurian strata. The vertically stacked pay zones within this anticlinal structure are normally pressured, and each has distinctly different gas-water contact.

The Yingnan 2 condensate gas pool belongs to a sandstone reservoir. The reservoir physical properties are moderate or relatively poor in porosity and permeability. The caprock is not well developed and mainly comprises two types, i.e. argillic rock and consolidated sandstones. The argillic caprock is mainly distributed in the Tertiary and Cretaceous strata, while the direct caprock, calcite-cemented consolidated sandstone, a kind of abnormal and rare occurrence, is mainly distributed in the interval of 3410.0~3491.0 m (Lower Jurassic). The caprocks belong to consolidated lithic fine-grained sandstones in lithology, the porosity ranges from 7.0 to 12%, and the permeability is less than $1.0 \times 10^{-3} \mu m^2$, $0.519 \times 10^{-3} \mu m^2$ in mean,

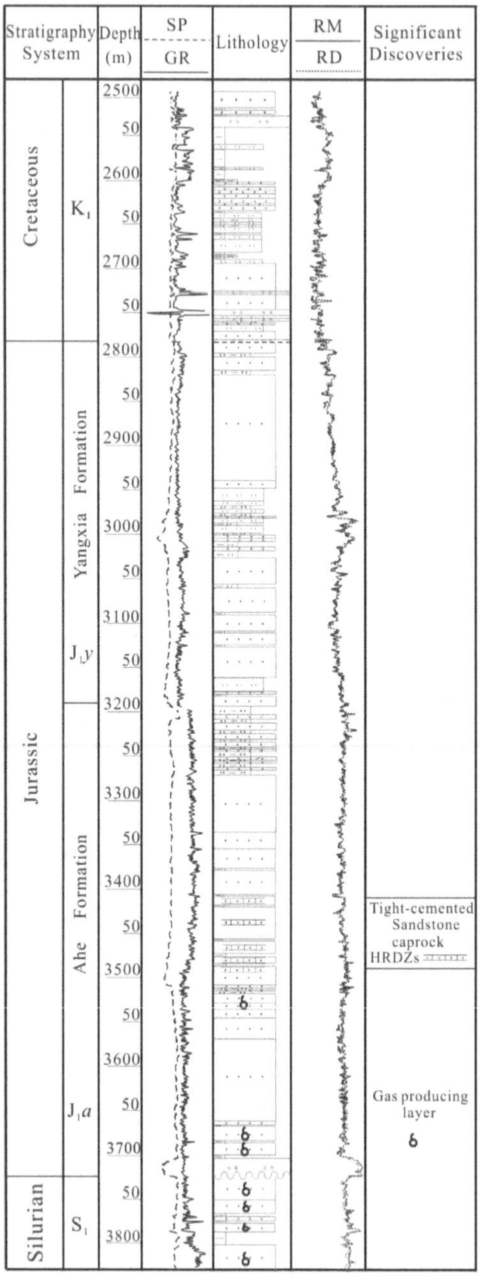

Figure 2. Stratigraphic section of the Yingnan 2 well in northeastern Tarim Basin, NW China.

which means that the abilities of sealing and capping are relatively poor. In one word, trap consolidation develops well in this gas pool mainly due to the interaction between organic and inorganic interaction during hydrocarbon filling and diffusing upwards.

The consolidated sandstone in the caprock is a unique feature in the Yingnan 2 gas pool. This paper demonstrates the relationship between tight sandstones and HRDZs by the study on the lithology and carbon and oxygen isotopes of the cement in tight sandstones of the Yingnan 2 condensate pool, trying to rebuild the gas pool forming process with the organic and inorganic interaction from the view of the history of hydrocarbon charging and tight sandstones development.

3 SAMPLES AND ANALYTICAL TECHNIQUE

35 core sandstones were sampled from the lower Jurassic and Silurian sediments in the Yingnan 2 well, and they are gray and gray-brown sandstones. The routine petrographic analysis, cathodoluminescence analysis and fluorescence analysis for fluid-inclusions were conducted on these samples. Stable carbon and oxygen isotope analyses were performed on carbonate-cement in these samples by two methods:

1) Bulk rock: For this routine analysis, approximately 200 g sample was finely ground in a miniature ball-mill and appropriate amount was reacted with 100 percent orthophosphoric acid at 50°C for 24 hours. After purification, the produced carbon dioxide gas was subjected to $\delta^{13}C$ and $\delta^{18}O$ analysis on a Finnigan-Mat 252 stable isotope mass spectrometer. The final results were reported in per mil notation relative to the V-PDB standard.

2) Laser micro-pyrolysis: The analysis set was made by Qiang Zitong et.al. (1996). A laser was fired at the carbonate-cement surface on a thin section and the carbonate was decomposed to produce CO_2 gas which was analysed by mass spectrometry. The method focus on collecting CO_2 generated from carbonate-cement of tight sandstones by laser heating and can basically distinguish different stage of carbonate-cement of sandstones. Because of the imperfection of laser micro-pyrolysis technique, the method is still at exploring stage at present (Greenwood et al., 1998; Zhong Ning-ning et al., 2001) .

According to the above-mentioned method, 10 samples were chosen to conduct stable carbon and oxygen isotope analyses on the carbonate-cements by laser micro-pyrolysis method, while 4 of them were also subjected to stable carbon and oxygen isotope analyses on the carbonate-cements by routine bulk rock acid dissolving method.

4 RESULTS AND DISCUSSION

4.1 *Cementation and isotopes of tight sandstone*

Cementation of sandstone reservoirs in the Yingnan 2 gas pool is dominantly made of authigenic carbonate mineral – calcite. From Silurian to lower Jurassic, calcite cement is extremely abundant (6.3~26%) and cement grains are in poikilitic form. Calcite occurs as mosaic to poikilitic crystal and develops compressed and deformed twin crystal with increasing burial depth.

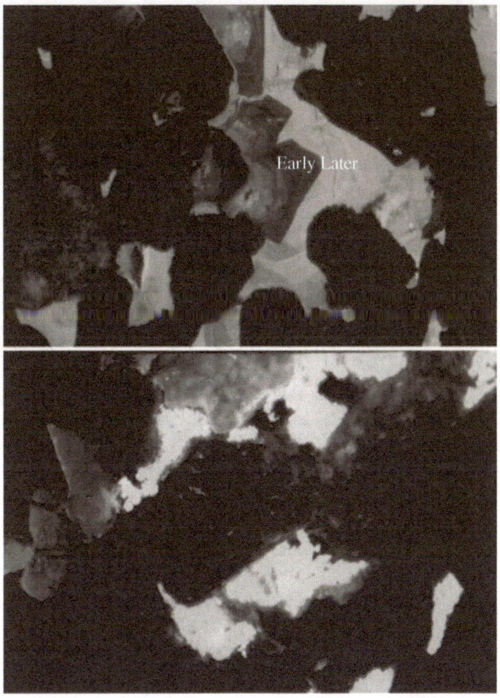

Figure 3. Cathodoluminescent characteristics of carbonate-cemented sandstone in the Yingnan 2 well; a. The earlier-phase nacarat calcite is dissolved, then it is replaced by later saffron calcite with sharp euhedral crystal face.(J_1a, 3336.2 m, × 125); b. The earlier-phase nacarat calcite is dissolved and replaced by the later saffron calcite. (J_1a, 3685.7 m, × 125).

Calcite of the tight-cemented sandstone shows two luminescence phases under the cathodoluminescence microscope. According to the relation of replacement and paragenetic sequence, calcite cement occurs in two events. The former phase of calcite shows nacarat, and the latter shows saffron. The nacarat calcite that first fills intergranular pores and cement grains were later dissolved by pore fluid. The calcite that grows from clean grain surface has embayed and hackly remnant crystals and scatters in saffron calcite, it must have been precipitated before saffron calcite formation. The saffron calcite occurs commonly as a pore-filling poikilitic cement from small dissolved pore to large one that postdates nacarat calcite and replaces it (Fig. 3).

The stable carbon and oxygen isotope values on carbonate cements of 10 sandstone samples in the

Table 1. Stable carbon and oxygen isotope values of carbonate-cemented sandstones in the Yingnan 2 well.

No.	Depth (m)	Age	Lithology	Micro-pyrolysis sample (‰, PDB) $\delta^{13}C$	$\delta^{18}O$	Bulk sample (‰, PDB) $\delta^{13}C$	$\delta^{18}O$
1	2783.6	J_1y	light grey sandstone	−26.15	−20.24		
2	2890.82	J_1y	light grey fine-grained sandstone	−5.67	−19.96	−1.3	−10.3
3	3001.9	J_1y	light grey fine-grained sandstone	−3.79	−12.44	−1	−9.1
4	3336.3	J_1a	light grey medium-grained sandstone	−6.89	−15.62	−2.7	−10.6
5	3336.3	J_1a	light grey medium-grained sandstone	−7.94	−20.68		
6	3338.7	J_1a	light grey medium-grained sandstone	−9.06	−21.85		
7	3338.7	J_1a	light grey medium-grained sandstone	−8.57	−20.16		
8	3495.15	J_1a	light grey medium-coarse sandstone	−6.96	−18.86		
9	3691.4	J_1a	light grey coarse-grained sandstone	−5.7	−18.12	−3.4	−13.1
10	3814.5	S	light grey medium-grained sandstone	−7.17	−17.42		

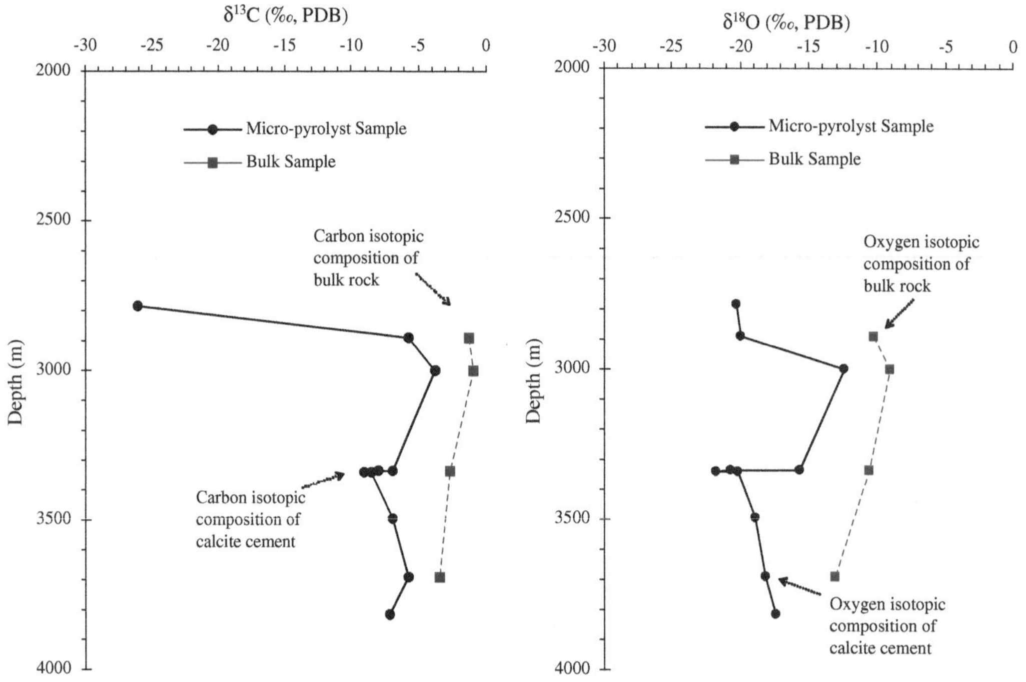

Figure 4. Comparison between $\delta^{13}C$ and $\delta^{18}O$ values of carbonate-cement in the Yingnan 2 well measured by laser micro-pyrolysis method and bulk rock acid dissolving method.

Yingnan 2 well are listed in Table 1. By routine bulk rock acid dissolving method, the $\delta^{13}C$ values of carbonate cements range from −1.0‰ to −3.4‰, while the $\delta^{18}O$ values range from −9.1‰ to −13.1‰, respectively. By the laser micro-pyrolysis method, the $\delta^{13}C$ values of carbonate cements range from −3.79‰ to −26.15‰, while the $\delta^{18}O$ values range from −12.44‰ to −21.85‰. Apparently, the isotopic values conducted by bulk rock acid dissolving method are totally heavier than those values conducted by laser micro-pyrolysis method (Fig.4), whatever for $\delta^{13}C$ or $\delta^{18}O$ values.

As shown in Table1 and Fig. 4, the $\delta^{13}C$ values of carbonate-cement of tight sandstones in the Yingnan 2 well mainly ranges from −7.17 to −9.06‰. However, an extremely depleted value (−26.15‰) occurs at the depth of 2783.6 m. Based on the stable carbon isotope fractionation mechanism (Hudson, 1977), it can

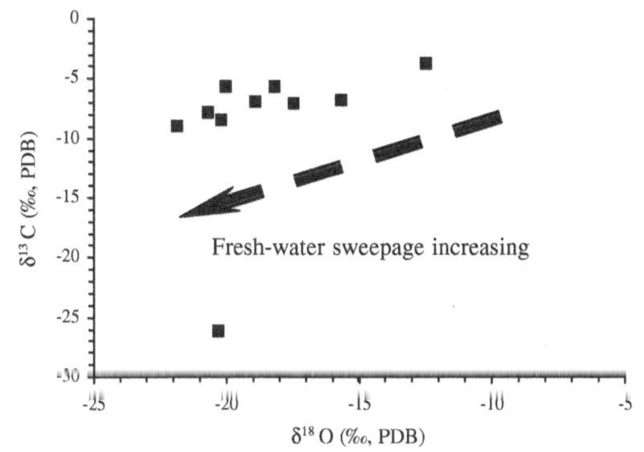

Figure 5. Covariant trend of stable carbon and oxygen isotopes of carbonate-cement sandstones in the Yingnan 2 well.

be seen that the $\delta^{13}C$ value of normal marine carbonate rock and carbonate cement with inorganic origin should range from −4‰ to +4‰, while the $\delta^{13}C$ value of organic matter with organic origin should range from −20‰ to −35‰ due to series of dynamic isotope fractionation. As for the methane gas, the $\delta^{13}C$ values with biogenic origin are less than −65‰ while those with thermogenetic origin are usually within −30‰∼−55‰ (Suess et al.,1989). Furthermore, the methane derived from thermal chemical decomposition can not directly lead to the precipitation of in-situ cements with quite low $\delta^{13}C$ value (Heydari et al., 1988; Sassen et al. 1987). Thus, the extremely negative excursion of stable carbon isotopic values indicates that the carbon source of the carbonate cement is from biological origin. In other word, the tight sandstones in the Yingnan 2 well is definitely a product of the hydrocarbon-related diagenetic zones.

O'Brien et al. (1995,1999) holds that the very intense hydrocarbon leakage and oxidation due to low integrity trap are probably inclined to form the carbonate cements with isotopic composition approximating with CH_4gas ($\delta^{13}C < -25‰$), while the relatively weak hydrocarbon seepage is probably apt to form the cements with isotopic composition approximating with normal carbonates ($\delta^{13}C > -4‰$). Thus, the extremely depleted isotopic composition of tight sandstones in the Yingnan 2 well definitely demonstrates that the HRDZs are caused by the slow leakage of hydrocarbon seep and then suffer from biodegradation.

In Fig. 5, the $\delta^{18}O$ values of carbonate-cements have a good relationship with the $\delta^{13}C$ values in the Yingnan 2 well. Especially, the $\delta^{18}O$ values of carbonate-cements seem to have a larger negative excursion (−12.44‰∼−21.85‰), while the main $\delta^{13}C$ values of carbonate-cements have a relatively smaller negative excursion (−3.79‰∼−9.06‰). These excursions might caused by two factors: 1) increase of temperatures of the pore water in the burial process; 2) a cushion of penetrated fresh water on the carbonate-cement in the forming process of tight sandstones.

The $\delta^{18}O$ values of carbonate-cement in the Yingnan 2 well (−12.44‰∼−21.85‰) are completely in the range of fresh water as a result of diagenesis of fresh water percolation (Lohman,1988). Meanwhile, the $\delta^{13}C$ values of carbonate-cement (−7.17‰∼−26.15‰) seem to approach those of CO_2 in the microbe metabolism process (Fig. 4). In other word, the hydrocarbon oxidation in the compact sandstones occurs in the instance of fresh water percolation. Therefore, it can be inferred that the HRDZs occur in the percolation zones of the atmosphere water, which just corresponds to the bacteria living zone (surface of deposits/water), as well as sulphate reduction zone (shallow stratum water distribution zone) and methane zymosis zone.

4.2 *Variations of natural gas compositions in the Yingnan 2 gas pool*

In the reservoir intervals (profile or section) in the Yingnan 2 gas pool, the percentage of methane content (C_1%) and the ratios of $C_1/(C_2+C_3)$ of natural gases decrease with depth decreasing(Fig. 6), but the ratio of iC_4/nC_4 increases with depth decreasing (Fig. 7). This kind of decreasing trend can be caused by the following factors: 1) hydrocarbon accumulation occurring in the late stage as well as lack of excellent caprock, an obvious gas leakage can be predicted. Thus, methane gases will preferentially migrate upwards to those heavier gaseous hydrocarbons, i.e, ethane, propane, butane and isobutene. 2) during the occurrence and development process of possible HRDZs, methane is consumed by bacteria or escapes, which can cause the gradual depletion of methane in shallower reservoir intervals.

Furthermore, in the reservoir intervals (profile or section) in the Yingnan 2 gas pool, the values of $\delta^{13}C_1$and $\delta^{13}C_2$of natural gases become gradually heavier with depth decreasing. From the bottom (3805 m) to the top (3470 m), the value of $\delta^{13}C_1$changes from −37.5‰ to −36.2‰, and the value of $\delta^{13}C_2$ varies from −34.7‰ to −30.9‰ (Fig. 8).

Generally speaking, with depth increasing organic matter becomes more and more mature and its $\delta^{13}C$ values should become heavier. Obviously, the more depleted trend of $\delta^{13}C$ values of methane and ethane with depth increasing is not the result of maturation fractionation.

Moreover, it is accepted that, to some extent, in the process of natural gas migration through pores and fractures of sediments, the stable carbon isotopic fractionation will not occur in ethane or heavier gases (Schoell,1983; Faber, 1984; Fuex, 1980), but will occur in methane (Zhang Tongwei, 1995). Thus, the more depleted trend of $\delta^{13}C$ values of ethane with depth increasing is also not the result of migration fractionation and leakage fractionation.

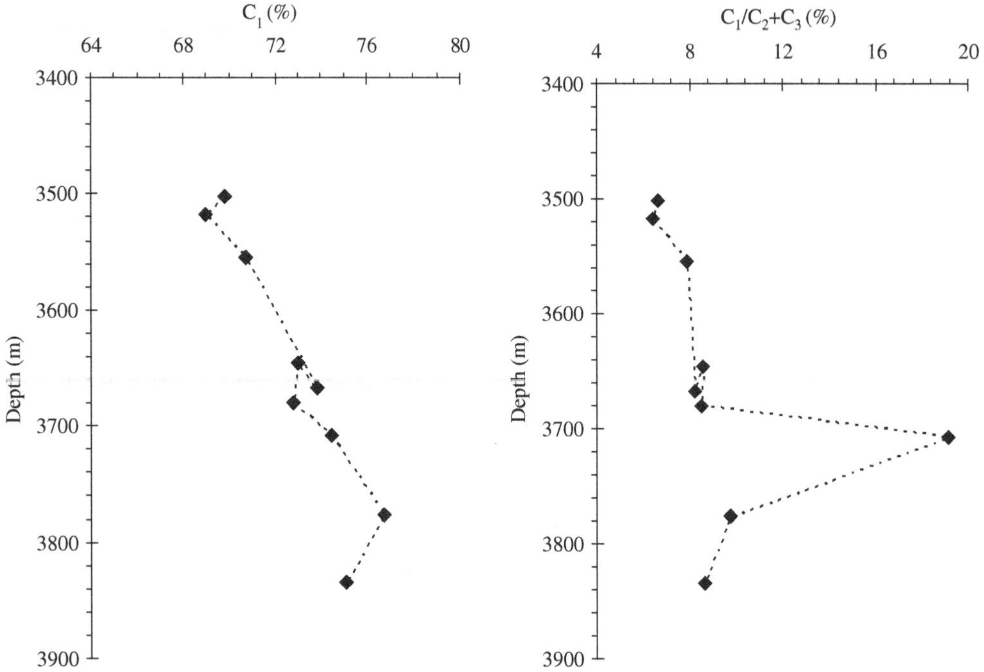

Figure 6. Variation of relative content of methane with depth in the Yingnan 2 gas pool.

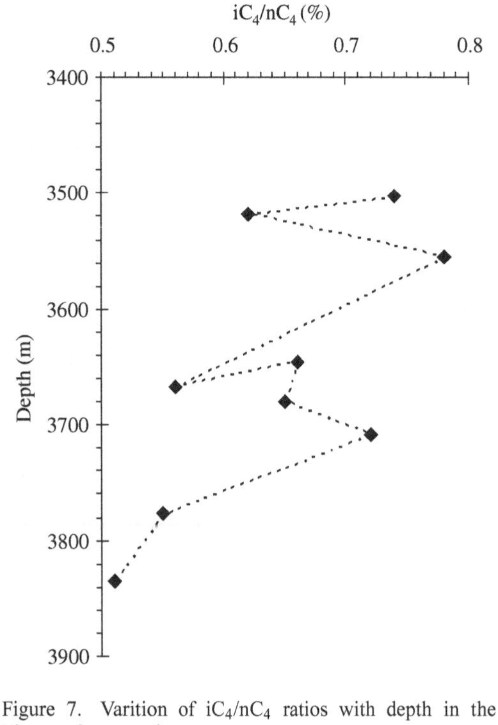

Figure 7. Varition of iC_4/nC_4 ratios with depth in the Yingnan 2 gas pool.

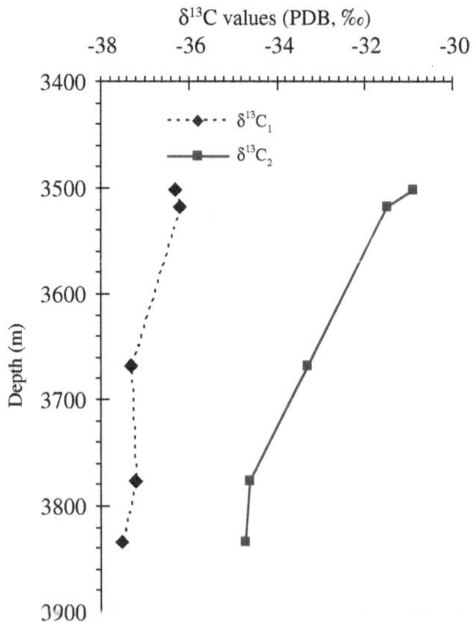

Figure 8. Varition of $\delta^{13}C_1$ and $\delta^{13}C_2$ values of natural gases with depth in the Yingnan 2 gas pool.

Therefore, the stable carbon isotope fractionation of natural gases in the Yingnan 2 gas pool is another geochemical record, and it should be related with the formation of tight-cemented sandstones. In other word, in the formation process of HRDZs, ^{12}C in upward-migrated gaseous hydrocarbons is preferentially consumed, utilized by microbe and transformed into the carbonate-cement, as a result, an unusually negative $\delta^{13}C$ value (−26.15‰) occurs in the calcite.

4.3 *Inorganic and organic interaction in HDRZs in the process of gas hydrocarbon accumulations*

After the early Caledonian orogeny in the Tadong area, great amounts of sediments were deposited in the Cambrian and Ordovician period. The marine source rock deposited from the Cambrian to early Ordovician period underwent through several stages: oil-generating window, wet-gas generation, and kerogen-cracking gas generation and an old reservoir formed in the Palaeozoic sediments. In the period of the Hercynian-Yanshanian orogeny, the Tadong area was exposed to erosion, and the former shallow reservoir was destroyed while some deeply-buried reservoirs might remain due to being separated from oxidation zone. In the early period of the Yanshanian orogeny, lower Jurassic sediments covered the Palaeozoic structure, and hydrocarbons in the Palaeozoic structure were adjusted and diffused upwards due to tectonic movement. During the middle-late Jurassic epoch, lower Jurassic sediments were exposed to leakage zone of meteoric water and were encountered by hydrocarbons migrating upward, which meets the prerequisites of the HRDZs formation (Fig. 9-a).

Besides the necessary methane with organic origin as a carbon source, appropriate conditions such as the leakage depth of meteoric water and formation temperature are needed in the formation of the HRDZs. The leakage depth of meteoric water is usually about 3000 m and a suitable formation temperature is in the range of 20∼75°C. Geothermal gradient in the Tadong area was about 2∼2.5°C/100 m in the late Mesozoic era, some deeply-buried gas hydrocarbons could migrate into the shallower sandstones in the sedimentation break of the middle-late Jurassic epoch. If oxygen is enriched in the pore water zone, methane might be oxidized by the aerobic bacteria as follows: $CH_4+2O_2=CO_2+2H_2O$.

The produced CO_2 by oxidization of aerobic bacteria can be combined with Ca^{2+} to form calcite and the light carbon isotope is transferred to calcite, which causes the negative excursion of carbon isotope in the calcite ($\delta^{13}C = -7.17\sim -26.15$‰). As a result, the $\delta^{13}C$ value of carbonate-cements reveals the characteristics of CO_2 derived from microbial metabolism, and the $\delta^{18}O$ value (−12.44‰ to −21.85‰) of carbonate-cements reveals the characteristics of meteoric water.

Formation water usually exists in a reducing environment, which is caused by complete consumption of oxygen by aerobic bacteria in pore water. Under the circumstance of exhaustion of oxygen, sulfate can be utilized by some bacteria as oxidant to produce CO_2, H_2S and H_2O as follows: $CH_4+SO_4^{2-}+2H^+=CO_2+H_2S+2H_2O$.

During the formation period of the HRDZs, two kinds of biochemical reactions are involved, aerobic bacteria (methane bacteria) and anaerobic bacteria (sulfate bacteria) are processing at all the times. Pyrite in the cement of calcite is the direct evidence of anaerobic sulfate bacteria's action (O'Brien et al., 1999; Ritger et al., 1987). With the occurrence of cements deposited in pores of rocks, tight cap-rocks come into being generated in the sandstone. So, the formation of carbonate-cements is the result of organic and inorganic interactions.

During the late period of the Yanshanian orogeny, a large amount of deeply-buried hydrocarbons in old reservoirs, in the form of crude oil-cracking gases or residual wet/dry gases, moved upwards into the HRDZs. As a result, about 90 m tight sandstone at depths from 3400 m to 3491 m was formed in the Jurassic sediment in the Yingnan 2 well. Especially when a red argillaceous rock (K_1kp) deposited as regional cap-rock in the Cretaceous period, a good trap for deeply-buried hydrocarbons was taken into shape in this region (Fig. 9-b).

In the early period of the Hercynian orogeny, the Yingjisu depression experienced rapid subsidence, and natural gases formed in the deep source rock migrated along faults and finally accumulated in the Jurassic sandstone. Due to specific characteristics of tight sandstone trap, the compactness of the Yingnan 2 structure was not perfect, which caused the leakage and diffusion of methane with priority. And the gaseous hydrocarbons became wet with obvious composition fractionation in the Yingnan 2 gas pool. When it came to the late stage of the Yanshanian orgeny, great amounts of gases were charged into the trap and the gas leakage formed a dynamic balance, and finally such a secondary and deeply-buried condensate gas pool formed in this region (Fig. 9-c).

5 CONCLUSION

The formation of tight sandstone as cap-rock in the Yingnan 2 gas pool is the result of organic and inorganic interactions. This kind of tight sandstone is characterized by obvious calcite cementation. And the $\delta^{13}C$ value of carbonate-cements has a distinct characteristic of biogenic origin with strongly negative excursion (from −7.17 to −26.15‰ shown in this paper), while the $\delta^{18}O$ value reveals the presence of sedimentary seepage of meteoric water (from −12.44‰ to −21.85‰ shown in this paper). All the evidence shows that tight sandstone is the product of the HRDZs. The data on chemical composition, ratios

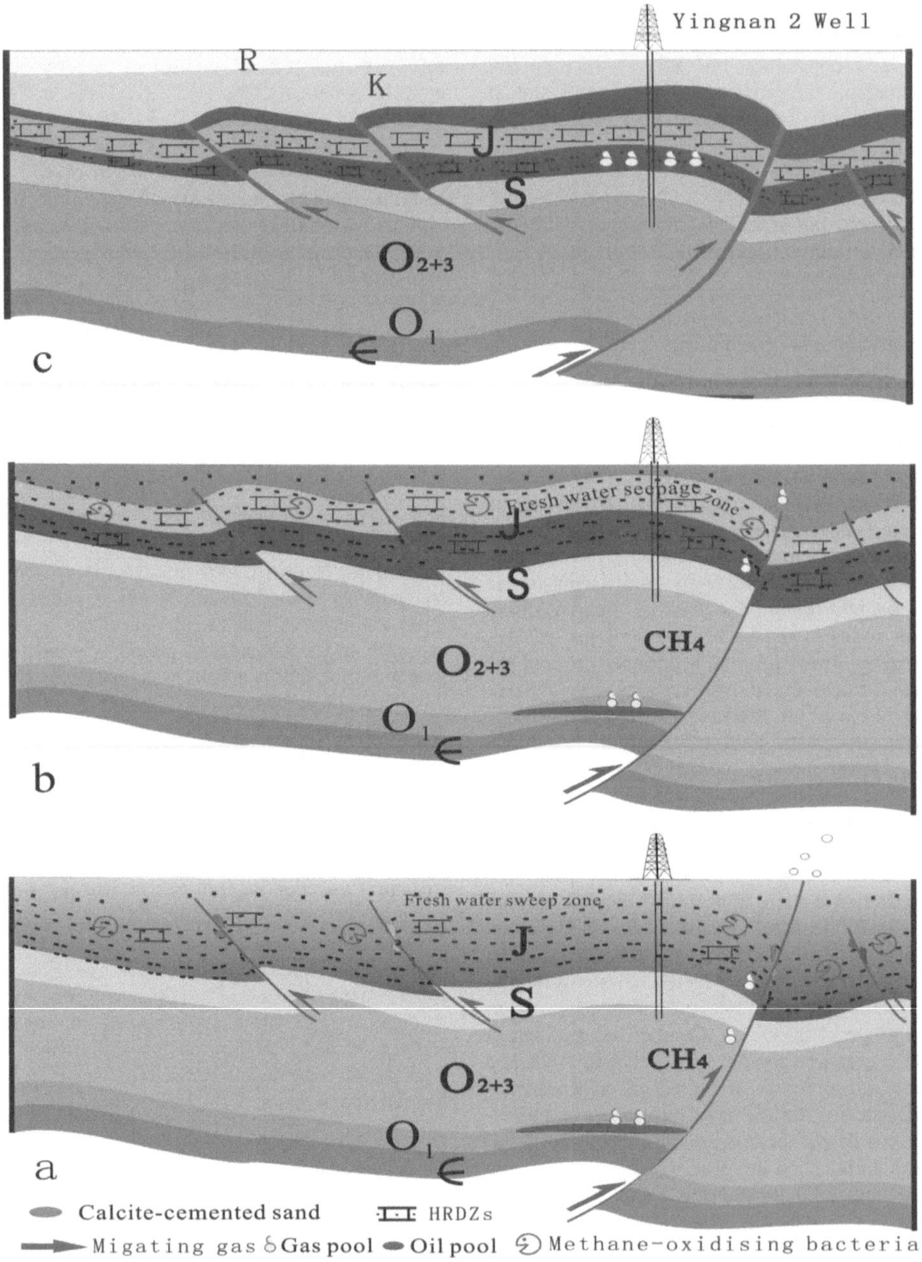

Figure 9. Formation model for the Yingnan 2 gas pool.

and stable carbon isotopes of fluid hydrocarbons reveal that the formation of the HRDZs in the Yingnan 2 gas pool is simultaneous with the process of hydrocarbon seepage and biodegradation, which cause the fractionation of natural gas compositions and stable carbon isotopes.

In the Yingnan 2 gas pool, an obvious hydrocarbon leakage takes place at the early stage and hydrocarbon accumulation occurs at the late stage. The compactness of tight sandstone as cap-rock develops gradually along with the gas hydrocarbon accumulation. Finally, a secondary condensate gas pool such as the Yingnan

2 reservoir is developed after the formation of the HRDZs in this region.

ACKNOWLEDGEMENTS

We thank Dr. Wang Lansheng, Technicians Zheng Ping and Zhang Zhonghua in the Institute of Southwest Sichuan Petroleum Corporation, PetroChina for providing laser micro-pyrolysis analysis on stable carbon and oxygen isotopes of carbonate cements. The research is financially supported by National Basic Research Program of China.

REFERENCES

Faber E, Stahl, W J. 1984. Geochemical surface exploration for hydrocarbons in North Sea. AAPG Bulletin, 68(3): 363~386.

Fuex A N. 1980. Experimental evidence against an appreciable isotopic fractionation of methane during migration. Organic Geochemistry, 12: 725~732.

Greenwood P F, George S C, Hall K. 1998. Applications of laser micropyrolysis–gas chromatography–mass spectrometry .Organic Geochemistry, 29(5–7):1075~1089.

Heydari E, Moore C H, Sassen R. 1988. Late burial diagenesis driven by thermal degradeation of hydrocarbons and thermochemical sulfate reduction: Upper Smackover carbonates, southeast Mississippi salt basin. Amer. Assoc. petrol. Geol. Bull., Abstr., 72:197.

Hudson J D. 1977. Stable isotopes limestone lithification. Jour. GeolSoc. London, 133: 637~660.

Lohman K C. 1988. Geochemical pattern of meteoric systems and their application of the stuy of Paleokarst, In N. P. James and P. W. Choquette, eds., Paleokarst: New York, Springer-Verlag, 58~80.

O'Brien G W, Lisk M, Duddy I, et al. 1999. Plate convergence, foreland development and fault reactivation: primary controls on brine migration, thermal histories and trap breach in the Timor Sea, Australia. Marine and Petroleum Geology, 16(6): 533~560.

O'Brien G W, Woods E P. 1995. Hydrocarbon-related diagenetic zones (HRDZs) in the Vulcan Sub-Basin, Timor Sea: recognition and exploration implications. APEA Journal, 35(1), 220~252.

Peckmann J, Reimer A, Luth U, et al. 2001. Methane-derived carbonates and authigenic pyrite from the northwestern Black Sea. Marine Geology, 177:129~150.

Peckmann J, Thiel V , Michaelis W , et al. 1999. Cold seep deposits of Beauvoisin(Oxfordian, southeastern France) and Marmoritio(Miocene, northern Italy):Microbially induced authigenic carbonates. International Journal of Earth Sciences, 88:60~75.

Qiang Zitong, Ma Deyan, Gu Dayong et al. 1996, Laser microsampling for isotopic analysis Natural Gas Industry, 16(6), 86~89.

Ritger S, Carson B, Suess E, 1987. Methane-derived authigenic carbonates formed by subduction-induced porewater expulsion along the Oregon/Washington margin. Geological Society of America Bulletin, 98: 147~156.

Sassen R, Grayson P, Cole G, et al. 1991. Hydrocarbon seepage and salt dome related carbonate reservoir rocks of the US Gulf Coast. AAPG Bulletin, 75(9) :1537~1538.

Sassen R, Moore C H, Meendsen F C, 1987. Distribution of hydrocarbon source potential in the Jurassic Smackover Formation. Org. Geochem., 11:379~383.

Schoell M. 1983, Genetic Characterization of natural gase. AAPG Bulletin, 67: 2225~2238.

Stakes D S, Orange D, Paduan J B, et al. 1999, Cold-seeps and authigenic carbonate formation in Monterey Bay, California .Marine Geology, 159 (1~4) : 93~109.

Suess E, Whiticar M J. 1989. Methane-derived CO2 in pore fluids expelled from the Oregon subduction zone. Palaeogeography, Palaeoclimatology, Palaeoecology , 71: 119~136.

Zhang Tongwei, Chen Jianfa, Wang Xianbin et al. 1995. Isotopic geochemical traces of gaseous hydrocarbons for natural gas migration. Acta Sedimentologica Sinica, 13:(2) 70~76(in Chinese with English abstract).

Zhong Ning-ning, Greenwood P. F., 2001, Approach on in situ laser micropyrolysis GC-MS analysis of sedimentary organic matter and its prospects. Geochimica, 30 (6) : 605~611 (in Chinese with English abstract).

2. Source rock evaluation, migration and reservoir geochemistry

Petroleum Geochemistry and Exploration in the Afro-Asian Region – Digang Liang, Darui Wang & Zhenxi Li (eds)
© 2008 Taylor & Francis Group, London, ISBN 978-0-415-44083-7

Structural change of extracted lignin by white rot and brown rot fungi and thermal alteration of the lignin after the microbial treatments

K. Toh & T. Murae
Department of Earth and Planetary Sciences, Faculty of Sciences, Kyushu University, Hakozaki, Fukuoka, Japan

ABSTRACT: The lignin was extracted from Oak wood (*Quercus serrata*). The brown-rot fungus (*Laetiporus sulphureus*) destroyed the polysaccharide part in the lignin preferentially. The white-rot fungus (*Coriolus versicolor*) destroyed the aromatic part in the lignin preferentially. The lignin samples obtained after the treatment by the rot fungi were heated stepwise at 200, 250, and 300°C in nitrogen gas. The carbohydrate-free lignin yielded by the brown-rot fungus gave a kerogen-like substance by heat of 200°C. The lignin treated with white rot fungi was decomposed without yielding the kerogen-like substance by the increase of heat temperature. Brown rot fungi can shorten the reaction pas of the transformation of lignin into terrestrial kerogen. A significant part of natural gas is generated by thermal cracking of terrestrial kerogen. So brown rot fungi may enhance generation of natural gas from plant resources.

1 INTRODUCTION

Lignin is a structural component of the cell walls of vascular plants (Sarkanen & Ludwig 1971), and the second most abundant naturally produced polymer in the biosphere (after cellulose). Lignin links through covalent bonds to noncellulosic wall polysaccharides, especially "hemicelluloses" (Isherwood 1965, Kirk & Farrell 1987, Stafford 1988). Various acids, including p-coumaric, ferulic, and diferulic acid, are linking to lignins through ester or ether bonds. Such bifunctional compounds, capable of forming both of ester and ether linkages, act as bridging units between lignin and "hemicelluloses" (Tanner & Morrison 1983, Scalbert et al. 1985, Bolwell 1988, Hartley et al. 1988). Lignins also link through covalent bonds to some proteins of cell walls (Whitmore 1982) too. Therefore, lignin essentially contains carbohydrate moiety in its structure. However, the role of the polysaccharide moiety in chemistry of lignin is unclear.

Structural units derived from lignin are widely distributed in natural environments as a result of their natural abundance, wide distribution, and resistance to microbial degradation (Hedges & Mann 1979). Lignin structural units also are present in humic substances extracted from soil and sediments (Ertel & Hedges 1984, 1985). Although lignin is not readily decomposed by most microorganisms (Crawford 1981), some type of microorganisms, such as wood-rotting fungi, can break down chemical structure of lignin. Wood-rotting fungi are divided into three groups, white-rot, brown-rot, and soft-rot fungi. These fungi can remove all of the cell wall components of wood, i.e., lignin, cellulose, and hemicellulose (Cowling 1961). White-rot basidiomycetes belong to one of the few groups of microorganisms that can decompose lignin (Otjen & Blanchette 1985). Brown-rot basidiomycetes can partially affect the lignin structure. Soft-rot fungi decompose lignin slowly.

The relative abundance of lignin and its relative resistance to microbial degradation during early diagenesis indicate that it can be selectively preserved and concentrated in some sedimentary deposits (Tegelaar et al. 1989). The lignin affected by microbes is further altered by geothermal heating according to the increase of burial depth in the sedimental environments. Thus, it is a large component of many coals, Type III kerogens, and other hydrocarbon precursors. Therefore, heating experiments of lignin have been carried out by several investigators (Nassar & Mackay 1984, Brezny et al. 1983, Hayatsu et al. 1984). A chemical kinetics model of the transformation of lignin-derived sedimentary organic matter under natural sub-surface conditions has been proposed to simulate a series of reactions that account for the observed structural transformations using geologically relevant kinetics (Payne & Ortoleva 2001). Although microbial activity can alter the reaction pathways of lignin by affecting the early structural transformations, the above model did not address these issues specifically. In this paper, we report on the thermal alteration of the extracted-lignin affected by rot fungi.

2 EXPERIMENTAL

2.1 *Preparation of lignin*

The extraction procedure of lignin in this study essentially followed those reported by Yokoyama et al. (1969a, b) for the preparation of lignin from milled wood. In our study, pieces of oak wood (*Quercus serrata*) were ground with a ball mill and filtered through a 100-mesh sieve. The wood powder was sonicated in a mixture of dioxane and water (8:2, V/V). The extraction was repeated until the extract became colorless. The residue obtained by evaporation of the solvent was dissolved in a mixture of pyridine, acetic acid, and water (9:1:4, V/V). Chloroform was added to the solution to make a 6:1 mixture of the final ratio of pyridine to chloroform. The structure of the lignin was examined using elemental analysis, NMR and IR spectra. The spectral data of our extracted-lignin were essentially the same as those reported for the lignin prepared from bark of Quercus by Yokoyama et al. (1969a, b).

2.2 *Treatments of the extracted lignin by rot fungi*

Decomposition of lignin by wood-rotting fungi was carried out using the strain of white-rot fungus (*Coriolus versicolor*; IFO 30340), and the strain of brown-rot fungus (*Laetiporus sulphureus*, IFO 30745). These fungi were obtained from the Institute of Fermentation, Osaka (IFO). Stock strains were maintained on a malt-extract agar medium (ATCC Culture Medium 325), which contains malt extract (20.0 g), glucose (20.0 g), peptone (1.0 g), agar (20.0 g), and distilled water (1 l). The culture medium for lignin treatments was prepared using glucose (l0 g/l) and ammonium tartrate (1.2 mM) according to the method described by Kirk et al. (1978). Lignin was dissolved in DMSO (480 mg/6 ml) and was added into the liquid culture (300 ml) in a Erlenmeyer flask. The stock strain in the agar medium was inoculated into the liquid medium containing lignin. The liquid medium was maintained at 24°C in the dark. 10 ml of the culture medium was collected at 10 day intervals and was freeze-dried.

2.3 *Thermal alteration of the lignin*

Thermal alteration was examined on the lignin treated by rot fungi for 90 days and the lignin not treated by fungi. The fungi-untreated lignin was freeze-dried and then put in five Pyrex tubes. The culture solution containing lignin treated by rot fungi was freeze-dried. Each 0.4 ml of the DMSO solution was transferred into a Pyrex tube, and freeze-dried. The Pyrex tubes were sealed after replacing the inner air with nitrogen. Five sealed tubes were prepared for each sample, and the tubes were heated stepwise at 150, 200, 250°C and then at 300°C. The samples were kept for 15 min at each temperature. One tube of each sample was cooled after heating at each temperature. IR spectra of the samples after the thermal treatments were measured by KBr disk methods.

2.4 *Analytical instruments*

NMR spectra were taken with a JEOL JNM-GX400 (400 MHz) in DMSO-d_6. Chemical shift of the solvent (DMSO) was used as internal standards (^{1}H NMR: 2.5 ppm, ^{13}C NMR: 39.5 ppm). The concentration of the sample was 40 mg/ml. IR spectra were determined using a Perkin Elmer 1600 series FT-IR by KBr disk method. Vitrinite reflectance was determined using a Carl Zeiss MPM microscope photometer operated by photometric analysis program "photan".

2.5 *Preparation of vitrinites*

The vitrinite was prepared from the coaly fragments which were collected from the Shimanto and Setogawa Groups of Cretaceous and Tertiary accretionary prisms in the eastern end of the outer zone of the southwest part of Japan arc. Rock samples of sediments containing coaly fragments were crushed into powders between 20 and 100 meshes. The powders were fractionated using aqueous solution of zinc chloride having specific gravity of 1.6. The floating part was collected and was dried at 60°C. The dried powder was buried in a plastic block. The surface of the block containing vitrinites was polished into mirror face.

3 RESULTS AND DISCUSSION

3.1 *Alteration of the IR spectrum of lignin by microbial degradation*

The IR spectrum of the lignin extracted from *Quercus serrata* wood in this study (Fig. 1a) showed absorptions due to aliphatic C-H (2920 cm^{-1}), the ester carbonyl group (1730 cm^{-1}), the conjugated carbonyl group (1680 cm^{-1}), the aromatic C=C bond (1600 and 1500 cm^{-1}), the methoxyl C-O (1270 and 1220 cm^{-1}), and the aromatic C-H (830 cm^{-1}). The following spectral features are characteristic for angiosperm lignin; (1) absorption due to ester carbonyl is strong, (2) intensity of absorption due to syringyl C-O (1220 cm^{-1}) and that due to guaiacyl C-O (1270 cm^{-1}) are almost equal, (3) intensity of absorption at 1460 cm^{-1} is stronger than that at 1500 cm^{-1}.

Figure 1b shows the IR spectrum of the lignin being treated by brown-rot fungi (*L. sulphureus*) for 90 days. On the degradation by brown-rot fungus (*L. sulphureus*), the relative intensity of absorptions due to aliphatic C-H (2920 cm^{-1}) and ester carbonyl group (1730 cm^{-1}) decreased, and the relative intensity of absorption due to the conjugated carbonyl group (1685 cm^{-1}) increased. The IR spectrum of the lignin

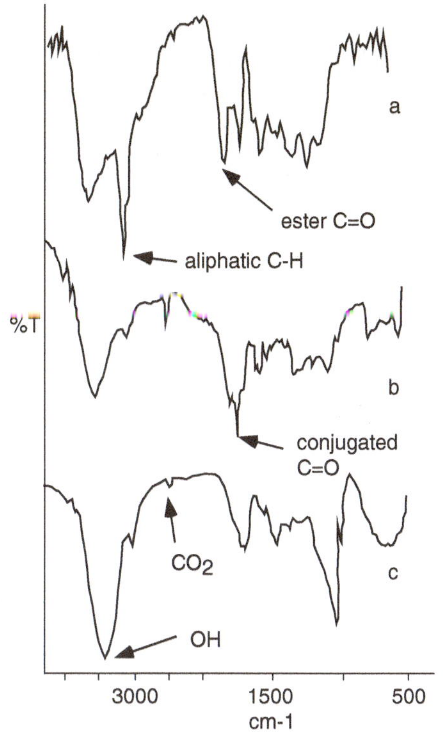

Figure 1. FT-IR spectra of (**a**) lignin extracted from oak wood, (**b**) lignin after 90 days of treatment by brown-rot fungus (*L. sulphureus*), and (**c**) lignin after 90 days of treatment by white-rot fungus (*C. versicolor*).

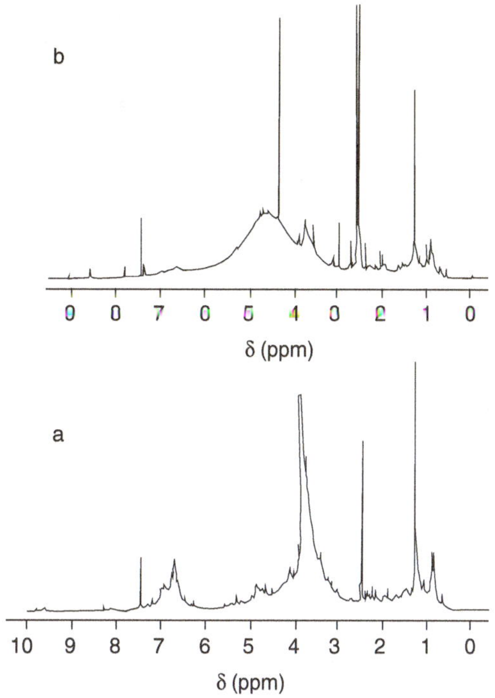

Figure 2. ^{1}H-NMR spectra (400 MHz in DMSO-d_6) of (**a**) lignin extracted from oak wood and (**b**) the lignin after 90 days of treatment by brown-rot fungi (*L. sulphureus*). In the spectrum **b** the overflowing peak at 2.7 ppm is due to water.

treated by white-rot fungus (*C. versicolor*) for the same 90 days showed the decrease of relative intensity of absorptions due to the aliphatic C-H (2920 cm^{-1}) and the ester carbonyl group (1730 cm^{-1}), and increase of the relative intensity of the absorption due to the conjugated carbonyl group (1685 cm^{-1}) was also observed in the spectrum. However, unlike the degradation by brown-rot fungus, the degradation by white-rot fungus very much enhanced the relative intensity of the absorptions due to hydroxyl group (3400 and 1100 cm^{-1}) as shown in Figure 1c. This remarkable difference of IR spectrum between the lignin affected by the brown-rot fungi and the lignin affected by the white-rot fungi suggests the followings. Although brown rot fungi decompose the carbohydrate part mainly, white rot fungi decompose the aromatic part prior to carbohydrate part.

3.2 *Alteration of the NMR spectrum of lignin by microbial degradation*

The feature of ^{1}H NMR spectrum of the lignin used in this study (Fig. 2a) is same as the spectrum which has been reported by Yokoyama et al. (1969b) on the milled-wood lignin. The signals at 0.8–2.7 ppm were assigned to the protons on aliphatic side chains in phenolic moieties, the signals at 3.65–3.95 ppm were assigned to the protons of aliphatic H-C-O, H-C-OH and methoxyl groups, those at 4.8–5.8 ppm were assigned to olefinic protons, and those at 6.5–8.0 ppm were assigned to aromatic protons. The intensive signal at ~3.8 ppm which was observed in the spectrum of the lignin not treated by microbe (Fig. 2a) diminished remarkably in the spectrum of the lignin treated by *L. sulphureus* for 90 days (Fig. 2b). This significant alteration of NMR spectrum is interpreted by preferential decomposition of carbohydrate part by the brown-rot fungus. This is compatible with the conclusion given on the basis of IR spectroscopy.

The selective decomposition of a carbohydrate part of the extracted lignin by the brown-rot fungus is suggested from ^{13}C NMR data too. Many peaks at 50–90 ppm in the ^{13}C NMR spectrum of lignin being not treated by microbe (Fig. 3a) are assignable to aliphatic carbon having a hydroxyl group such as carbohydrate (Silverstein and Webster, 1998). Most of these peaks disappeared in the ^{13}C NMR spectrum of lignin obtained after 90 days of treatments by

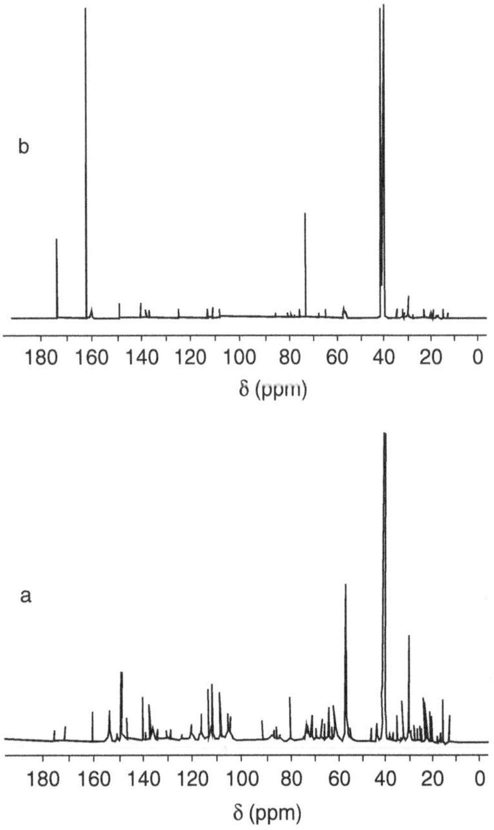

Figure 3. ^{13}C-NMR spectra (100 MHz in DMSO-d_6) of lignin extracted from oak wood (**a**) and lignin after 90 days treatment by brown-rot fungus (*L. sulphureus*) (**b**). The strong peak in (**a**) and (**b**) is assigned to DMSO, and was used as the standard for chemical shift (39.5 ppm).

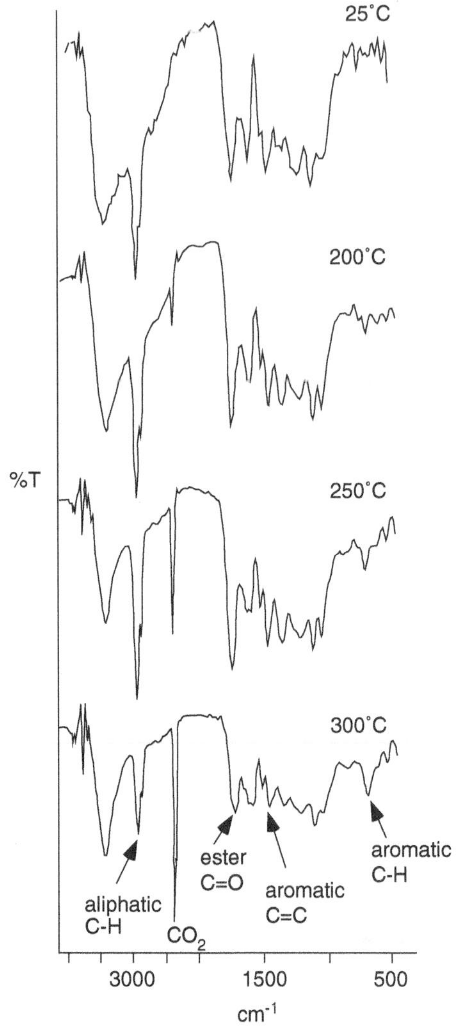

Figure 4a. Thermal alteration of the IR spectrum (after heating for 15 minutes at each temperature) of lignin extracted from oak wood.

L. sulphureus (Fig. 3b). This indicates that the brown rot fungus decomposed preferentially the carbohydrate part of the extracted lignin.

3.3 *Thermal alteration of lignin affected by microorganisms*

Figure 4 shows IR spectra obtained by step-wise heating at 150°C, 200°C, 250°C, and 300°C for 15 min at each temperature about the lignin extracted from Oak wood (*Quercus serrata*) (Fig. 4a), lignin treated by the white-rot fungus (Fig. 4b), and lignin treated by the brown-rot fungus (Fig. 4c). The lignin not treated by microbe did not show obvious change in the spectral feature until the heating temperature reached 300°C. However, on heating at 300°C, the lignin not treated by microbe showed remarkable decrease of intensity of absorption due to aliphatic C-H (2920 cm^{-1}) and ester carbonyl group (-C(=O)-O-) (1730 cm^{-1}). And intensity of absorption due to the aromatic portion (830 cm^{-1}) increased (Fig. 4a).

The intensity of absorption due to carbohydrate moiety of lignin treated by white rot fungus (*C. versicolor*) decreased on heating at 200°C (Fig. 4b). But a part of the peak still remained, and it disappeared on heating at 250°C. Although that spectrum changed on heating at 300°C, its feature was different from that obtained on heating at 300°C from the lignin being not treated by fungi. In the process of thermal metamorphism of lignin, the carbohydrate moiety, which left after decomposition by white-rot fungi, probably disturbs the formation of aromatic structure which is resistant against thermal alteration.

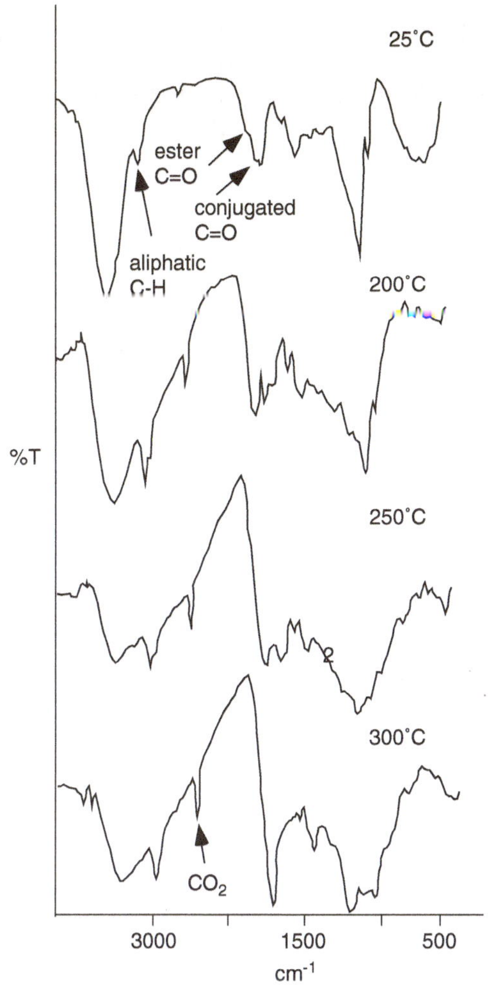

Figure 4b. Thermal alteration of the IR spectrum (after heating for 15 minutes at each temperature) of the lignin after 90 days of treatment by brown-rot fungus (*L. sulphureus*).

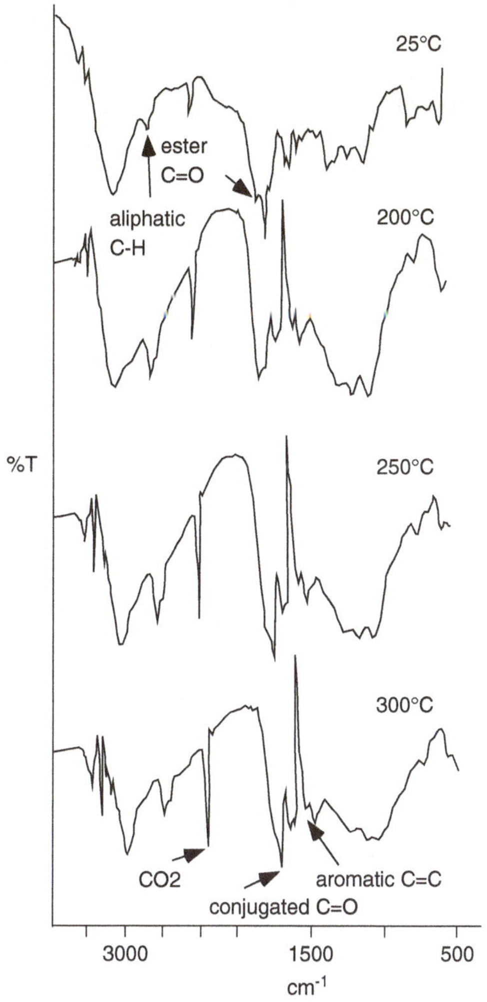

Figure 4c. Thermal alteration of the IR spectrum (after heating for 15 minutes at each temperature) of lignin after 90 days of treatment by white-rot fungus (*C. versicolor*).

The lignin treated by the brown rot fungus (*L. sulphureus*) showed remarkable change of the IR spectral feature by heating at 200°C (Fig. 4c). By further heating at higher temperature, the IR spectrum of that lignin did not show remarkable change. Removing the carbohydrate part by brown rot fungi made it possible to change the lignin structure into a thermally stable aromatic structure at low temperature. The feature of the IR spectrum of the above thermally stable aromatic compound showed high similarity to the feature of the spectrum given by heating biologically unaffected lignin at much higher temperature. This fact indicates that the lignin being not treated by rot fungi requires thermal decomposition of the carbohydrate part prior to the formation of heat stable aromatic structure. The IR spectrum of a vitrinite having reflectance of Ro(%) = 1.2 was also closely related to that of the lignin which was heated at 200°C after treatments by brown rot fungi. Therefore, brown rot fungi may greatly enhance the yield of type III kerogen by maturation of organic matter in natural underground conditions. These processes are summarized in Figure 5.

4 CONCLUSION

Natural gas is generated from terrestrial kerogen by heating at high temperature in the course of being buried in the deep ground. For natural gas generation,

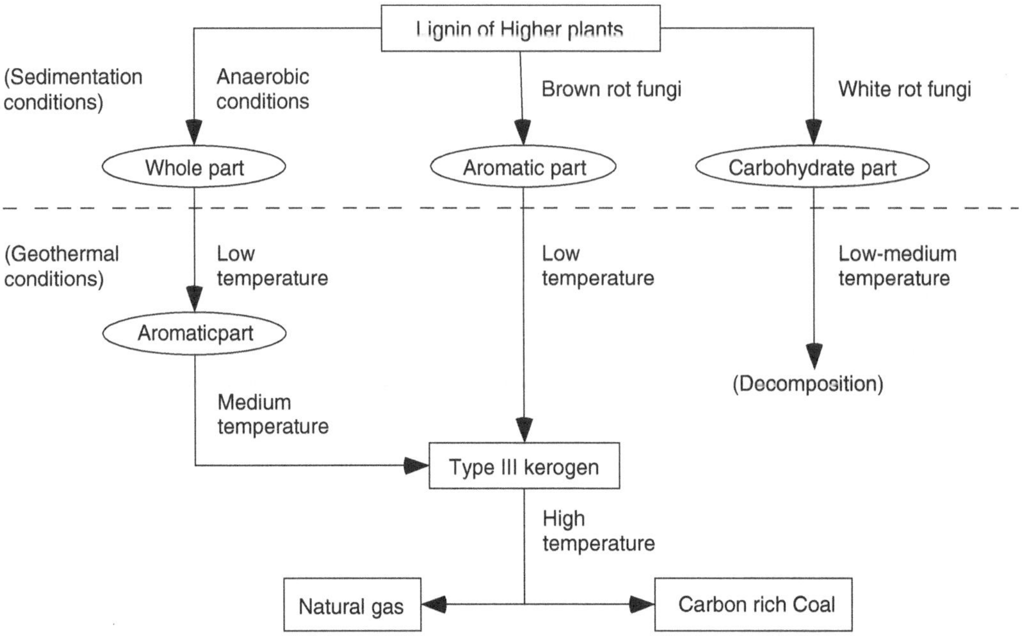

Figure 5. Scheme of alteration path of lignin in sedimental and geothermal conditions. Sedimental conditions include the case under the effects by brown rot fungi, white rot fungi and no effect by fungi. Words in □ mean the starting material and products. Words in ○ mean intermediates.

phenolic structure of lignin is necessary to be kept until the temperature to be enough high level. During the thermal alteration of natural lignin to form terrestrial kerogen, selective removal of polysaccharide moieties in the lignin structure is necessary. White rot fungi decompose aromatic moiety in the lignin at room temperature prior to removal of polysaccharide moiety and prevent kerogen formation. Brown rot fungi remove the polysaccharide moiety in lignin selectively at room temperature, and omit various pyrolysis reaction steps of polysaccharide degradation to form kerogen under geothermal conditions. Therefore, brown rot fungi probably enhance the speed and the yield for generation of natural gas.

REFERENCES

Bolwell, G.P. 1988. Synthesis of cell wall components: Aspects of control. *Phytochemistry* 27: 1235–1253.

Brezny, R. et al. 1983. Low temperature thermolysis of lignins: I. Reactions of β-O-4 model compounds. *Holzforshung* 37: 199–204.

Cowling, E.B., 1961. Comparative biochemistry of the decay of sweet gum sapwood by white-rot fungi and brown-rot fungi. *U. S. Department of Agriculture Technical Bulletin* 1258. U.S. Department of Agriculture, Washington, D. C.

Crawford, R.L. 1981. *Lignin biodegradation and transformation*. New York: John Wiley & Sons, Inc.

Ertel, J.R. & Hedges, J.I. 1984. The lignin component of humic substances: Distribution among soil and sedimentary humic, fulvic, and base insoluble fractions. *Geochim. Cosmochim. Acta* 48: 2065–2074.

Ertel, J.R. & Hedges, J.I. 1985. Sources of sedimentary humic substances: Vascular plant debris. *Geochim. Cosmochim. Acta* 49: 2097–2107.

Hartley, R.D., et al. 1988. 4,4'-Dihydroxytruxillic acid as component of cell walls of *Lolium multiflorum*. *Phytochemistry* 27: 349–351.

Hayatsu, R et al. 1984. Artificial coalification study: preparation and characterization of synthetic macerals. *Org. Geochem.* 6: 463–471.

Hedges, J.I. & Mann, D.C. 1979. The characterization of plant tissues by their lignin oxidation products. *Geochim. Cosmochim. Acta* 43: 1809–1818.

Isherwood, F.A. 1965. Biosynthesis of lignin, in: Biosynthetic Pathways. In Pridham, J.B. & Swain, T. (Eds.), *Higher Plants*: 133–146. London: Academic Press.

Kirk, T.K. et al. 1978. Influence of culture parameters on lignin metabolism by *Phanerochaete chrysosprjum*. *Arch. Microbiol* 117: 277–285.

Kirk, T.K & Farrell, R.L. 1987. Enzymatic "combustion": The microbial degradation of lignin. *Annu. Rev. Microbiol* 41: 465–505.

Nassar, M.M. & Mackay, G.D.M. 1984. Mechanism of thermal decomposition of lignin. *Wood and Fiber Science* 16: 441–453.

Otjen, L. & Blanchette, R.A. 1985. Selective delignification of aspen wood blocks in vitro by three white rot basidiomycetes. *Appl. Environ. Microbiol.* 50: 568–572.

Payne, D.F. & Ortoleva, P.J. 2001. A model for lignin alteration – part I: a keinetic reaction-network model. *Organic Geochem.* 32: 1073–1085.
Sarkanen, K.V. & Ludwig, C.H. 1971. *Lignins, Occurrence, Formation, Structure and Reactions.* New York: John Wiley & Sons, Inc.
Scalbert, E. et al. 1985. Ether linkage between phenolic acids and lignin fractions from wheat straw. *Phytochemistry* 24: 1359–1362.
Silverstein, R.M. & Webster F.X., *1998. Spectrometric identification of organic compounds, 6th ed.* New York: John Wiley & Sons, Inc.
Stafford, H.A. 1988. Proanthocyanidins and the lignin connection. *Phytochemistry* 27: 1–6.
Tanner, G.R. & Morrison, I.M. 1983. Phenolic-carbohydrate complexes in the cell walls of *Lorium perrene. Phytochemistry* 22: 1433–1439.
Tegelaar, E.W. et al. 1989. A reappraisal of kerogen formation. *Geochim. Cosmochim. Acta* 53: 3103–3106.
Whitmore, F.W. 1982. Lignin-protein complex in cell walls of *Pinus ellotii*: Amino acid constituents. *Phytochemistry* 21: 3.
Yokoyama, S. et al. 1969a. High resolution nuclear magnetic resonance spectra of model compounds related to lignin with dimethyl sulfoxide as solvent. *Kogyo Kagaku Zasshi* 72: 346–352 (in Japanese: *C.A.* 70: 116386e).
Yokoyama, S. Et al. 1969b. High resolution nuclear magnetic resonance spectra analysis of the structures of milled wood lignin and thiolignin. *Kogyo Kagaku Zasshi* 71: 353–358 (in Japanese: *C.A.* 70: 116384e).

Petroleum Geochemistry and Exploration in the Afro-Asian Region –
Digang Liang, Darui Wang & Zhenxi Li (eds)
© 2008 Taylor & Francis Group, London, ISBN 978-0-415-44083-7

Application of reservoir geochemistry in Jiuxi Depression

Yong Hu
Department of Energy, China University of Geosciences, Beijing, PR China

Jiang Da, Dajiang Zhang, Keming Cheng & Ying Xong
Research Institute of Petroleum Exploration and Development, PetroChina, Beijing, PR China

ABSTRACT: This paper, with theory and research methods of reservoir geochemistry, discussed the petroleum populations and families, hydrocarbon-generating history of source rocks and the petroleum migration direction in Jiuxi Depression. The crude oils from the six oil fields of Jiuxi Depression all belong to the same petroleum population, and can be divided into three oil families, namely maturity, high maturity and low maturity families. The three oil generating sub-sags in Jiuxi Depression, Qingnan Sub-sag, Shibei Sub-sag and Dahongquan Sub-sag, have different hydrocarbon-generating threshold depth, i.e. 4150 m, 3000 m and 3300 m, respectively. The property parameters and nitrogen compound parameters of the crude oil suggested that the oil migration of Jiuxi Depression mainly is from west to east, i.e. from the Qingxi oil field to the Yaerxia oil field, then to the Laojunmiao oil field and to the Shiyougou oil field.

1 INTRODUCTION

Reservoir geochemistry, a sub-discipline of geochemistry, thriving in recent years, is an interdisciplinary subject between geochemistry and petroleum engineering. It takes the elements of geochemical analysis to make a powerful comprehension for the hydrocarbon accumulation and reservoir development. It involves the study of reservoir fluids, reservoir fluids correlation, hydrocarbon migration and reservoir fluids with reservoir rocks. As the study of reservoir fluids in an effort to account for their spatial and compositional variations, reservoir geochemistry deals with the alteration of oil and gas in the reservoir and helps predict the distribution of hydrocarbons and understand reservoir continuity. With more and more frequent applications in petroleum industry, reservoir geochemistry has showed its obvious advantage and importance (Cubitt J M and England W A, 1997; Lin, R., 1998). In this article, the theory and research methods of reservoir geochemistry were adopted to study the division of oil populations and families, hydrocarbon generation history and petroleum migration in Jiuxi Depression.

2 GEOLOGICAL BACKGROUND

Jiuxi Depression, with the area of 2700 Km^2, is in the western Jiuquan basin of northwest China. It has been explored more than 70 years. Up to now, six oilfields

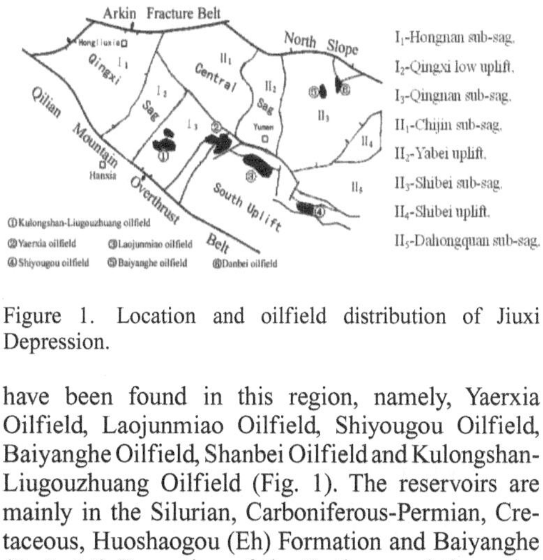

Figure 1. Location and oilfield distribution of Jiuxi Depression.

have been found in this region, namely, Yaerxia Oilfield, Laojunmiao Oilfield, Shiyougou Oilfield, Baiyanghe Oilfield, Shanbei Oilfield and Kulongshan-Liugouzhuang Oilfield (Fig. 1). The reservoirs are mainly in the Silurian, Carboniferous-Permian, Cretaceous, Huoshaogou (Eh) Formation and Baiyanghe (K, L, M) Formation of the Tertiary, and the early-Cretaceous lacustrine mudstone and muddy limestone (Chijinbao Formation, Xiagou Formation and Zhonggou Formation) are the main source rocks of this depression (Chen, J. et al, 2001; Huo, Y., 1998).

3 SAMPLE PREPARATION AND ANALYSIS

The samples of this study include crude oils and hydrocarbon source rocks. The 33 crude oil samples were collected from the six oilfields in Jiuxi Depression, namely Yaerxia Oilfield, Laojunmiao Oilfield, Shiyougou Oilfield, Baiyanghe Oilfield, Shanbei Oilfield and Kulongshan-Liugouzhuang Oilfield. The samples of hydrocarbon source rocks include cores, cuttings and outcrops. The cores and cuttings were collected from Wells Qing 2-12, Qing 2-1, Bainan 3, Dacan 2, Long 105, Long 8, Long 1, Long 103, Liu 104, Liu 102 and Liu 4 in Qingnan Sub-sag, Shibei Sub-sag and Dahongquan Sub-sag. The cutting were collected at interval of 20 meter. The outcrops samples were collected from the Hongliuxia and Hanxia profiles.

For the crude oil samples, the stable carbon isotope analysis, GC/MS, GC/MS/MS, concentration quantization analysis of biomarkers and neutral nitrogenous compound analysis were performed. All the analyses were done under the normal conditions. The rock samples were analyzed with the method of rock pyrolysis. The collected samples were extracted, and the GC/MS and biomarkers analyses were applied to the extracts.

4 RESULTS AND DISCUSSION

4.1 *Division of oil populations and families*

Petroleum population means the same kind of crude oils, which has the definite and common source rock, and its geochemistry characters are not only alike but also clearly different from other oils. Due to the difference of maturity, migration effect and biodegradation etc, the same petroleum population can be divided into different oil families with different geochemical characteristics. So, one petroleum population could include one or more oil families. In oil and gas exploration, the discrimination and division of oil populations and families are very important.

31 crude oil samples from the 6 oilfields in Jiuxi had been analyzed. All the analysis results suggested that all the crude oils in Jiuxi Depression belong to the same population, originating from the source rock with the same sedimentary environment, which is deposited in the weak-moderate salt water and moderately reducing terrestrial lacustrine environment, with the mixed source materials of the terrestrial higher plants and algae, while the crude oil belongs to three families, namely maturity, high maturity and low maturity families.

4.1.1 *Division by stable carbon isotope composition*

According to the $\delta^{13}C$ distribution of compositions for individual n-alkanes of the 31 oil samples, the carbon isotope of saturates, arene, non-hydrocarbons and asphaltene showed the decreasing tendency, while the distribution of individual n-alkanes was smooth except for weak abnormity. The two methods all showed the intensive distribution of $\delta^{13}C$ values, suggesting the same origin for the oils.

4.1.2 *Division by saturated hydrocarbon GC*

All the n-alkanes are distributed similarly, showing a singlet pattern and odd carbon predominance with the prominent carbon peak of C_{23} and minor C_{21}. The GC parameters of the crude oils also were greatly alike, CPI was about 1.2, OEP was about 1.13 and Pr/Ph ranged from 0.74 to 1.05 with the mean of 0.86. This evidence suggested that the crude oils in this region are generated in the normal stage. Furthermore, the predominance of phytane indicated the source rock is deposited in a reducing environment.

4.1.3 *Division by molecular geochemistry method*

Through analyzing the m/z191, m/z217 and m/z231 mass chromatograms for terpanes, steranes, methyl steranes and dinosteranes, respectively, of the crude oil samples obtained from different oilfields in this region, we found all the fingerprints of biomarkers had a similar distribution, implying that the crude oils in this region have the same origin (Fig. 2).

Terpanes: Tricyclic terpanes and tetracyclic terpanes were not rich, the Ts of C_{27}-trisnorhopane was more than Tm of that, C_{29}Ts, C_{30} rearranged hopane

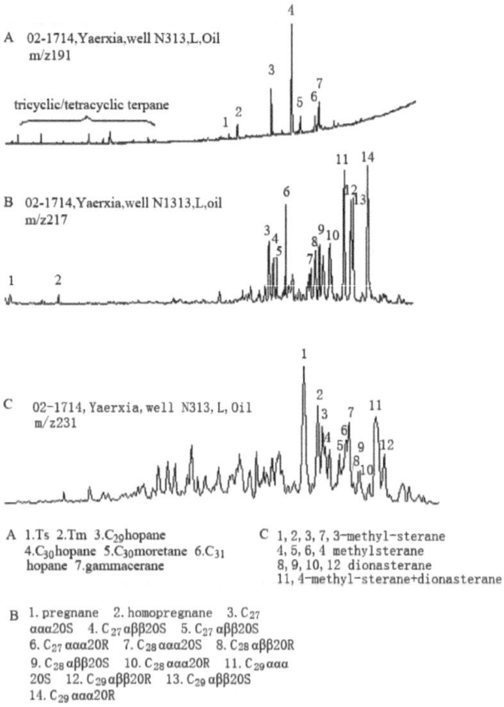

Figure 2. Mass Chromatograms (m/z191, m/z217 and m/z231) for terpanee, steranes and methyl steranes /dinosteranes of the Jiuxi oils.

and C_{29} moretane were not rich either while gammacerane was relatively rich with the content more than C_{31} hopane. This evidence indicated that the source rock is deposited in a reducing environment with brackish water.

Steranes: Pregnanes and homopreganes were found with minor contents, and rearranged steranes were not rich either, indicating that the crude oils in this region are normally mature. The anti-L type distribution of C_{27}, C_{28} and C_{29} regular steranes with the predomination of C_{29} suggested that organic matter of the source rock is lower planktonic algae and higher plants.

Methyl steranes and dinosteranes: Crude oils in this region were rich in C_{30} methyl sterane and dinosterane. C_{30} methyl sterane mainly comes from homonemeaes and dinosterane is formed by dinophyceae, indicating that all the original matter of the source rock is derived from lower planktonic algae.

4.1.4 *Division by multi-parameters*

According to the carbon isotope composition, gas chromatograms (GC) of saturated hydrocarbons and fingerprints of biomarkers, all the crude oils in Jiuxi Depression are generated by the same source rock. We applied ten parameters obtained from concentration quantitative analysis of biomarkers to further divide the oil populations and families. From the star map of the ten parameters (Fig.3), we found that the curves of $\alpha\alpha\alpha$ (R) sterane C_{27}/C_{29}, $\alpha\alpha\alpha$ (R) sterane C_{28}/C_{29} and gammacerane/C_{30}hopane (1–3) which are usually used to refer to original matter of source rocks and sedimentary environments are the same, suggesting a common original matter and sedimentary environment for the source rock. In the same environment, the parameters 1–4 could be referred to maturity. Most of the oil samples have the same curves, except for Sample A (from Well Qing 2–7), Sample B (from Well Long 4) and Sample C (from Well Long 8). Thus, according to the values of parameters 4–10, we found that the values of parameters 4–9 of Sample A is lower

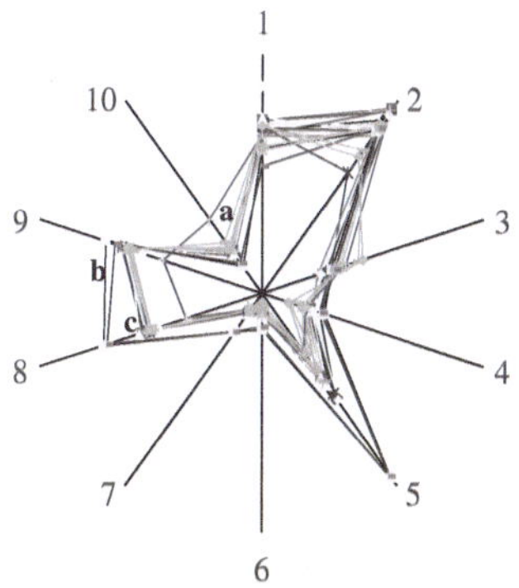

Figure 3. Star plot of biomarker parameters of the Jiuxi oils.

than the normal value while the value of parameter 10 (C_{30}Mo/H) is above the normal, indicating that it belongs to a low maturity oil, which also has been proved by the drilling result. The characters of parameters 4–10 of Samples B and C is opposite to Sample A, so they have high maturity, and for other samples, the values indicated they belong to a normal mature oil.

Consequently, according to the results obtained from the above methods, we determined that the crude oils in Jiuxi Depression belong to the same population, the source rock is deposited in the lacustrine reducing environment with brackish water, and original matter of the source rock is higher plants and algae. The crude oils are featured into three familiars, namely the high mature oil of Well Long 4 and Well Long 8, the low maturity thick oil of Well Qing 2–7 and the normal mature oil of other wells.

4.2 *Hydrocarbon generation history*

The single layer serial profiles, source rock pyrolysis profiles and geochemical profiles of crucial wells were made to determine the hydrocarbon-generating threshold depth for the lower Cretaceous source rock in the three hydrocarbon generating sub-sags of Jiuxi Depression, i.e. Qingnan Sub-sag, Shibei Sub-sag and Dahongquan Sub-sag. The results suggested that the threshold depth for the lower Cretaceous source rock in these three sub-sags is 4150 m, 3000 m and 3300 m, respectively.

4.2.1 *Serial profile analysis of a single layer*

The method is as follows. Taking Qingnan Sub-sag as the research object, source rocks of the same layer were sampled and analyzed from the edge to center of the K_1g_2 lower Cretaceous Xiagou Formation from eight wells (Well Long 8, Well Qing 2–12, Well Long 1, Well Long 103, Well Qing 2-1, Well Liu 104, Well Liu 102 and Well Liu 4) (Table 1) and a relationship of hydrocarbon conversion rations and depth is shown in Fig. 4. From which it can be seen that from the edge to center, the thermodynamic activity is reinforced with the increase of depth and so does the hydrocarbon-generating activity, the hydrocarbon conversion ration continuously increases as well. The hydrocarbon conversion ratio is up to 3% as the Ro reaching to 0.7%. Consequently, the corresponding depth is 4150 m, which can be regarded as the threshold for the Cretaceous source rock of Qingnan Sub-sag.

4.2.2 *Analysis of source rock pyrolysis profile*

The rock pyrolysis profile is also used by geologists to rapidly and economically analyze the total longitudinal organic matter (TOC), original matter type and maturity so as to obtain geochemical information for further exploration (Wang, Q. and Cheng, J., 1998; Wu, L. et al., 1998).

Table 1. Geochemical data of the $K_1g_2^2$ source rock of Qingnan Sub-sag.

Profile	Well	Depth (m) or location in the structure	Strata	1,8-/1,2-DMC	1,8-/1,3-DMC	1,8-/1,6-DMC	1,8-/1,7-DMC	1,8-/2,3-DMC	1,8-/2,4-DMC	1,8-/2,5-DMC	1,8-/2,7-DMC	1,8-/3,4-DMC
Qingxi	Liu104	4300–4352	K1g	2.173	0.530	0.704	0.947	3.847	1.235	2.356	1.122	5.620
	Long4	4519–4520	K1g	2.863	0.756	0.907	1.174	6.966	1.615	3.065	1.648	6.546
Yaerxia	Y700	Middle	E(K)	3.257	0.670	0.878	1.672	7.312	2.250	5.118	2.373	7.094
	Y503	South		2.911	0.818	0.917	1.142	8.717	1.920	2.760	2.466	8.347
Laojunmiao	D204	North	E(L)	3.352	0.912	1.026	1.298	9.477	1.896	3.648	2.585	7.596
	365	South	E(K)	3.632	0.881	1.098	1.478	9.849	2.027	4.036	2.938	8.486
Shiyougou	274	South	E(M)	4.733	1.273	1.535	1.616	13.616	2.804	5.251	3.882	7.494
	N662	North	E(M)	3.461	0.967	1.279	1.504	8.951	1.765	3.226	2.400	6.579
Baiyanghe	116	North	E(E_{3h})	2.837	0.759	0.958	1.104	8.687	1.599	2.703	2.307	6.551
	67	Top	E(E_{3h})	2.946	0.767	0.877	1.183	8.151	1.672	2.906	2.383	7.428

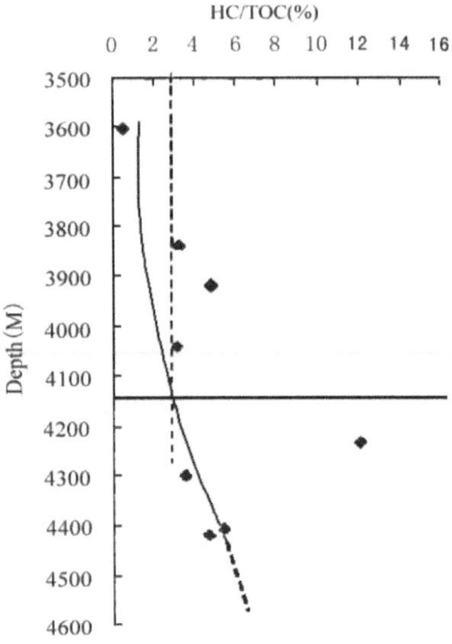

Figure 4. Hydrocarbon conversion ratio of the $K_1g_2^2$ source rock in Qingnan Sub-sag.

Figure 5 shows the pyrolysis profile of the lower Cretaceous source rock of Well Long 105, Qingnan Sub-sag. According the TOC, $S_1 + S_2$ and I_H data, the good source rock in this well is mainly in Xiagou Formation with the depth of below 4150 m. The highest value of Tmax is more than 435°C, suggesting that the source rock is already mature. The index of production and hydrocarbon indexes longitudinally increase with depth. Furthermore, below 4150 m, these parameters increase more quickly, suggesting that in this well 4150 m is the threshold for large generation of hydrocarbons.

The same method was used to analyze the source rock from Well Qing 2–12 of Qingnan Sub-sag, Well Bainan 3 of Shibei Sub-sag and Well Dacan 2 of

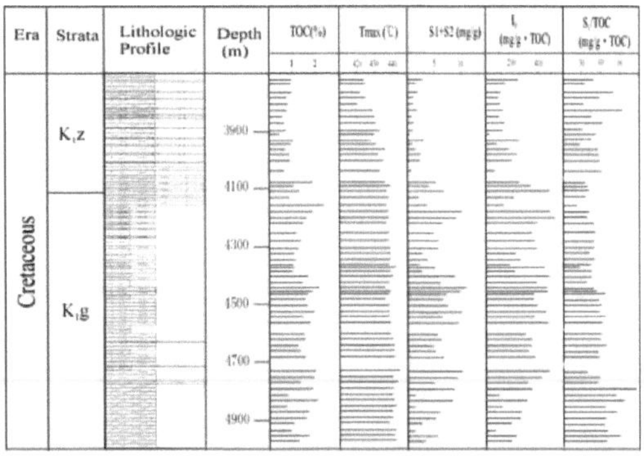

Figure 5. Pyrolysis profile of the lower Cretaceous source rock in Well Long 105 of Qingnan Sub-sag. TOC % refers to total organic content, T_{MAX} to the highest pyrolysis temperature of kerogen, $S_1 + S_2$ to hydrocarbon generative potential, IH to hydrogen index and S_1/TOC to hydrocarbon index.

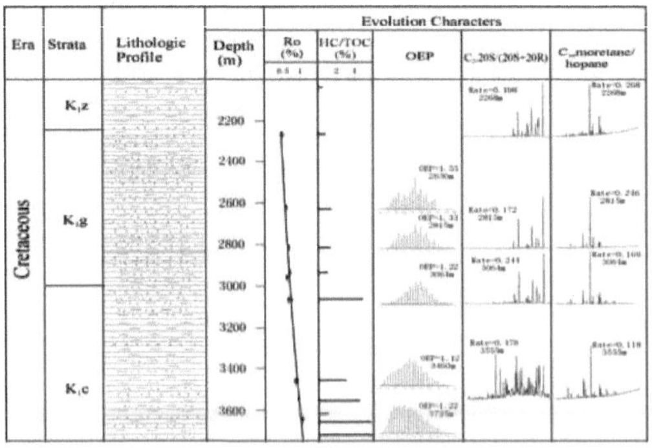

Figure 6. Evolution profile of lower Cretaceous source rock in Well Bainan 3 of Shibei Sub-sag. Ro = vitrinite reflectance, HC/TOC = hydrocarbon conversion ratio, OEP = odd-even predominance of n-alkanes, C_{29}20S/(20S+20R) = isomerization parameter of steranes, C30 moretane/hopane = isomerization parameter of hopanes.

Table 2. Dimethylcarbazole parameters of oil in the Jiuxi oilfield.

Well Number	Depth (m)	Strata	Chloroform Bitumen A (%)	Saturated Hydrocarbon + Arene (%)	TOC (%)	A/ TOC (%)	HC/ TOC (%)	Hydrogen Index (mg/g. TOC)	Oxygen Index (mg/g. TOC)	Degradation Ratio (%)	Type	Ro (%)
Long 8	3604	$K1g_2^2$	0.014	31.40	0.80	1.76	0.55	133	84	12.04	II_2	0.73
Qing 2–12	3840	$K1g_3^1$	0.107	43.26	1.41	7.57	3.27	394	40	34.97	II_1	0.67
Long 1	3917	$K1g_2^2$	0.141	26.50	0.77	18.36	4.87	125	66	12.18	II_2	0.75
Long 103	4417	$K1g_1^2$	0.044	57.37	0.54	8.17	4.69	1225	176	13.83	II_2	0.77
Qing 2–1	4040	$K1g_3^1$	0.108	31.85	1.07	10.12	3.22	192	42	18.46	II_2	0.72
Liu 104	4300	$K1g_2^2$	0.084	43.54	1.02	8.31	3.62	166	44	17.17	II_2	0.83
Liu 102	4232	$K1g_2^2$	0.178	67.48	0.99	17.96	12.12	210	91	24.40	II_2	0.8
Liu 4	4406	$K1g_2^2$	0.119	54.91	1.21	9.86	5.41	262	64	25.38	II_2	0.83

Dahongquan Sub-sag. The results suggested that the thresholds for Well Qingnan 2–12, Well Bainan 3 and Well Dacan 2 are 4050 m, 3000 m and 3300 m, respectively. Because Zhonggou Formation was eroded about 1000 m in Shibei Sub-sag and Dahongquan Sub-sag, the threshold depth in these two sub-sags is shallower.

4.2.3 *Analysis of geochemical profile*

The variations in Ro, TOC, saturated hydrocarbon chromatograms (MC) and biomarkers in geochemical profile can be used to analyze the hydrocarbon generation history (Wu, L. et al., 1998).

According to Fig. 6, the vitrinite reflectance (Ro), hydrocarbon conversion ratio and $C_{29}20S/(20S+20R)$ steranes of Well Bainan 3 increase with depth, on the contrary, the OEP and C_{30} moretane /hopane decrease with depth, indicating that with the increase of thermodynamic activity, the source rock matures gradually. Below 3000 m, the source rock (Chijinbao Formation) reach the threshold with the Ro more than 0.7%, HC/TOC more than 3%, OEP about 1.22, $C_{29}20S/(20S+20R)$ sterane around 0.244 and C_{30} moretane/hopane around 0.169, which is accordant with the threshold depth determined by pyrolysis profile.

In addition, the same parameters were used to analyze the lower Cretaceous geochemical profile of Well Long 105 of Qingnan Sub-sag. The results indicated that the threshold in this well is 4200 m, which is also close to the depth determined above.

It is notable that in depths of 3000 m, 3300 m and 4200 m, the vitrinite reflectance of Well Bainan 3, Well Dacan 2 and Well Long 105 is above 0.7%, suggesting that the hydrocarbon-generating process of the lower Cretaceous source rock is obviously depressed by the lacustrine carbonate environment and thus during the hydrocarbon generation the isomerization was slowed down. It can be concluded that in the corresponding depth of 0.4–0.7% Ro, the source rock of Jiuxi Depression is in immature-low maturity stage.

4.3 *Direction of petroleum migration*

The oil samples of Jiuxi Depression were selected to analyze nitrogenous compounds, which are generally found in the non-hydrocarbon distillates of oils and are mainly methylcarbazole, dimethylcarbazole and trimethylcarbazole. Carbazoles with methyl at both 1 and 8 ring carbons, at one of 1,8 ring carbons or no methyl at either 1,8 ring carbon are called covered, half-exposed type and exposed type, respectively. In general, during petroleum migration, the sorption between covered-nitrogen isomer and water or minerals is relatively weak, whereas the sorption to exposed-nitrogen and half-exposed nitrogen isomers is stronger. So the ratio of covered-nitrogen isomers to exposed-nitrogen or half-exposed-nitrogen isomers will rise with the increase of oil migration distance.

9 parameters of nitrogenous compounds were selected to discuss the petroleum migration direction of Jiuxi Depression (Table 2). According to Table 2, the dimethylcarbazole parameter (DMC) of the Kulongshan and Liugouzhuang oils is less than that of oils from other oilfields, indicating the oil of the Yaerxia, Laojunmiao and Shiyougou oilfields migrates from Qingnan Sub-sag. Thus the petroleum migration direction is from west to east, namely from the Qingxi oilfield to Yaerxia oilfield, to Laojunmiao oilfield and to Shiyougou

oilfield (Fig. 7). This kind of migration regularity is in accordance with the property characters of the oils (Table 3) (Chen, J., et al, 2001; Yang, Z., et al, 2003). Although the geochemical features of the Baiyanghe oil and Danbei oil in Shibei Sub-sag are greatly similar to those of the oil in the Laojunmiao anticline belt, it is proposed that according to regional geological features, the oil of the Baiyanghe oilfield and Danbei oilfield can not migrate from Qingnan Sub-sag but migrate from the lower Cretaceous source rock of Shibei Sub-sag. The dimethylcarbazole parameter (DMC) of the two Baiyanghe oil samples is greater than that of the Qingxi oil but less than the values of the Laojunmiao oil and Shiyougou oils (Fig. 7), indicating that the migration distance is relatively short, and the oil of the Baiyanghe oilfield migrates from Shibei Sub-sag.

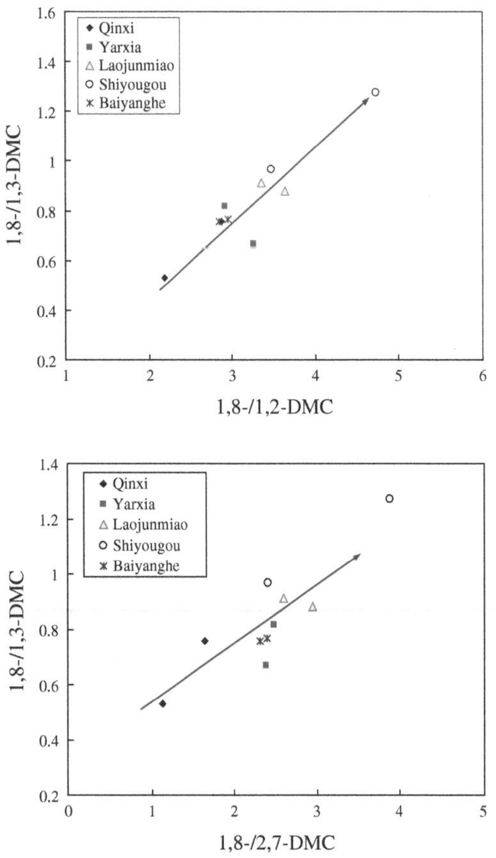

Figure 7. Direction of petroleum migration in Jiuxi Depression indicated by dimethylcarbazole parameters.

5 CONCLUSION

All the crude oils in Jiuxi Depression belong to the same population, originating from the source rock with the same sedimentary environment, which is deposited in the weak-moderate salt water and moderately reducing terrestrial lacustrine environment with mixed source materials of terrestrial higher plants and algae. Furthermore, the crude oil in Jiuxi Depression can be divided into three families, maturity, high maturity and low maturity families.

The hydrocarbon-generating threshold depth for the lower Cretaceous source rock in the three hydrocarbon-generating sub-sags of Jiuxi Depression, i.e. Qingnan Sub-sag, Shibei Sub-sag and Dahongquan Sub-sag is determined to be 4150 m, 3000 m and 3300 m, respectively. Because the lower Cretaceous in Jiuxi Depression is a lacustrine carbonate environment which has a negative influence on the maturity of organic matter, it is suggested that in Jiuxi Depression the source rock with Ro of 0.4%–0.7% should be immature or low maturity.

According to the parameters of neutral nitrogenous compounds and the property regularity of crude oils, it is suggested that the petroleum migration in Jiuxi Depression is mainly from west to east, i.e. from the Qingxi oil field to the Yaerxia oil field, to the Laojunmiao oil field, and finally to the Shiyougou oil field.

ACKNOWLEDGEMENTS

Funds for this work are gratefully acknowledged from Yumen Oilfield Company, PetroChina. We thank Yang Zhiming for his kind help in sample collection

Table 3. Oil property of Qingxi-Shiyougou in Jiuxi Depression.

Oilfield	Strata	N-alkane (%)	Arene (%)	Non-hydrocarbon + asphaltene (%)	Density (g/cm^3)	Wax content (%)	Freezing point (°C)	Note
Shiyougou	L	75.50	8.46	10.12	0.8611	13.00	−0.94	
	M	74.24	10.52	11.21	0.8647	14.50	−1.90	
	K	74.96	11.90	11.92	0.8630	13.83	0.10	
Laojunmiao	L	71.21	11.91	12.20	0.8698	16.56	7.96	↑
	M	68.72	13.30	12.77	0.8725	12.26	10.21	Migration
	L	69.25	13.33	12.62	0.8757	14.73	14.80	direction
Yaerxia	K1	67.88	11.83	17.46	0.8738	18.86	21.56	
	S	65.99	11.98	15.33	0.8859	16.03	19.26	
Qingxi	K1	64.48	13.24	20.84	0.8878	20.35	21.13	

and research. We are grateful to research team of Key Lab of Petroleum Geochemistry, CNPC, for the suggestions and help.

REFERENCES

Chen, J., Chen, J., Zhang, L., Zhong, N. 2001. New opinions for the oil and gas occurrence and future exploration in Jiuxi basin. *Petroleum Exploration and Development* V28, P1–22

Connan J. 1984. Biodegradation of crude oils in reservoirs. In: Brooks J. & Welte D.H. (eds) *Advances in Petroleum Geochemistry*. Academic Press, London

Cubitt J M, England W A. 1997. *Reservoir Geochemistry. Translators*: Wang T., Zhang Z. Petroleum Industry Press, Beijing

Haven H.L. 1996. Application and limitations of Mango's light hydrocarbon parameters in petroleum correlation studies. *Organic Geochemistry* 24, N10/11 P957–976

Henry I. Halpern. 1995. Development and applications of light-hydrocarbon-based star diagrams. *AAPG Bulletin* V79 N6, P801–815

Huo, Y. 1998. *History of China Petroleum Industry, volume 13: Yumeng oilfield*, Petroleum Industry Press, Beijing

Larter, S.R. & Aplin, A.C. 1995. Reservoir geochemistry methods, application and opportunities, *In the Geochemistry of Reservoirs*. Geological Society of London, Special publication No86

Li, M., Larter, S.R., Stoddart D and Bjoroy M. 1995. Fractionation of pyrrolic nitrogen compounds in petroleum during migration: derivation of migration related geochemical parameters. *In the Geochemistry of Reservoirs*. Geological Society of London, Special publication No 86

Lin, R. 1998. Petroleum *exploration and reservoir geochemistry*, Petroleum Industry Press, Beijing

Milner C.W.D, Rogers M.A. & Evans C.R. 1977. Petroleum transformations in reservoirs. *Journal of Geochemical Exploration* N7 P101~153

Seifert W.K. 1987. Sterancs and terpancs in kerogens phrolysis for correlation of oils and source rocks. *Geochem Cosmochim Acta* V42, N5

Tissot B.P., Pelet R. and Ungerer, PH. 1987. Thermal History of Sedimentary Basin, Maturation Indices, and Kinetics of Oil and Gas Generation. *AAPG Bull* V71, N12, P1445~1466

Wang, Q., Cheng, J. 1998. *Oil and gas geochemistry*, Press of China University of Geosciences, Wuhan

Wu, L., Gu, X., Sheng, Z., et al, 1986. *Rapid quantitative evaluation of source rock by pyrolysis*, Sciences Press, Beijing

Yang, Z., Wang, Q., Shi, J., Li, M., Wang, L. 2003. Geological and geochemical analysis for the reason of oil richness in Qingxi sag of Jiuxi depression, *Acta Sedimentological Sinica* V21, 695~701

Zhang, M., Zhang J. 1998. Application of Mango's light hydrocarbon parameters. *Petroleum Exploration and Development* V25, N6, P26~28

*Petroleum Geochemistry and Exploration in the Afro-Asian Region –
Digang Liang, Darui Wang & Zhenxi Li (eds)*
 ISBN 978-0-415-44083-7

Solid residue of thermal simulation experiment on *Laminaria* and its petroleum geological significance

Wenbin Liu[1,2], Xiancai Lu[1], Jianzhong Qin[2], Kai Hu[1], Zhiming Li[2], Lizeng Bian[1] & Mingxia Hu[2]

[1] *State Key Laboratory for Mineral Deposits Research, Department of Earth Sciences, Nanjing University, Nanjing, Jiangsu, P. R. China*

[2] *Wuxi Research Institute of Experimental Geology, SINOPEC, Wuxi, Jiangsu, P. R. China*

ABSTRACT: As an important bioprecursor of hydrocarbons in source rocks, Seaweeds (showing as macro-algae) have been attracting more and more attention from petroleum geologists. *Laminaria*, as a typical species of *Phaeophyta*, was used in the thermal simulation experiments in this study. FT-IR, SEM, diffuse reflectance spectrum, reflectivity and microscopic observation were employed to study the thermal evolution, potential of hydrocarbon generation and changes in pore structure of *Laminaria*. It is discovered that *Laminaria* has significant potential of hydrocarbon generation even under relatively high maturity of organic matter. The change in pore structure of *Laminaria* in geological environment is important for the preservation and expulsion of gaseous hydrocarbons. The results will highlight the exploration and evaluation of marine petroleum system.

Marine macro-algae are an important member in temperate and tropical marine communities worldwide. Seaweeds, called as marine forest (Campebell *et al.*, 2001), are not only the primary producers acting as the lowest member in the marine food chains, but also an important constructor (as reef-builders) providing substrate and shelter for other organisms. *Laminaria*, a common large *Phaeophyte*, is a primary seaweed in the sea ecosystem (Campebell N.A. *et al.*, 2001). In recent years, many studies provided various microscopic observation and geochemical data supporting that macro algae are a nonnegligible bioprecursor of hydrocarbons in marine source rocks (Liu Qing *et. al.*, 2001; Zhao Mengjun *et. al.*, 1999; Cui Panfeng *et. al.*, 2000; Wang Feiyu *et. al.*, 2001; Bernard Durand, 1980). The potential of hydrocarbon generation should be emphasized in the evaluation, exploration and exploitation of marine petroleum systems, although the ability of hydrocarbon generation and the primary expulsion behavior of hydrocarbons from marine macro-algae need to be further studied.

By integrating thermal simulation experiments with spectrum measurement and microscopic observation, the ability of hydrocarbon generation of *Laminaria* and the significance of hydrocarbon expulsion are investigated preliminarily in this paper.

1 SAMPLES AND EXPERIMENTS

Modern *Laminaria* (*Phaeophyta)* was selected as the experimental samples. Thermal simulation experiment was carried out in a tubular muffle equipped with a Pt-Lr-Pt thermoelectric couple, nitrogen flux 100SCCM (Education Ministry Key Laboratory for Mesoscopic Material, Nanjing University). The sample was heated with temperature programmed at 6°C/min and preserved on a constant temperature for 4 hours followed by cooling naturally. The constant temperatures were set as 150°C, 200°C, 250°C, 300°C, 350°C, 400°C and 500°C, respectively. Lignite, as maturity reference, was heated under the same conditions.

The reflectivity of the solid residue was measured on MPV-III photometer in Science and Technology Center Laboratory of China University of Mining and Technology (Beijing). Topography was observed on Philips XL30 SEM in Wuxi Research Institute of Experimental Geology. FT-IR spectrum was measured on Nicolet NEXUS870 spectrometer in Analyzing Center of Nanjing University, and diffuse reflection spectrum was measured on Lambda900 UV/VIS/NIR spectrometer in Surface Geochemistry Research Institute of Nanjing University.

2 RESULTS

2.1 *Microscopic observation of the solid residue*

As observation on transmission optical microscope, the color of *Laminaria* changes gradually from sandy beige to pea green(150°C), wine(200°C), maroon (250°C, 300°C), brown (350°C), and black (400°C, 500°C) as heating temperature increasing (Plate I). The shape of *Laminaria* samples also changes regularly. The biotexture of the residue can be identified before 300°C (included), but it disappears gradually after 350°C, and becomes into various irregular conformation. As it is heated up to 400°C, the solid residue is homogeneous and amorphous.

2.2 *Reflectivity*

Reflectivity of solid residue of *Laminaria* is shown in Table 1 and Fig.1. It rises regularly from 0.13% to 2.48% while temperature increasing from 150°C to 500°C, which is lower than that of lignite heated to the same temperature on the whole. The changes can be fitted by an exponent equation, which suggests that the reaction can be described using Arrhenius Equation. Furthermore, the reflectance of solid residues has a linear relation to that of lignite with correlation coefficient of 1.0769 (Fig.2). The reflectivity of solid residue of *Laminaria* at 500°C is nearly equal to that of lignite.

2.3 *Diffusing reflection spectrum*

Figure 3 shows the diffusing reflection spectrum of solid residue of *Laminaria*. Four aspects could be concluded below. (1) The diffusing reflectivity decreases between 400 nm and 2500 nm with the temperature increasing. The decreasing extents vary with different wavelength; (2) In visible light scope, the reflectivity intensity changes regularly. It increases firstly (150°C∼200°C), then decreases (200°C∼350°C) and increases again (350°C∼500°C). The red/blue ratio increases in the beginning, then decreases persistently (Table 2). (3) In the wavelength range of 750 nm∼2500 nm, the diffusing reflectance of the solid residue decreases with the temperature increasing. The shape of curve changes from rightward incline to smooth horizon. (4) When temperature is lower than 300°C, there are absorption vales of O-H bend vibration at 1350 nm and 1450 nm and of O-H stretch vibration at 1800 nm and 1950 nm, indicating that the samples may contain H_2O, protein, starch or cellulous (Pu *et al.*, 2000). These vales disappear gradually while the temperature increasing, suggesting that the content of O-H bonds decreases gradually. In addition, the absorption vale at 2300 nm changes significantly, and its assignment needs further investigation to affirm.

2.4 *FT-IR spectra*

Figure 4 shows the FT-IR spectra of the solid residue of *Laminaria* in thermal simulation experiment. As previous studies indicated, the absorption vale shows the occurrence of H_2O or O-H bonds in the sample at 3300 cm^{-1}, asymmetry bend and stretch vibration of $-CH_2$ at 2920 cm^{-1} and 1460 cm^{-1}, asymmetry bend and stretch vibration of $-CH_3$ at 2860 cm^{-1} and 1380 cm^{-1}, outwards bend vibration of aromatic nucleus C-H at 810 cm^{-1} and 750 cm^{-1} (Yao Suping *et al.*, 1996), stretch vibration of carbonyl at 1730 cm^{-1}, C=C of benzene at 1640 cm^{-1} and 1610 cm^{-1}, -NH- bend at 1535 cm^{-1}, methyl ketone or carbonyl at 1460 cm^{-1} and 1380 cm^{-1} (Yao Suping *et al.*, 1996), respectively. Based on FT-IR spectra of the samples, the following conclusions can be obtained. (1) The absorption vale of aliphatic $-CH_2$ (at 2920 cm^{-1} and 1460 cm^{-1}) still exists at 400°C, even up to 500°C. (2) The vale of $-CH_3$ asymmetry bend and stretch vibration (at 2860 cm^{-1} and 1380 cm^{-1}) is not obvious and disappears at 350°C. (3) The vale relation to aromatic nucleus (at 1640 cm^{-1}, 1610 cm^{-1} and 1600 cm^{-1}) enhances and it shows that the extent of aromatization increases gradually. (4) The vale relation to -NH- (at 1535 cm^{-1}) enhances remarkably, which may be caused by the reaction with N_2 flow as the samples are heated.

2.5 *SEM observation*

The change process of pores of *Laminaria* in the thermal simulation experiment is presented in Plate II, SEM photos of the thermal simulation experiment samples. The biological fabric pores of the sample change into secondary ones.

The original tabular fabric pore in the straight-cut lamina is enveloped with thick cell wall (with thickness of 1–2 μm). There are dispersion particles between walls, which may be pieces of organelle or pipeline. A few fabric pores in the cell wall are observed (Plate II-1). The original biological fabric of *Laminaria* changes scarcely when samples are heated at 150°C and 200°C, except that the pore wall becomes coarse (Plate II-2) and cracks occur (Plate II-3). The sample surface becomes more smooth when it is heated up to 250°C and 300°C. Original biological fabric begins to disappear gradually. The pore wall is deformed, and secondary micropores are produced (Plate II-5). While laminaria is heated to 350°C, all of the original fabric disappears and it is characterized dominantly by abundant secondary pores, which are round or ellipse with uniform radius of 0.2–1.0 μm. The distribution of micropores is homogeneous in the whole sample, and does nothing with its original fabric (Plate II-6). When it is heated at 400°C, the number of secondary pores increases obviously, and the radius increases slightly. Some of the pores are connected together into

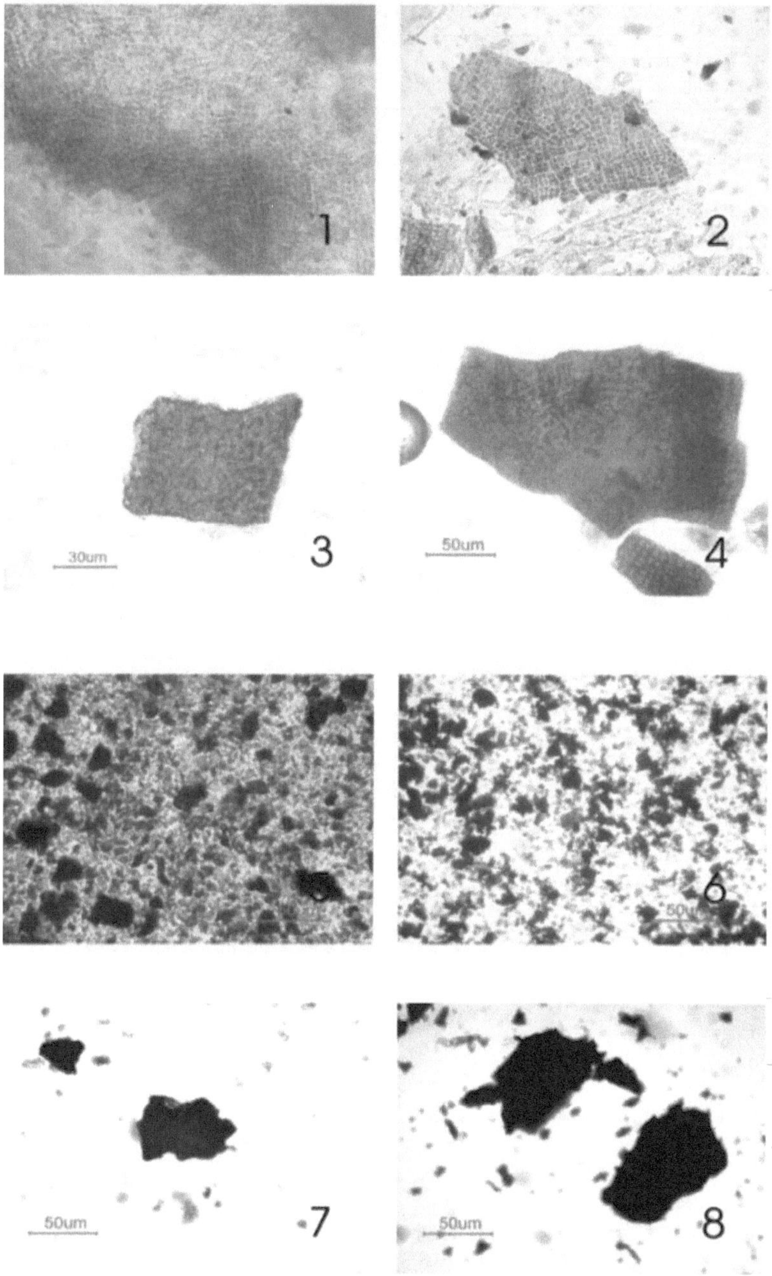

Plate I. Transmitted light Micrograph of Thermal Experiment Samples 1. Original, 2. T150, 3. T200°C 4. T250, 5. T300°C, 6. T350, 7. T400°C, 8. T500°C.

irregular pores (Plate II-7). While it is heated at 500°C, the original biological pores are enlarged into round cavity with radius of about 20 μm, and simultaneously more micropores are formed. In the high-resolution SEM image, the micropores are present in irregular shapes with radius of about 100 nm. Statistical data of SEM images suggest that the volume of micropores takes about 1/3 of the whole sample (Plate II-8). It is proposed that during the thermal evolution the pore structure of Laminaria varies along in two different

Table 1. Reflectance of solid residues of thermal simulation experiment.

Temperature Sample	150°C	200°C	250°C	300°C	350°C	400°C	500°C
Lignite	0.37	0.37	0.50	0.74	1.06	1.25	2.46
Laminaria	0.13	0.23	0.26	0.76	0.67	1.00	2.48

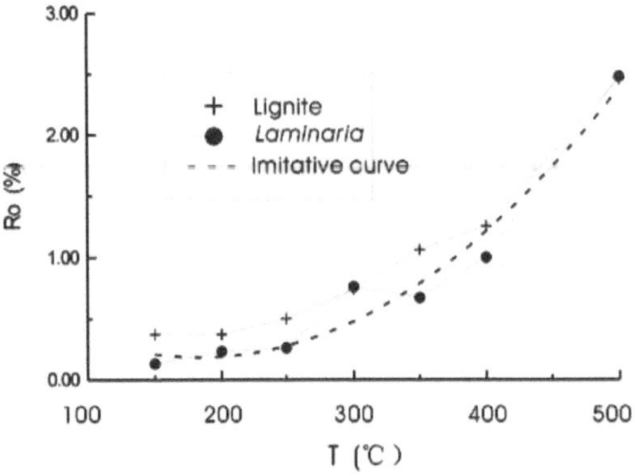

Figure 1. Reflectance of thermal simulation experiment samples related with temperature.

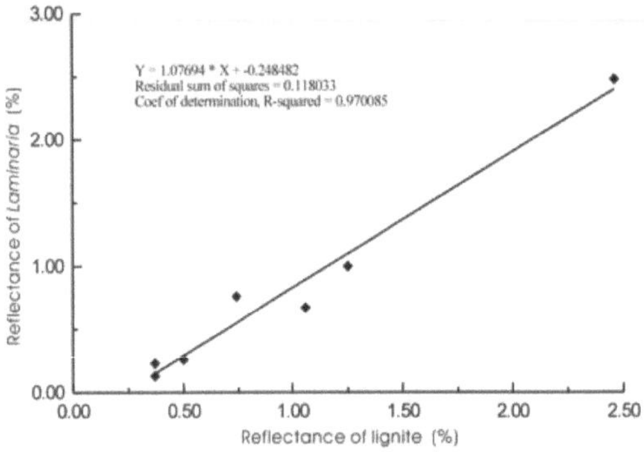

Figure 2. Reflectance of solid residues of *Laminaria* and lignite in thermal simulation experiment.

directions, including pore enlargement and formation of micropores at molecular level, which is consistent to the previous observation of coals (Zhang Hui, 2001).

3 DISCUSSION

3.1 *Potential of hydrocarbon generation*

In the present study, 350°C is proved as an important temperature point in the thermal process of Laminaria because the absorption vale of aliphatic groups changes greatly and secondary pores come forth at this stage, which is attributed to intensive hydrocarbon generation and expanding effect caused by gas production. The fact that the absorption vale of aliphatic groups till remains at 400°C shows that *Laminaria* still has a ability of hydrocarbon generation at this temperature point. These two temperature points are much higher than those for pelagic algae. Wang K.F. *et al.*, (1994) found that *Pediastrum* generates hydrocarbon extensively at 300°C. Whereas, both temperature points are similar to those for spore-pollen, which generates CO_2 and N_2 at 300°C and methane/ethane at 400°C. In a thermal simulation experiment on coals from Hailaer Basin, Fu Li (2003) found that the amount of gas generated from coals increases obviously at 359°C, which is similar to *Laminaria* in this study. Li T.Z *et al.*, (2002) considered that sapropelic coal begins with the second hydrocarbon generation stage at 380°C. It is notable that our experiment is designed as an open system without aqueous phase. According to the basic principle of chemical kinetics, the temperature for hydrocarbon generation in the open system is somewhat lower than that in the close system. In other words, the threshold temperature of *Laminaria* is higher than that of pelagic algae and sapropelic coal, but similar to spore-pollen. It may be related to its unique cell structure as *Laminaria* has a kind of rubber-like cell walls (Campebell N.A., 2002).

The reflectivity of solid residue of *Laminaria* is lower than that of vitrinite at same temperature. Sometimes the reflectivity of vitrinite is much higher (Wang Z.Y.*et al.*, 1995). This may be caused by two factors, retardation of hydrocarbon generation and abundant secondary pores.

3.2 *Hydrocarbon expulsion and pore evolution*

Pore structure is a crucial parameter in coal petrology because the pore system in coals with high porosity provides channels for hydrocarbon expulsion and preservation space of hydrocarbons (Liu Xiaoyan *et al.*, 1997). Only when the absorption on coal surface is saturated or the adsorption energy is conquered, can hydrocarbons be expulsed out (Liu D.H. *et al.*, 1987). Porosity and specific surface area are proposed as determining factors influencing hydrocarbon expulsion. As the porosity of coal is higher than 10% and the

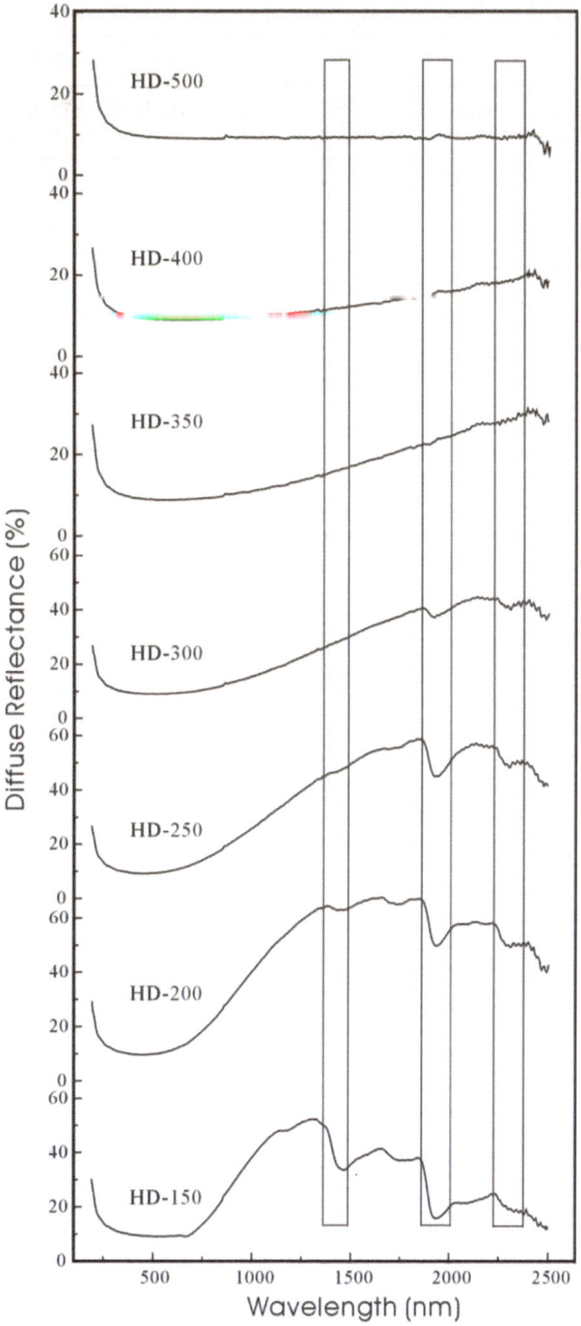

Figure 3. Diffuse reflective spectrum of solid residue of thermal simulation experiment.

hydrocarbon content higher than 86 mg/g coal, hydrocarbon expulsion occurs easily (Luo Xia *et al.*, 2003). On the other hand, the pores of coal are the main space for hydrocarbon preservation (Bi Hua *et al.*, 1997). Therefore, it is of importance to study pore structure of kerogen (Fu Xuehai *et al.*, 2001). Previously substantive studies mainly focused on the pore structure and its classification. There are different classifications, such as original pore, secondary pore and inner pore (Bi Hua *et al.*, 1997), or original pore,

Table 2. Statistics of diffusing reflectance of samples in visible light range.

T	150°C	200°C	250°C	300°C	350°C	400°C	500°C
Σ VIS	727.145	890.367	787.7576	706.5289	676.3044	685.2727	699.9221
R/B*	1.115	1.561	1.342	1.091	0.995	0.960	0.950

*: R: 750~630 nm, G: 570~490 nm B: 460~430 nm

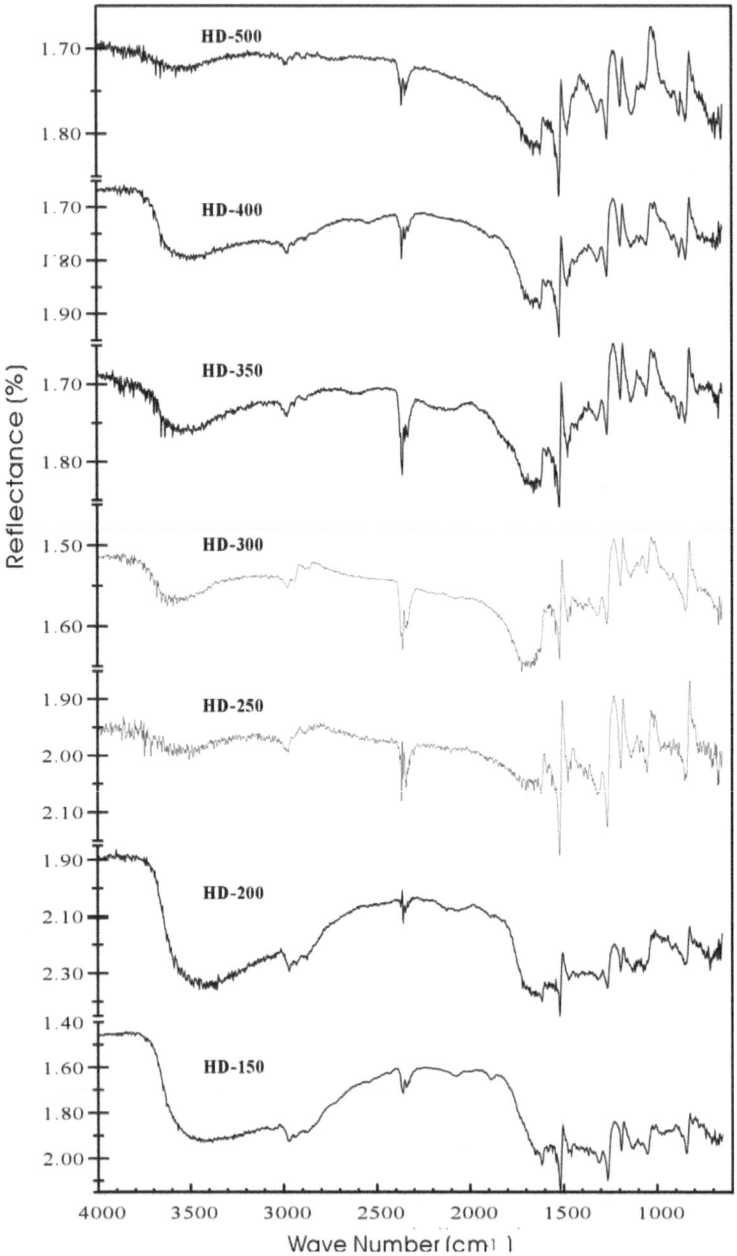

Figure 4. FT-IR spectra of the solid residue in thermal simulation experiment.

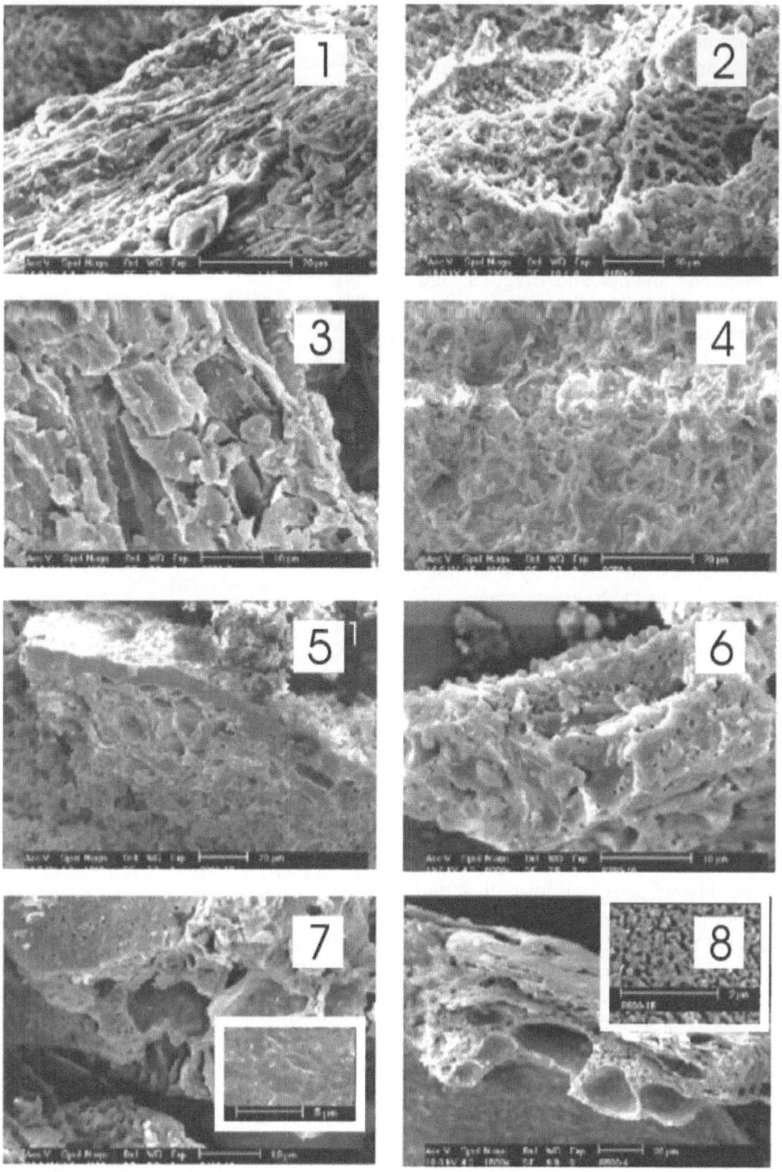

Plate II. SEM of Thermal Experiment Samples 1. Original, 2. T150, 3. T200°C, 4. T250°C, 5. T300, 6. T350°C, 7. T400°C, 8. T500°C.

metamorphose pore, exogenous pore, and mineral inner pore (Zhang H., 2001; Zhang H.*et al.*, 2003). Some studies investigated pore shape (Chen P. *et al.*, 2001; Zhu Z.B. *et al.*, 1994; Zhong L. W. *et al.*, 2002) and pore width distribution (Zhao Z.G. *et al.*, 2001), and summarized its evolution rules. It is found that coal type, gasification media and temperature determine pore structures (Xiang Y.H. *et al.*, 2002), During the heating and coking processes, the generation of CO_2 opens the closed pore firstly, then continues with a pore-enlargement process caused by gasification of H_2O. Therefore, lowering the concentration of vapor or raising the activation temperature will be favor for the development of micropores. Original pores in coals are distorted, dwindle, close and disappear with the enhancement of metamorphism. As temperature increases, the metamorphose pores become larger pores and molecular pores gradually (Zhang H., 2001).

The solid residue of *Laminaria* has two kinds of pores, original fabric pores and secondary pores. The original fabric pores are dominant before 300°C and disappear gradually at higher temperatures. Then secondary pores become into a dominant type of pores. Because the experiment is carried in an open environment and water vapor and CO_2 are expulsed in time, which is responsible for the lack of macropores and dense occurrence of micropores. When the sample is heated up to 500°C, the surface is obviously coked, and the pores get smaller and irregular, and connect together (Plate II-8). The dense micropores of macro algae have double effects on hydrocarbon expulsion. One is that they offer good channels for hydrocarbon expulsion. On the other hand, they can hold gas and obstruct hydrocarbon expulsion. It is necessary to impersonally and synthetically evaluate source rocks containing abundant organic matter derived from macro algae.

Furthermore, micropores affect the reflectivity more or less. The surface of the solid residue is smooth at 300° C, corresponding to high reflectivity that equals to that of lignite (Table 1, Fig. 1). When micropores are developed, the reflectivity of the solid residue reduces. But it is not close to that of lignite until 500°C.

4 CONCLUSION

Two viewpoints are reached by the present systematical study on the solid residue of thermal simulation experiment on *Laminaria*. (1) Threshold temperature for hydrocarbon generation of macro algae is higher than that of pelagic algae. The reflectivity is lower than that of vitrinite at the same temperature. Attentions must be paid to the maturity of marine source rocks in evaluation. (2) Pore structure of *Laminaria* changes regularly from original biological ones to smooth surface, then the generation of secondary micropores occurs and pore radius decreases. The development of micropores and variation in pore width affect the hydrocarbon expulsion and the measurement of reflectivity.

ACKNOWLEDGEMENTS

We are grateful to Dr. Ji Jufeng, Dr. Yao Suping, Department of Earth Sciences, Nanjing University and anonymous reviewers for their critical comments on the manuscript. Great thanks go to Dr. Fang Jiahu, China University of Mining & Technology, for reflectance measurement. Dr. Liu Weixin and Dr. Liu Guangxiang, Wuxi Research Institute of Experimental Geology, are thanked for their great help in the study.

REFERENCES

Bernard Durand, 1980, Kerogen – insoluble organic matter from sedimentary rocks, Graham & Trotman Ltd.

Bi Hua, Peng Gelin, Yang Minhui, 1997, The Characteristics of Gas-generating Rock and Resources Assessment of Coalbed Mathane of Carboniferous Permian System in Lianyuan Basin, Middle Hunan, Natural Gas Industry, 17(4):10–14

Campebell N.A., Reece J.B., 2002, Essential biology (Photocopy), Higher Education Publishing Company, Beijing, P:322

Chen Ping, Tang Xiuyi, 2001, The research on the adsorption of nitrogen in low temperature and micropore properties in coal. Journal of China Coal Society, 26(5): 552–556

Cui Panfeng, Li Jian, Hu Haitao et. al., 2000, Research Significance of Macro Algae Fossil in the Study of Sedimentary Facies in Ansai Oil Field, 21(5):36–40

Fu Li, 2003, Hydrous Thermal Simulating Experiment on Hydrocarbon Generation of Coal in Hailaer Basin, Geology-Geochemistry, 31(2): 100–104

Fu Xuehai, Qin Yong, Xue Xiuqian et.al., 2001, Research on Fractals of Pore and Fracture- Structure of Coal Reservoirs, Journal of China University of Mining & Technology, 30(3):225–228

Li Tianzhong, Qing Yong, Zhang Yousheng, et.al., 2002, Blue Shifting Phenomena in Hydrocarbon Re-generation, Chinese Journal of Geology, 37(1):62–69

Liu Dehan, Fu Jiamo, 1987, Primary Study on the Stage and Characteristic of Gas and Oil from Coal, Beijing, Petroleum Industry Publishing Company

Liu Qing, Hu Kai, Zhang Linye et. al., 2001, Organic Geochemistry of the Lower Permian Source Rocks in the South Shengli-Hetian Exploration Region, Xinjiang, Western China, Journal of Nanjing University(Natural Science), 37(1):114–121

Liu Xiaoyan, Zhu Fuxiang, Chen Xin et.al., 1997, Oil Generation and Migration Characteristics of Experimental Simulation from Coal and Geological Significance, Journal of Daqing Petroleum Institute, 21(2):5–8

Luo Xia, Li Jian, Hu Guoyi et.al., 2003, An Experiment of Generation and Expulsion of Oil from the Jurassic Coals and its Implication of Coal Generated Oil in Ordos Basin, Petroleum Geology & Experiment, 25(1): 76–80

Pu Ruiliang, Gong Peng, 2000, High Spectrum Remote Sensing and its Application, Higher Education Publishing Company, Beijing, P:185–228

Wang Feiyu, Bian Lizeng, Zhang Shuichang et.al., 2001, Two Kinds of Hydrocarbon-generating Source material in Ordovician Marine Source Rock in Tarim Basin, Science in China, 31(2):96–102

Wang Hongmei, Zhang Wenhuai, Xie Shuchen et.al., 2000, The Organic Geochemistry Characteristics of Jinya Gold Deposit, Guangxi, Acta Petrologica Sinica, 16(4): 602–608

Wang Zhaoyu, Chen Kemin, Zhang Baisheng, 1995, The Study on the Characteristics and Evolution Regularity of Product of Gas under Pyrolysis Simulation Experiments, Petroleum Exploration and Development, 22(3):36–40

Wu Guoxuang, Wang Kaifa, He Chenquan et.al., 1994, Research on Thermal Effects of Modern Pediastrum and its Geologic Significance, Petroleum Exploration and Development, 21(1):53–56

Xiang Yinhua, Wang Yang, Zhang Jianmin, Dong Zhongbing, Li bin, Study on structural properties and their affecting factors during gasification of chars. Journal of Fuel Chemistry and Technology, 2002, 30(2): 108–112

Yao Suping, Jin Kuili et.al., 1996, Study the Thermal Evolution Rule of Maceral Using Micro Fluorescence and FT-IR Spectrum, Acta Sedimentologica Sinica, 14(3): 1–10

Zhang Hui, 2001, Genetical Type of Pores in Coal Reservoir and its Research Significance, Journal of China Coal Society, 26(1):40–44

Zhang Hui, Li Xiaoyan, Hao Qi, 2003, SEM Observation on Chinese coal, Beijing, Geology Publishing Company, P70

Zhao Mengjun, Zhang Baomin, Bian Lizeng et.al., 1999, Ordovician III-resembling Kerogen Source Rock and Characters of its Gas, Chinese Science Bulletin, 44(21): 2333–2336

Zhao Zhigen, Tang Xiuyi, Study of micropore in coal by low-temperature nitrogen adsorption method and its significance. Coal Geology & Exploration, 2001, 29(5):28–30

Zhong Lingwen, Zhang Hui, Yun Zhengrong et.al., 2002, Influence of Specific Pore Area and Pore Volume of Coal on Adsorption Capacity, Coal Geology & Exploration, 30(3):26–29

Zhu Zhibin, Ma Zhihua, Lin Shiying, Mitsuho Hirato, Masayuki Horio, Characteristics of coal char gasification at high temperature (II): the effect of pore structure on coal char gasification. Journal of Chemical Industry and Engineering (China), 1994, 45(2):155–161

Petroleum Geochemistry and Exploration in the Afro-Asian Region – Digang Liang, Darui Wang & Zhenxi Li (eds)
© 2008 Taylor & Francis Group, London, ISBN 978-0-415-44083-7

Algorithm for production allocation in multilayer commingled production: Research and application

Yuewen Wang & Shuangfang Lu
Daqing Petroleum Institute, Heilongjiang, PR China

Wei Fang & Juhe Zhang
Daqing Oil Field Exploration and Development Institute, Heilongjiang, PR China

ABSTRACT: Chromatographic fingerprint techniques were in hot water ever in the field of multilayer commingled production of oil wells. We applied a set of algorithm that combines partial least-square method (PLS) and artificial nerve networks (ANN) in this paper and got hold of a solution. PLS was used to select the parameters automatically and ANN was used to predict the result of production allocation. The algorithm is non-linear, highly precise and adaptive for multilayer commingled production. It can reduce difficulty of acquiring a single oil sample in certain degree. We applied chromatography fingerprint data of a single layer oil, artificial commingled oil in laboratory and oil commingled production in the test area of the Daqing and Liaohe oilfields. The results computed with the algorithm were acceptable and close to laboratorial proportion and MFE test results. The computation technology will give an economical and applicable way to production allocation of commingled production of oil wells with geochemistry methods.

1 INTRODUCTION

Kaufman (1987) computed production allocation of two layers commingled production with crude oil chromatographic fingerprints in the North Sea oil field. He established two layers plate with peak high ratio. The technique brought development to dynamic geochemistry monitor of oil reservoirs (Kaufman et al. 1987, 1990, 1997). Afterwards, many geochemists in China were studying the feasibility of this method in nonmarine oilfields (Feng Zihui et al. 1999, Fu Xiaotai et al. 2000, Wang Zhenping et al. 2000, Jin Xiaohui et al. 2001, and Zou Yuzheng et al. 2001). Compared with present production log methods, chromatographic fingerprint technique is advantageous in short cycle, low cost, big controlling area. It is suitable to thick or unconventional oil reservoirs and it does not delay normal production, etc.. However, it has not been applied widely to production of oil fields until now. Many constraints exit in the technique.

Less appropriate algorithm is the first constraint. It is nonlinear superimposition relations between peak high ratio of chromatogram fingerprints and production allocation. Predecessors once used internal standard method and partial least-square method to solve the problem, the precision, however, was insufficient for three layers or multilayer problems, which are usual in nonmarine oilfields. Negative output appeared sometimes. It was difficult to choose suitable fingerprints, which is the second constraint. Only a small number of fingerprints are suitable. Suitable parameters are characteristic fingerprints. Selection of characteristic fingerprints was a troublesome job. The job had to be done by experts, therefore, it was time-consuming and not objective. It is difficult to acquire all single oil samples and the characteristics of oils would change with water flooding, which is the third constraint. Most of oil wells in nonmarine oilfields are exploited in multilayer commingled production.

This paper studies the feasibility applying partial least-square method (PLS) and artificial nerve networks (ANN) algorithm to production allocation in multilayer commingled production.

2 ARITHMETIC MODELS

2.1 *Simple partial least squares regression model*

2.1.1 *Fundamental*

Wold S. and Albano C. presented partial least squares method for the first time in 1983. The production allocation in multilayer commingled production is a problem of simply dependent variable. Parameters of single layer oil are independent variables while parameters of mixed oil are dependent variables. Simple partial least squares regression model (PLS1) is applicable.

Generally, chromatogram fingerprints are highly multiple correlations. The correlations make error enlarge, parameter estimation lapse and the model lose robustness. In addition, the regression coefficient is difficult to explain when applying LS at discretion. As a result, least square method (LS) is inapplicable. PLS1, however, can solve the difficulties effectively. Firstly, PLS1 is effective even when multiple correlations exist. Secondly, it can still obtain solution when the number of sample points is less than number of independent variables. Thirdly, all independent variables are included in the model. Fourthly, it can reduce random noise effectively. Fifthly, the regression coefficients are easy to explain practical problems.

A simplified PLS1 is presented here to the specific problem. If F_0 is the standardized variable of dependent variable y, E_0 is standardized matrix of independent variable X, then

$$w_1 = \frac{E_0' F_0}{\|E_0' F_0\|}, \quad t_1 = E_0 w_1,$$

$$p_1 = \frac{E_0' t_1}{\|t_1\|^2}, \quad E_1 = E_0 - t_1 p_1' \tag{1}$$

The program will check across validation variable Q_1^2. The program continues to perform the second step if $Q_1^2 > 0.0975$ or $Q_1^2 = 0.0975$. One component t_1 is selected if $Q_1^2 < 0.0975$. When $h = 2, \ldots, m, E_{h-1}$ and F_0 are known , then

$$w_h = \frac{E_{h-1}' F_0}{\|E_{h-1}' F_0\|}, \quad t_h = E_{h-1} w_h,$$

$$p_h = \frac{E_{h-1}' t_h}{\|t_h\|^2}, \quad E_h = E_{h-1} - t_h p_h' \tag{2}$$

The program will check across validation variable Q_h^2. The program continues to perform the $h+1$ step if $Q_h^2 > 0.0975$ or $Q_h^2 = 0.0975$. M components $(t_1, \ldots, t_m)$ are selected if $Q_h^2 < 0.0975$. Next step, F_0 is regressed in $t_1, \ldots, t_m$, *that is*

$$\hat{F}_0 = r_1 t_1 + \cdots + r_m t_m \tag{3}$$

where $t_1, \ldots, t_m$ is linear combination of E_0, which means

$$t_h = E_{h-1} w_h = E_0 W_h^*,$$

thus, the left of expression (3) can be shown as linear combination of E_0, then replace the left of expression (3) with y and $t_1,\ldots,t_m$ with $x_1,\ldots,x_p$

$$\hat{y} = a_0 + a_1 x_1 + \cdots + a_p x_p \tag{4}$$

Where

$$Q_h^2 = 1 - \frac{PRESS_h}{SS_{h-1}},$$

$$PRESS_h = \sum_{i=1}^{n} (y_i - \hat{y}_{h(-i)})^2,$$

$$SS_h = \sum_{i=1}^{n} (y_i - \hat{y}_{hi})^2,$$

The variable with "hi" is the fit value of the sample point i when regressing with components $t_1, \ldots, t_h$ and all sample points, the variable with "$h(-i)$" is the fit value of sample point i when regressing with components $t_1,\ldots,t_h$ and all sample points but i.

Generally, lesser variables were selected in order to avoid multiple correlation before PLS method is invented, which would lose information and harm precision. The arithmetic included principally principal component regression and ridge regression. They are highly complex when large numbers of variables exist. The former is difficult to use in accepting or rejecting major constituents and the latter is difficult in estimating ridge parameters. PLS is similar to principally principal component regression in vercoming collinearity but conquering its shortcoming of ignoring some independent variables. Moreover, PLS is much less complex in computation.

2.1.2 *Selecting parameters automatically*

Chromatogram fingerprints of oils are complex, and they include validation, repeatability, and noise. In the past, characteristic fingerprints were selected based on the geochemistry characteristics of compounds. The process was time-consuming and subjective. Now, PLS1 can be used to select characteristic fingerprints. Instead of selecting parameters directly, PLS1 disassembles and recombines the parameters, then picks up synthetic variables, which interpret the system better, and wipes off redundant messages and noise. PLS1 should be able to acquire results more reliable by working like this.

To be specific, firstly, we picked up the common parameters of each sample and standardized each of them as individual vector. Standardization means orthogonalization and unitization. Secondly, we performed expressions (1) to (4), the coefficients $a_1, \ldots, a_p$ were main elements to accept or reject parameters. Thirdly, the exceptional points were deleted, which was kept away from averaging. For instance, t_1/t_2 ichnography and T^2 ellipse could be drawn when two major components were selected. The equation T^2 ellipse is

$$\frac{t_{1i}^2}{s_1^2}+\frac{t_{2i}^2}{s_2^2}=c \quad (5)$$

where t_{1i},t_{2i} are major components, s_1, s_2 are sample variance, c is associated with sample F distribution. If number of sample points is n and number of major components is m, then

$$c=\frac{m(n-1)(n^2-1)}{n^2(n-m)}F_{0.95}(n-m,m)\text{, When}$$

$$T_i^2=\frac{1}{n-1}\left(\frac{t_{1i}^2}{s_1^2}+\frac{t_{2i}^2}{s_2^2}+\frac{t_{mi}^2}{s_m^2}\right)\geq\frac{c}{n-1},$$

sample point i has a oversize contribution to $t_1, t_2,\ldots,t_m$ at 95% limitation and it may deflect analytical results. Therefore, sample point i is an exceptional point and should be deleted. Expression

$$\frac{t_{1i}^2}{s_1^2}+\frac{t_{2i}^2}{s_2^2}+\frac{t_{mi}^2}{s_m^2}=c$$

is a T^2 ellipse when $m=2$ and an ellipsoid of m dimension spaces when $m>3$ or $m=3$. The exceptional points are the points outside of the ellipse or ellipsoid. For example, points 13 and 18 are exceptional and should be deleted in Figure 1.

2.2 *Artificial nerve networks model*

2.2.1 *Fundamental*

Artificial nerve network is a kind of intelligent algorithm. We selected error back propagation algorithm (EBP) in this paper (Hagan, M.T. et al. 2002). The algorithm was proposed firstly by Rumelhart in 1986.

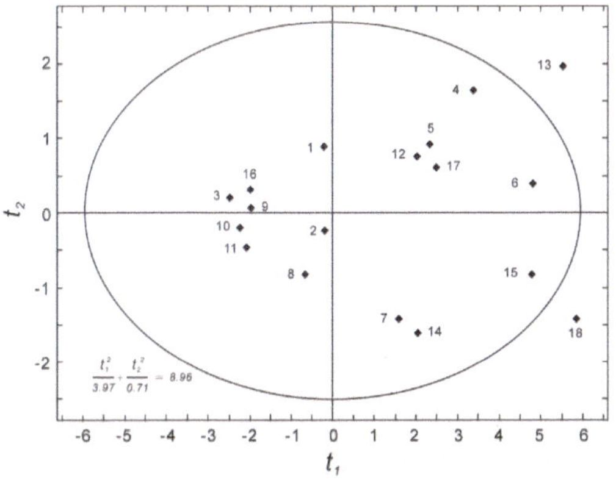

Figure 1. t_1/t_2 ichnography and T^2 ellipse.

To explain the fundamental, we took ANN with two hidden layers as an example (Fig. 2). Firstly, we input each set of selected parameters $X_K(K=1,2,\ldots,N)$ one by one. The parameters were passed through a series of sigmoid functional transformations and a series of matrix operations with a random weight matrix. The results were applied to the first hidden layer $H_{1,K}(K=1,2,\ldots,P)$. They were passed through a series of similar transformations and operations, then the results were applied to the second hidden layer $H_{2,K}(K=1,2,\ldots,Q)$, then they were passed through a series of similar transformations and operations, then the results were applied to the output layer $Y_K(K=1,2,\ldots,M)$. The outputs were production allocation percentage. The errors between outputs and true values were fed back according to the original inverse transformations and operations one by one layer. The network adjusted weight of each node according to the magnitude of error in process of feedback. Secondly, we repeated the first step with the new weight matrix; a new weight matrix was obtained. The process did not stop until the errors between outputs and true values were small enough. The weight matrix was saved in a template when the process was finished.

Sigmoid function is a nonlinear function. Unsymmetrical sigmoid function is $f(x)=1/(1+e^{-x})$ and the scope of function value is $0<f(x)<1$. Symmetrical sigmoid function is $f(x)=(1-e^{-x})/(1+e^{-x})$ and the scope of function value is $-1<f(x)<1$, Unsymmetrical sigmoid function is used in this paper because the production allocation percentage is in 0%–100%.

Characteristic fingerprints of commingled production oils were input and the proportion allocation would be achieved with the template. Some shortcomings exist in EBP. For example, study rate parameter should be small enough in order to diminish total errors, but the process to make template would be slow. The other way rounds, the process to make template would be accelerated if study rate increases, but the process would be instable and might be in local minimum point or stagnate in the error function

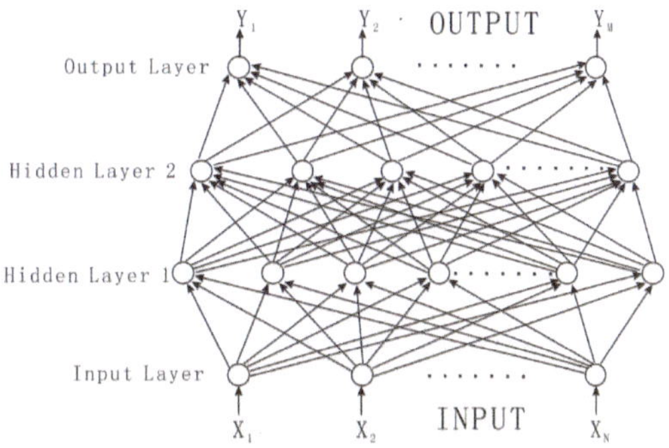

Figure 2. Schematic diagram of artificial nerve network with two hidden layers.

Table 1. Relative errors of model in different subsections.

Output percentage (%)	<1	1–5	5–10	10–25	25–50	50–75	75–100
Relative error (%)	negligible	15	10	8	6	5	3

Table 2. Contrast between MFE tests and the computation of double layer oils of Well 1-42-563 in Daqing Saertu Oil Field, China.

Name of oil layer	MFE test (%)	Results with PLS1 & ANN(%)	Absolute error (%)	Relative error (%)
SI	8.4	8.7	0.3	3.57
PII	91.6	91.3	−0.3	−0.33

Table 3. Contrast between laboratory proportion and computational allocation of three layer oils with different parameters in Daqing Lamadian Oil Field, China.

Name of Oil layer	Actual proportion (%)	Results with peak high ratio (%)	Absolute error (%)	Relative error (%)	Results with density (%)	Absolute error (%)	Relative error (%)
SI	40.02	40.39	0.37	0.92	40.81	0.79	1.97
SII	39.72	40.26	0.54	1.36	37.96	−1.76	−4.43
PI	20.26	19.35	−0.91	−4.49	21.23	0.97	4.79

steady section. Pure EBP algorithm, therefore, was inefficient to satisfy the actual requirement. Two main improvements are enumerated below.

2.2.2 *Adjustment of study rate*

Adjustment of connection weight is governed by study rate and gradient in EBP algorithm. In principle, the adjustment should be possibly biggest value in each step on condition of stabilization. A heuristic adjustment to study rate was applied. The rule is: (1) study rate would increase when total error E reduces; (2) study rate would reduce when total error E increases to a certain degree; (3) study rate would keep invariable when total error E increases a little. The rule is described in Expression (6).

$$a_k = \begin{cases} a \cdot a_{k-1} & E_k < E_{k-1} \\ b \cdot a_{k-1} & E_k > C \cdot E_{k-1} \\ a_{k-1} & others \end{cases} \tag{6}$$

Generally, $a = 1.05$, $b = 0.7$, and $C = 1.04$. Where, a_k and a_{k-1} are current and recent study rates, E_k and E_{k-1} are current and recent errors.

2.2.3 *Appending of the inertia item*

Traditional EBP algorithm based on standard gradient descent method is too slow to solve actual problems. Convergence rate of Newton algorithm is quick, but it requires rigorous initial values, and the inverse of a matrix has to be calculated in the iterative formula, so the computational complexities are much larger than standard gradient descent method.

We improved EBP algorithm based on Newton algorithm in this paper. A simple and effective method named inertia item method was chosen. The last results were appended to the results with a suitable ratio. It could avoid effectively the possibility that the results fell into the partial superior solution, thus the usability was enhanced.

$$W_k = W_{k-1} + \Delta W_k + \mu \cdot \Delta W_{k-1} \tag{7}$$

Where the last term is the inertia item, its coefficient is the inertia factor, generally we let the inertia factor $\mu = 0.45$.

2.2.4 *Practicability questions*

It is more appropriate to control relative errors. There are a series of errors in the process, which are instrumental system errors and artificial errors etc. Therefore, it is unnecessary to arrive at high precision to practical questions. Relative errors are controlled in different subsection (Table 1).

In addition, it is likely to stagnate during the process of making the template of multi group data. We evaluated roundly and selected preferentially the data. By doing so, the process is accelerated, the rounding error is reduced, and endless loop is avoided.

Table 4. Contrast between MFE tests and the computation of four layer oils of Well 1-50-J562 in Daqing Saertu Oil Field, China.

Name of oil layer	MFE test (%)	Results with PLS1 & ANN (%)	Absolute error (%)	Relative error (%)
SI	28.69	25.87	−2.82	−9.83
SII	32.66	30.08	−2.58	−7.90
PII2-5	22.53	25.65	3.12	13.85
PII8-10	16.12	18.40	2.28	14.14

Table 5. Contrast between laboratory proportion and computational allocation of five layer oils of Well Jin16-19-411 in Liao Oil Field, China.

Depth of oil layer (m)	Actual proportion (%)	Results with PLS1 & ANN (%)	Absolute error (%)	Relative error (%)
1322	16.89	17.33	0.44	2.61
1330	16.57	17.25	0.68	4.10
1355.5	41.81	38.52	−3.29	−7.87
1365	15.75	16.56	0.81	5.14
1461	8.98	10.34	1.36	15.14

3 PRACTICAL APPLICATIONS

We selected single layer and commingled production crude oil samples in some wells in the Daqing and Liaohe oilfields. Chromatogram fingerprints were obtained by gas chromatography (GC) using the reference of predecessor's researches (Feng Zihui et al., 1999, Fu Xiaotai et al. 2000, Wang Zhenping et al. 2000, Jin Xiaohui et al. 2001, and Zou Yuzheng et al. 2001). Artificial and instrument errors were reduced by repetitious experiments. Gas chromatography and mass spectrography analyses (GC-MS) are necessary in a new area for finding out the chemical characteristics of oils. We optimized 10–20 characteristic chromatogram fingerprints with PLS1, then computed 2–5 layers and proportioned problems with ANN algorithm.

It is observed that the errors increase with the advance of layers. On the part of algorithm, the errors are associated with computational rounding errors. These parts of errors are lesser. On the part of chemistry, the errors are associated with systematic and laboratorial errors, especially; they are associated mainly with the compound relations among single layer oils. The relative errors increase with decreasing the proportion of the oil. The computational results are practicable compared with the values of proportioning in lab and actual MFE test.

Difficulties would increase with the advance of the number of oil layers. The difficulties are not only in acquiring single layer oil samples but also in selecting the common steady parameters. The oil samples in other wellheads could replace the single layer oil, but the wells must be adjacent and the oil proportion in each layer be known.

4 CONCLUSIONS

The errors obtained from PLS and ANN algorithm are applicable to practical problems compared with the values of proportioning in lab and actual MFE test. The algorithm has four characteristics: (1) it could select parameters automatically; (2) PLS method could compute the results solely; (3) theoretically, the number of layers, which could compute, is infinite, as long as there are differences among crude oils and appropriate parameters could be selected; and (4) the results would never be negative.

This technique presents an economical and applicable way to production allocation of commingled production in oil reservoirs by geochemistry means. Moreover, the geochemistry means could monitor wells simultaneously in wide area with less production logs. As a result, the cost would be cut and the monitoring efficiency would be enhanced.

Some demerits still remain in this technique. (1) There must be differences among oils, which is to say no connectivity among oil layers. (2) The technique would not replace all production logs due to existing limitations, but it would decrease production logs by a large degree.

REFERENCES

Feng Zihui, Zhang Juhe & Guan Qiuhua. 1999. Production Allocation Calculations for Commingled Oils by the Oil Composition Analysis and Gas Chromatographic Method. *Petroleum Instruments.* April, 40–42 (in Chinese).

Fu Xiaotai, Wang Zhenping & Bi Lifei. 2000. Prediction of Output Fraction from Single Layer to Commingling Oil Well by High Performance Liquid Chromatography (HPLC). *Chinese Journal of Chromatography.* Vol.18, No.4, 318–321 (in Chinese).

Hagan, M.T. 2002 *Never Network Design.* Translated by Dai Kui et al., Xi'an Electron Scientific & Technical University Publishing House, Xi'an , China, pp. 197–255 (in Chinese).

Jin Xiaohui, Zhu Dan & Lin Renzi. 2001. Mathematical Modeling of Dynamic Chromatography Fingerprint Monitoring in the Oilfield Development. *Petroleum Exploration & Development,* Vol.28, No.3, 86–89 (in Chinese).

Jin Xiaohui, Zhu Dan & Lin Renzi. 2003. Partition Experiment of GC Fingerprint of Crude Oil. *Petroleum Exploration & Development.* Vol.25, No.1, 53–57 (in Chinese).

Kaufman R. L., Ahmed A. S. & Hempkins W. B. 1987. A New Technique for the Analysis of Commingled Oils and Its Application to Production Allocation Calculations, *16th Annual Indonesian Petroleum Assoc.*, IPA 87–23/21.

Kaufman R. L., Ahmed A. S. & Elsinger R. J. 1990. Gas Chromatography as a Development and Production Tool for Fingerprinting Oils from Individual Reservoirs: Applications in the Gulf of Mexico, in *GCSSEPM Foundation Ninth Annual Research Conference Proceedings*, 263–282.

Kaufman R. L., Dashti H., Kabir C. S., Pederson J. M., Moon M. S., Quttainah R. & Al-Wael H. 1997 Characterizing the Greater Burgan Field: Use of Geochemistry and Oil Fingerprinting: *SPE Paper* No. 37803, 385–394.

Wang Huiwen. 1999. *Partial Least-Squares Regression-Method andApplications.* National Defence Industry Publishing Company. Beijing, China. pp. 184–185; 200–234 (in Chinese).

Wang Zhenping, Fu Xiaotai, Lu Shuangfang, Pang Yanming, Xu Yunting & Dong Zhonglin. 2000. A Monitoring Technique for Production Allocation of Commingled Production Well with Liquid Chromatography, *Geochemistry.* Beijing, Vol. 29, No.3, 297–301 (in Chinese).

Zou Yuzheng, Cai Yuanming, Ma Ting & Mei Bowen. 2001. How to Eliminate the Defects of GCF Applied to the Production Partitioning Calculation of Crude Oil in Coproducing Wells. *Petroleum Exploration & Development.* Vol.23, No.2, 213–220 (in Chinese).

Petroleum Geochemistry and Exploration in the Afro-Asian Region –
Digang Liang, Darui Wang & Zhenxi Li (eds)

Oil-source rock correlation with multivariate statistical analysis

Zhenguang Li
Exploration and Development Research Institute, Daqing Oil Field Company Ltd. Daqing, China

ABSTRACT: 34 representative oil samples of the Heidimiao reservoir in the northern Songliao Basin are selected to perform a cluster analysis and study the relation of the oil property. The variables used in the analysis are the ratios of Pr/Ph, Ph/nC_{18}, Pr/nC_{17}, $\sum nC_{21}^{-}/\sum nC_{22}^{+}$, OEP, nC_{21}+nC_{22}/nC_{28}+ nC_{29} and the peak area of nC_{14} ~nC_{37} in saturated hydrocarbon chromatograms, there are altogether 30 parameters. The crude oil of the Heidimiao reservoir can be divided into two types and three subtypes. The cluster analysis of biomarkers shows that terpanes can be divided into two types and steranes can be also divided into two types. The factor analysis uses chromatographic parameters to make oil-source rock correlation, such as Pr/Ph, Pr/nC_{17}, Ph/nC_{18}, $\sum nC_{21}^{-}/\sum nC_{22}^{+}$, OEP and the ratio of nC_{21}+ nC_{22}/nC_{28}+ nC_{29}. The Type I oil in the Heidimiao reservoir is mainly derived from the oil source rock in Q1 Member and Type II oil comes mainly from that in Nen1 and Nen2 Members.

1 PREFACE

In recent years, with the deepening of exploration activities, some oil and gas fields such as Puxi, Putaohua, Aonan, Longnan and Xinzhao have been discovered in the Heidimiao reservoir of the northern Songliao Basin. These discoveries indicate that the Heidimiao reservoir has high hydrocarbon potential, thus, someone proposed again a necessity of studying its oil and source rock correlation. The author thinks that the previous studies have some referennce value (Liu et al., 1993), but they may have some limits because of the low discovery rate in the production wells in the reservoir at that time. Up to now, there have been nearly 80 wells with oils discovered in the Heidimiao reservoir. Nenjiang Formation (Nen 3,4 Members), in which the Heidimiao reservoir is located, distributes widely in an area of 120,000 km^2 in the northern Songliao Basin, with a vertical interval of 300–400 m. Because of the wide horizontal and vertical distribution range of the formation, one or two oil samples could not represent the whole reservoir. A study on the oil property and oil-source rock correlation is performed with multivariate statistical analysis.

The method of mathematical multivariate statistical analysis has developed rapidly and been applied in many disciplines in the past decade. Therefore, the multivariate statistical analysis has been introduced into other disciplines naturally. A multivariate approach can separate oils derived from different organic matter in nonmarine shales, marine shales, and marine carbonates (Peters et al., 1986). The multivariate statistical analysis has been used well in the differentiation of organic matter source input, depositional environments and the level of thermal maturity as well as in the prediction of source-rock characters from oil compositions (Irwin et al., 1990; Zumberge, 1987).

Cluster analysis is used to divide types of oils and source rocks (Cheng et al., 2002; Xiong et al., 2004). The use of multivariate statistics improves evaluation of large, complex data sets (Peters et al., 2002).

The study on Mahakam Delta in Indonesia by Peters et al. is an example of how to combine geochemistry with multivariate statistics and sequence stratigraphy to produce a new and successful model for exploration in a mature basin. This new geochemical-stratigraphic model for Mahakam Delta is based largely on oil-oil and oil-source rock correlation, the model upgrades the potential of outer shelf, where it influenced drilling that resulted in significant deepwater oil discoveries (Peters et al., 2002).

The content of multivariate statistical analysis has been included in the book, Biomarker Guide, Second Edition (Peters et al., 2005).

As a recommended method for oil and rock source correlation, Peters et al. valued the multivariate statistical analysis highly in their review paper (Peters et al., 2002). However, there are only a few of papers that have reported the use of the multivariate statistical analysis in oil-source correlation because of the available limitation of the analysis software and lack of samples.

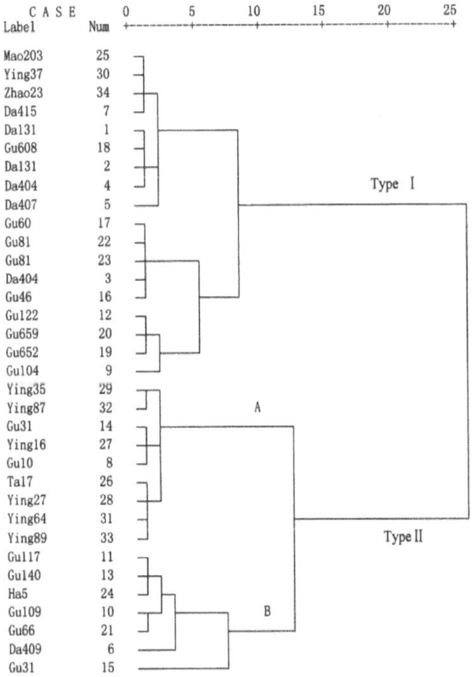

Figure 1. Hierarchical cluster analysis dendrogram based on multivariate statistical analysis of thirty chromatographic parameters for thirty-four oils from the Heidimiao reservoir in the northern Songliao Basin, China.

2 OIL TYPE CLASSIFICATION OF THE HEIDIMIAO RESERVOIR

2.1 *Oil type classification with chromatographic parameters*

34 representative oil samples of the Heidimiao reservoir were selected to perform a cluster analysis and study the relation of oil properties. Cluster analysis can avoid artificial determination and is beneficial for mathematical n-dimensional space classification when someone does not know how to classify a group of samples. SPSS11 software was adopted to perform cluster analysis. The variable parameters used in the analysis are the ratios of Pr/Ph, Ph/nC_{18}, Pr/nC_{17}, $\sum nC_{21}^- / \sum nC_{22}^+$, OEP, $nC_{21} + nC_{22}/nC_{28} + nC_{29}$ and the peak area of nC_{14}~nC_{37} in saturated hydrocarbon chromatograms, there are altogether 30 parameters.

The crude oil of the Heidimiao reservoir can be divided into two types shown in the dendrogram (Fig. 1). The middle & upper parts of Well Mao 203 (No.25) and Well Gu 104 (No.9) have the Type I oil; the bottom of Well Ying 35 (No.29) and Well Gu 31 (No.15) is regarded to have the Type II oil. The latter type is further classified into Type A (in Well Ying 35 and Well Ying 89) and Type B (in Well Gu117 and

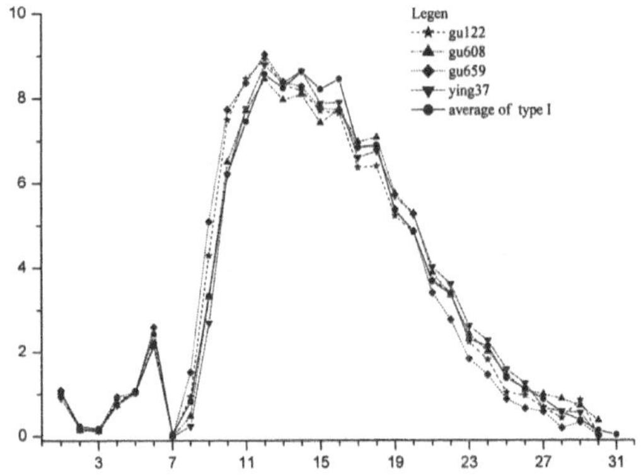

Figure 2. Chromatographic fingerprint characteristics of the Type I crude oil.

This map selects the samples with different space location (No.12, 18, 20 and 30) from the cluster dendrogram (Fig. 1) and shows the chromatographic fingerprint characteristics taken from the average value of 18 samples in the Type I crude oil to observe different space location samples and the distribution of the average chromatographic fingerprint value. 1 Pr/Ph, 2 Pr/nC_{17}, 3 Ph/nC_{18}, 4 $\sum nC_{21}^-/\sum nC_{22}^+$, 5 OEP, 6 $nC_{21}+ nC_{22}/nC_{28}+ nC_{29}$, 7-31 $nC_{14}- nC_{38}$ peak area, respectively.

Well Gu31) oils. The mathematical method was used to classify types of the studied oils, in which n parameters of m samples were taken as m points in n-dimensional space. The relations between samples were determined by distance, samples close to each other were classified to the same group, and the classification was not clear in geochemistry meaning.

To study the property of different types of crude oils, the spatial distance of the cluster analysis should be presented using the original parameters. In order to directly observe the characteristics of different types of crude oils, the analyzed Type I, Type II A and Type II B crude oils are involved to demonstrate chromatographic fingerprints, respectively, according to their parameter ratios and peak area values (Fig. 2, 3 and 4). From these figures one can see that the fingerprint characteristics of the every type of crude oils are close, which can be presented with the average value curve. Figure 5 shows the comparison among the different crude oils. Table 1 demonstrates the first 6 ratios of the 31 parameters for 34 oil samples and the data of their main peak carbon (MPC).

From Table 1, one can see that the Pr/nC_{17} and Ph/nC_{18} ratios of the Type I crude oil are lower than those of Type II, i.e. the ratios are 0.23 and 0.19 for the Type I oil while those are 0.48 and 0.34 for the Type II A crude oil and 0.58 and 0.51 for the Type II B crude oil, respectively. The ratio of $\sum nC_{21}^-/\sum nC_{22}^+$ of Type I crude oil is higher than that of Type II and at the same time the ratio of $nC_{21} + nC_{22}/nC_{28} + nC_{29}$ is also higher than that of the Type II crude oil, thus,

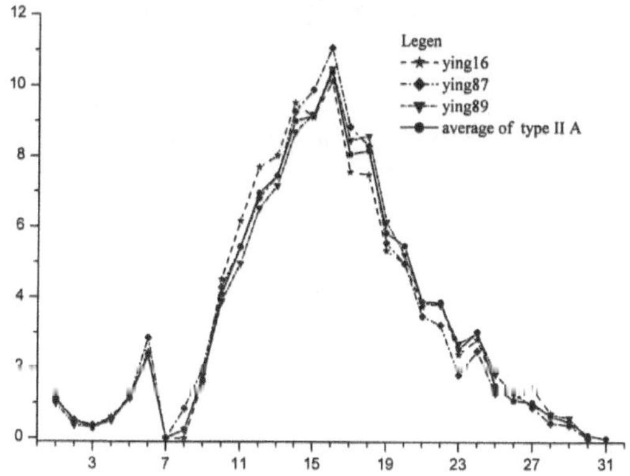

Figure 3. Chromatographic fingerprint characteristics of the Type IIA crude oil.

This map selects the samples with different space location (No. 27, 32 and 33) from the cluster dendrogram (Fig. 1) and shows the chromatographic fingerprint characteristics taken from the average value of 9 samples in the Type IIA crude oil to observe different space location samples and the distribution of the average chromatographic fingerprint value. The abscissa is the same as in Fig. 2.

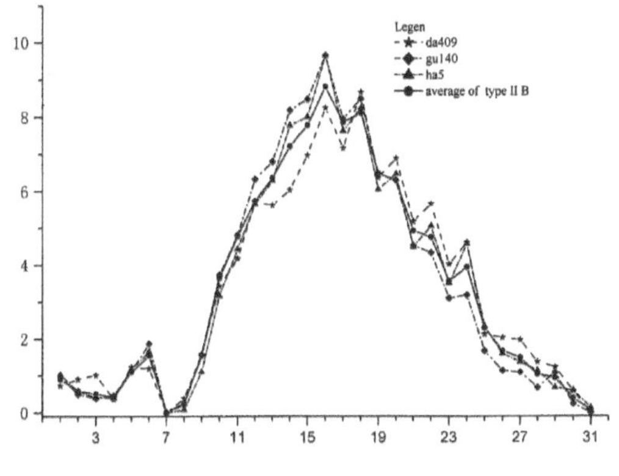

Figure 4. Chromatographic fingerprint characteristics of the Type 2B crude oil.

This map selects the samples with different space location (No. 6, 13 and 24) from the cluster dendrogram (Fig. 1) and shows the chromatographic fingerprint characteristics taken from the average value of 7 samples in the Type 2B crude oil to observe different space location samples and the distribution of the average chromatographic fingerprint value. The abscissa is the same as in Fig. 2.

it shows that the fraction less than nC_{22} in the Type I crude oil is superior.

The contrast in Fig. 5 obviously shows that the peak strength between $nC_{14}\sim nC_{20}$ of the Type I crude oil is the highest, that of the Type II A is the second, and that of the Type II B is the least. The same value within $nC_{26}\sim nC_{37}$ shows that the highest strength is the Type II B, the second one is the Type II A, the least one

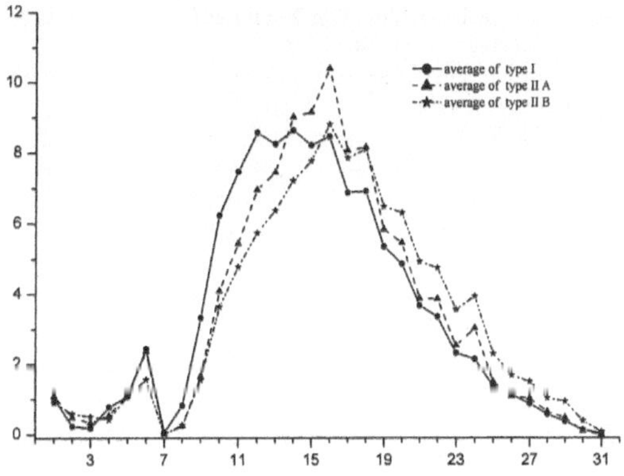

Figure 5. Comparison of chromatographic fingerprint characteristics among the Types I, 2A and 2B crude oils. The abscissa is the same as in Fig. 2.

is the Type I. The peak shape clearly responds to the chromatography fingerprint characteristics of different type of crude oils. In order to further study the difference in types of crude oils, statistical analysis of group-type fraction and the density of the crude oils should be conducted.

The result shows that there are some differences. The average value of the group-type fractions of the Type? crude oil is 74.9% for saturates, 12.8% for aromatics and 13.1% for nonhydrocarbons and asphaltene. The density of the Type I oil is 0.85 g/cm^3. That of the Type II A crude oil is 62.91% for saturates, 16.24% for aromatics and 20.85% for nonhydrocarbons and asphaltene, its density is 0.86 g/cm^3. That of the Type II B crude oil is 62.91% for saturates, 15.49% for aromatics and 21.59% for nonhydrocarbons and asphaltene. The density is 0.87 g/cm^3. Thus, it can be seen that the Type I crude oil has a higher content of aromatics but a lower content of nonhydrocarbons and asphaltene than the Type II A and Type II B oils. The Type II crude oil has similar group-type fractions. The density of the II B crude oil is a little more than that of the II A crude oil.

2.2 *Crude oil classification by parameters of biomarkers*

Usually, ratioing technique is used in the crude oil classification, i.e. adopting ratios of any two biomarkers in mass chromatograms, such as Ts/Tm, C_{27}/C_{29} sterane, 20S/(20S + 20R) of C_{29} sterane, etc.. However, there is a problem in adopting the rationing technique, when very high or low relative contents of two chemical compounds are adopted, its ratio may be the same, but its geochemical meaning may be absolutely different.

Adopting the cluster analysis method may sufficiently utilize the information of the biomarkers. At first, distributions of terpanes and steranes are normalized, respectively, then the relative contents of

Table 1. Chromatographic parameters of different types of crude oils in the Heidimiao reservoir in the northern Songliao Basin.

Parameters Type	Pr/Ph	Pr/nC_{17}	Ph/nC_{18}	$\sum nC_{21}^{-} / \sum nC_{22}^{+}$	OEP	$nC_{21+22}^{*} / nC_{28+29}$	MPC*
Type I	1.03	0.23	0.19	0.77	1.08	2.44	19–23
Type II A	1.09	0.48	0.34	0.54	1.16	2.36	23
Type II B	0.91	0.58	0.51	0.42	1.10	1.56	23–25

*stands for $nC_{21} + nC_{22}/nC_{28} + nC_{29}$.
MPC*stands for main peak carbon.

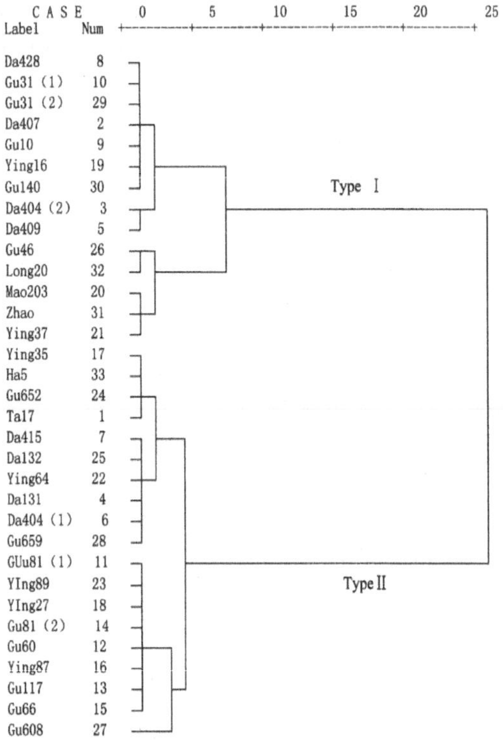

Figure 6. Hierarchical cluster analysis dendrogram based on multivariate statistical analysis of 26 terpane biomarkers for 34 oils from the Heidimiao reservoir in the northern Songliao basin.
This sketch shows that the studied oils can be divided into two types according to their terpane distribution.

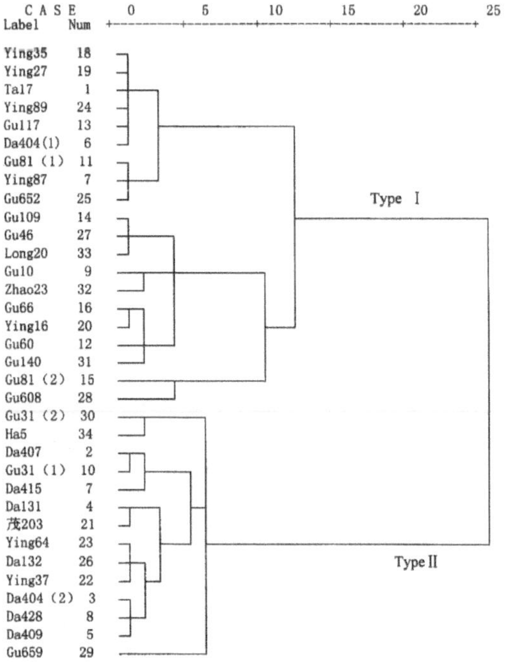

Figure 7. Hierarchical cluster analysis dendrogram based on multivariate statistical analysis of 14 sterane biomarkers for 34 oils from the Heidimiao reservoir in the northern Songliao basin.
This sketch shows that the studied oils can be divided into two types according to their sterane distribution.

each determined biomarker, which are regarded as cluster parameters, are employed in cluster analysis. The adopted parameters of biomarkers are a series of tricyclic terpanes from C_{19} to C_{29} and pentacyclic terpane series including hopanes, norhopanes, homohopanes, gammacerane etc. 26 biomarker parameters were adopted in the cluster analysis of the present study and 14 parameters were used in the cluster analysis on C_{27} to C_{29} steranes. The analysis results are shown in Fig. 6 and Fig. 7.

Fig. 6 shows that the crude oil can be classified into two types by terpane biomarkers, one is Type I from Well Da 428 (No.8) to Well Ying 37 (No.21), and the other is Type II from Well Ying 35 (No.17) to Well Gu 608 (No.27).

Fig. 7 shows that the crude oil can be classified into two types according to the characteristics of steranes, one is Well Ying 35 (No.18) to Well Gu 608 (No.28), and the other is Well Gu 31(No.30) to Well Gu 659 (No.29).

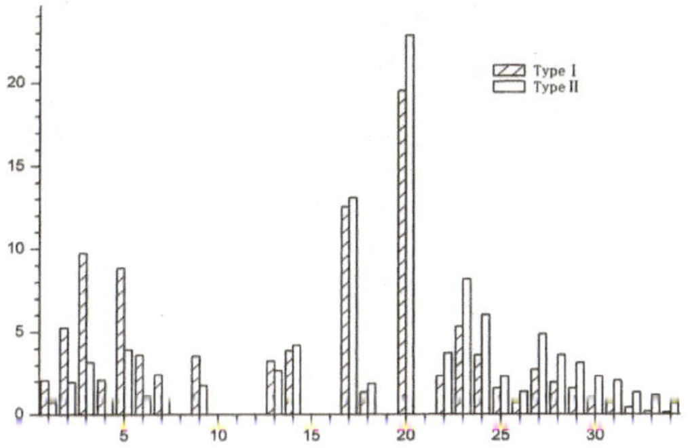

Figure 8. Comparison of terpane–fingerprint characteristics of the Heidimiao reservoir in the northern Songliao basin 1-7 refer to C_{19}-C_{25} tricyclic terpanes, 8 to C_{24} tetracyclic terpane, 9-12 to C_{26}-C_{29} tricyclic terpanes, respectively. 13 stands for Ts, 14 for Tm, 15 for C_{30} tricyclic terpane, 16 for 17β(H)-22,29,30- trisnorhopane, 17 for C_{29}-hopane, 18 for C_{30}-normoretane, 19 for 18α(H)-oleanane, 20 for $C_{30}\alpha\beta$-hopane, 21 for $C_{29}\beta\beta$-hopane, 22 for C_{30}-moretane, 23 for $C_{31}\alpha\beta$22S-hopane, 24 for $C_{31}\alpha\beta$22R-hopane, 25 for gammacerane, 26 for $C_{30}\beta\beta$-hopane, 27 for C_{32} $\alpha\beta$22S-hopane, 28 for $C_{32}\alpha\beta$22R-hopane, 29 for C_{33} $\alpha\beta$22S- hopane, 30 for $C_{33}\alpha\beta$22R-hopane, 31 for C_{34} $\alpha\beta$22S-hopane, 32 for $C_{34}\alpha\beta$22R-hopane, 33 for $C_{35}\alpha\beta$22S-hopane and 34 for $C_{35}\alpha\beta$22R-hopane, respectively.

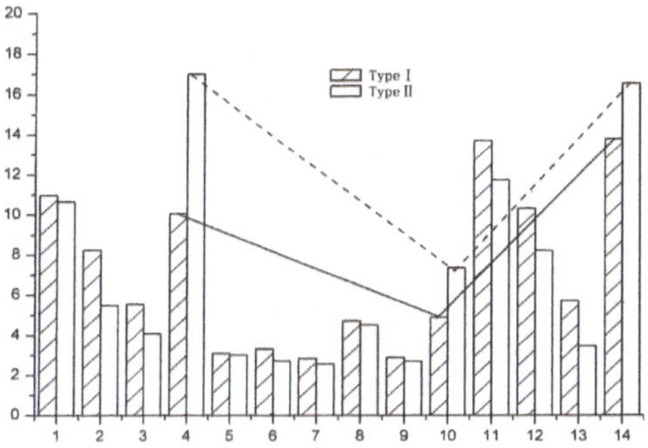

Figure 9. Comparison of sterane–fingerprint characteristics of the Hedimiao reservoir in the northern of Songliao basin. 1 refers to $C_{27}\alpha\alpha\alpha$20S sterane, 2 to $C_{27}\alpha\beta\beta$20R sterane, 3 to $C_{27}\alpha\beta\beta$20S sterane, 4 to $C_{27}\alpha\alpha\alpha$20R sterane, 5 to $C_{29}\beta\alpha$20R diasterane, 6 to $C_{29}\alpha\beta$20S diasterane, 7 to $C_{28}\alpha\alpha\alpha$20S sterane, 8 to $C_{28}\alpha\beta\beta$20R sterane, 9 to $C_{28}\alpha\beta\beta$20S sterane, 10 to $C_{28}\alpha\alpha\alpha$20R sterane, 11 to $C_{29}\alpha\alpha\alpha$20S sterane, 12 to $C_{29}\alpha\beta\beta$20R sterane, 13 to $C_{29}\alpha\beta\beta$20S sterane and 14 to $C_{29}\alpha\alpha\alpha$20R sterane, respectively.

In order to further study the difference of biomarkers among crude oils, each parameter of any type of crude oils was averaged and then compared with the help of the bar graphs (Fig. 8 and Fig. 9).

According to terpane property in Fig 8, it is seen that C_{19}~C_{25} tricyclic terpanes in the Type I crude oil are over two times higher than those in the Type II crude oil. Among which the difference between the ratio of

Table 2. Relative abundance and ratios of terpanes in different type of crude oils.
Peaks No. in Table 2 is the same as in Fig. 8.

Peak No.	I	II	I /II
1	2.04	0.73	2.82
2	5.25	1.93	2.72
3	9.73	3.16	3.07
4	2.07	0.96	2.15
5	8.83	3.95	2.24
6	3.60	1.11	3.24
7	2.40	0.97	2.47
8	3.53	1.75	2.02
13	3.24	2.65	1.22
14	3.87	4.20	0.92
17	12.52	13.05	0.96
18	1.33	1.84	0.72
20	19.52	22.83	0.86
22	2.32	3.72	0.62
23	5.32	8.17	0.65
24	3.60	6.03	0.60
25	1.58	2.28	0.69
26	0.72	1.38	0.52
27	2.73	4.86	0.56
28	1.91	3.59	0.53
29	1.56	3.13	0.50
30	0.99	2.28	0.43
31	0.71	2.03	0.35
32	0.41	1.33	0.31
33	0.16	1.14	0.14
34	0.09	0.66	0.14

peak No.3 (C_{21} tricyclic terpane) and No.6 (C_{24} tricyclic terpane) is the highest (Table 2). Peaks No.13 and No.14 (Ts, Tm) are similar. Peaks No.15~34 mainly belong to pentacyclic terpane series. The Type II crude oil has abundant pentacyclic terpane series. Biomarker ratios of Type II and I crude oils are shown in Table 2. C_{21} tricyclic terpane has the highest peak in tricyclic terpane series of the Type I crude oil while C_{23} tricyclic terpane has the highest peak in those of the Type II crude oil. Tricyclic terpane has a biogenetic derivation of microbe (bacteria or algae) and pentacyclic terpane has a biogenetic derivation of prokaryote (bacteria). Relative abundance and distribution of various kinds of terpanes show that the crude oil in Heidimiao is generated from two sets of source rocks.

Sterane is one of the often used biomarkers in oil source correlation, and also a common parameter to reflect organic matter input. From Fig. 9 it can be seen that two types of crude oils have C_{27}20R, C_{28}20R and C_{29} 20R steranes distributed in different V shape (Nos.4,10, 14 peaks), C_{29}20R sterane in the Type I crude oil is in superiority while C_{27}20R sterane in the Type II crude oil is in superiority, reflecting they come from different source rocks. The relative abundance of $C_{29}\alpha\alpha\alpha$20S and $C_{29}\alpha\alpha\alpha$20R steranes (No.11 and 14 peaks) indicates that the maturity of the Type I crude oil is higher than that of the Type II crude oil.

3 OIL – SOURCE ROCK CORRELATION

Oil and source rock correlation is one of the important subjects investigated in the oil and gas prospecting. Correlation is a geochemical comparison among oils or between oils and extracts from prospective source rocks in order to determine whether there exists a genetic relationship (Peters et al., 1993). Oil-source rock correlation is based on the concept that certain compositional parameters of migrated oils do not differ significantly from those of bitumen remaining in the source rocks. It is conducted based upon some similar characteristics among oils and source rocks of various types, such as stable carbon isotopes, characteristics of biomarkers and chromatography fingerprints etc. Sometimes it is hard in practice because the similarity is difficult to define. For example, there are some similarities among source rocks from a certain formation, yet it is impossible to seek out the most similar ones if correlations are not conducted among the entire source rocks from a certain horizon. In other words, it could not be concluded that the horizon does not generate the oil even though there is not a correlation between the rock sample from a certain horizon and the oil. Because properties of source rocks from the same horizon are not homogeneous, one or two samples cannot present the properties of the whole horizon. In order to avoid such limitation, it is better to introduce distribution scatter and multivariate statistical analysis to the oil and source correlation. Factor analysis, cluster analysis and discriminant analysis supply many effective ways for oil and source rock correlation. In order to avoid such limitation, the oil and source rock correlations in the study were conducted among total 109 mudstone and 34 oil samples. Among witch 11 samples were taken from 9 wells in Member IV of Nenjiang Formation symbolized by Nen 4 with depth ranging from 1043 m to 1400 m, 14 samples from 10 wells in Nen 3 with depth ranging from 1019 m to 1633 m, 14 samples from 11 wells in Nen 2 with depth ranging from 1174 m to 1681 m, 19 samples from 11 wells in Nen 1 with depth ranging from 1234 m to 1885 m, 17 samples from 8 wells in Members 2 and 3 of Qingshankou Formation symbolized by Qn 2 + 3 with depth ranging from 1657 m to 2359 m and 34 samples from 22 wells in Qn1 with depth ranging from 1595 to 2466 m. The oil and source rock correlation was conducted with factor analysis, which is multivariate statistical technique that creates new independent variables (i.e. factors). They are linear combinations of the original variables. The primary objective of the statistical analysis is to reduce the

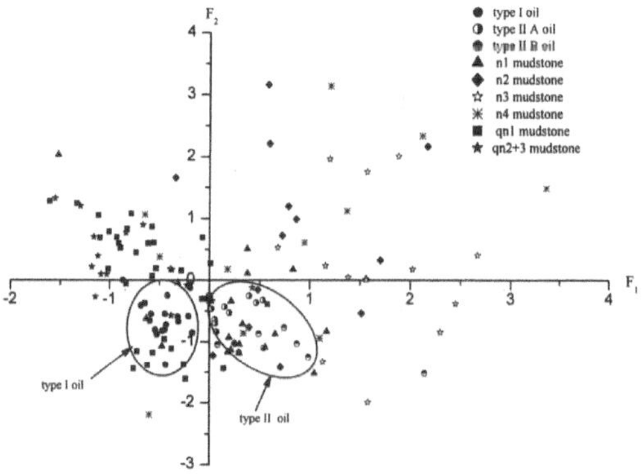

Figure 10. Factor analysis of oil-source rock correlation for the Hedimiao reservoir in the northern Songliao basin.

dimensionality of the data to a few important components or factors that can best explain the variation in the data. The combined factors reserve, to a great extent, the information of many original data.

Principal component analysis was also adopted for factorial analysis. The factors with the eigenvalues over 1 were chosen with varimax rotation. With which the factor analysis was employed to chromatographic parameters, such as Pr/Ph, Pr/nC_{17}, Ph/nC_{18}, $\sum nC_{21}^{-}/\sum nC_{22}^{+}$, OEP and the ratio of $nC_{21} + nC_{22}/nC_{28} + nC_{29}$. Fig. 10 was made based on factor analysis. Factors 1 and 2 account for 65 percent of the variation in the data set. Factor 1 is composed primarily of Pr/nC_{17}, Ph/nC_{18} and OEP while factor 2 consists of Pr/Ph and $\sum nC_{21}^{-}/\sum nC_{22}^{+}$. It can be seen from the factor figure (Fig. 10) that the crude oil in Heidimiao can be divided into two types, which are just corresponding to the two types of oils defined by chromatographic parameter cluster analysis. The Type I oil can be related to the source rock in Qn1 Member and the Type 2 to the source rock in Nen 1 Member and Nen 2 Member partially. Therefore, it can be suggested that the Type I oil in the Heidimiao reservoir is mainly derived from the source rock in Qn1 Member and the Type II oil comes mainly from the source rock in Nen 1 and Nen 2 Members.

4 CONCLUSIONS

With multivariate statistical method multiple parameters can be applied to conduct the classification of crude oils. This kind of classification is especially available for more samples because artificial effects can be avoided and the manpower of classification can be reduced. The correlation of multiple samples and parameters of oil source rocks can be used for the research on the relationship between source rocks and crude oils in each layer and then the problem arising from using less samples, which are sometimes unrepresentative, can be avoided.

The cluster analysis of 34 representative oil samples from the Heidimiao reservoir in the northern Songliao basin shows that there are two different types of crude oils in this reservoir which can be divided by their different characteristics and the factor analysis shows that the Type I crude oil mainly comes from the Qn1 Member source rock while the Type II crude oil is mainly derived from the N1 Member source rock and has a certain contribution by the N2 source rock.

REFERENCES

Cheng Keming, Xiong Ying, Zeng Xiaoming et al., 2002. Research on coal-derived hydrocarbon in Turpan-Hami Basin. Acta Petrolei Sinica 23, 13–17 (in Chinese).

Irwin, H. and Meyer, T. 1990. Lacustrine organic facies. A biomarker study using multivariate statistical analysis. Organic Geochemistry 16, 176–120.

Liu Xiaoyan, Wang Xiwen and Chen Xin 1993. Geochemical characteristics of Heidimiao crude oil. Journal of Daqing Petroleum Institute 17, 24–29 (in Chinese).

Peters, K. E., Moldowan, J. M., Schoell, M. and Hempkins, W. B. 1986. Petroleum isotopic and biomarker composition related to source rock organic matter and depositional environment. Organic Geochemistry 10, 17–27.

Peters, K. E. and Moldowan, J. M. 1993. The Biomarker Guide: Interpreting Molecular Fossils in Petroleum and Ancient Sediments. Prentice Hall, Englewood Cliffs, Nj.

Peters, K. E. and Fowler, Martin G. 2002. Application of petroleum geochemistry to exploration and reservoir management. Organic Geochemistry 33, 5–36.

Peters, K. E., Walters, C. C. and Moldowan, J. M. 2005. The Biomarker Guide. Volume 2: Biomarkers and Isotopes in Petroleum Exploration and Earth History, pp 479–482, Second Edition Cambridge University Press

Xiong Ying, Cheng Keming, Yang Zhiming, et al., 2004. Advances in the correlation of oils and source in the Jiuxi depression, Northwest China, Petroleum Exploration & Development 31, 36–39 (in Chinese).

Zumberge, Hohn, E. 1987. Prediction of source rock characteristics based on terpane biomarkers in crude oils: a multivariate statistical approach. Geochimica et Cosmochimica Acta 51, 1625–1637.

Petroleum Geochemistry and Exploration in the Afro-Asian Region – Digang Liang, Darui Wang & Zhenxi Li (eds)
© 2008 Taylor & Francis Group, London, ISBN 978-0-415-44083-7

Interaction between basalts and oil source rocks in rift basins: CO_2 generation

Qiang Jin, Li Xu & Yi Zeng
China University of Petroleum, Dongying, Shandong, China

Congli Wan
Geological Institute of Petroleum Transportation Company of the Shengli Oilfield, Ltd. SINOPEC, China

ABSTRACT: Basalts interbedded with oil source rocks are discovered frequently in rift basins of eastern China, where CO_2 is found in reservoirs around or within basalts such as in the Bingnan reservoir of Dongying Depression. In the reservoirs, CO_2 with heavy carbon isotope compositions ($\delta^{13}C > -10‰$ PDB) is often over 40% of natural gases, and is believed to come from the degassing of basaltic magma from the mantle. Having investigated the Binnan reservoir, the authors suggest that the CO_2 is resulted from interactions between the source rocks and basalts. As the source rocks around basalts are rich in carbonate minerals, volcanic minerals, transitional metals and organic matter, during their burial history some of the transitional metals catalyzed the thermal degradation of organic matter into hydrocarbons and the decomposition of carbonate minerals into CO_2, which is reproduced in thermal simulations on the source rock with transitional metals (Ni and Co). This kind of CO_2 occupies 55–85% of the total gases generated in the simulation, and its $\delta^{13}C$ ranges in $-11 \sim -7.2‰$ PDB, which is very similar to those of CO_2 found in the Binnan reservoir. The co-generation of CO_2 and hydrocarbon gases makes it possible to be accumulated together in one trap. In other words, if the CO_2 is degassed directly from basaltic magma or from mantle, it could not be accumulated with hydrocarbon gases because it came into the basin much earlier than hydrocarbon generation and much earlier than trap formation. Therefore, the source rocks around basalts generate hydrocarbons and CO_2 simultaneously by catalysis of Co and Ni transitional metals, which is useful for explanation on co-accumulations of hydrocarbon gases and CO_2 in rift basins in eastern China.

Keywords: Basalt, Source rock, Interaction, CO_2, Hydrocarbon gas.

1 INTRODUCTION

Volcanic rocks, especially basalts, are often distributed within oil source rocks in rift basins including the Songliao basin, Bahai Bay basin, North Jiangsu basin and so on (Jin, 1998a; Zhang, 1998; Du, 1998; Xiong and Lu, 1996; Zhai et al., 2004). Around the basalts, hydrocarbon gases and CO_2 are accumulated together. In the accumulation, the CO_2 is often over 40% and with heavy stable isotope composition ($\delta^{13}C$ generally greater than $-10‰$), and its origin is believed either from the degassing of basaltic magma or from mantle (e.g. Dai et al., 1994; Jeffery and Kaplan, 1988; Berner et al., 1983; Dai et al., 2002). In fact, the $\delta^{13}C$ of CO_2 discovered in the Binnan reservoir ranges in $-10‰ \sim -7‰$ (Dai et al., 2002), and hydrocarbons associated with CO_2 are characterized as organic origin because the $\delta^{13}C$ of methane is lighter than $-46‰$ or even lighter than $-55‰$ PDB (Jin et al., 1999a; Patrick et al., 2004).

Some of the authors of this paper conducted a series of experiments and recognized that olivine and zeolite in the basalt are interacted with organic matter in the source rocks to generate more hydrocarbons during thermal degradation of kerogen (Jin et al., 1999b). The interaction is applied to explanation of the origin of immature or low-maturity oils and gases found in rift basins (Wan et al., 2001; Wan and Jin, 2001; Jin, 2001). In recent thermal simulations, transitional elements were added into the source rock samples that contain high carbonate minerals, and much CO_2 was produced in addition to hydrocarbons generated in the simulation. $\delta^{13}C$ of the CO_2 ranges in $-11 \sim -7‰$, thus, the authors believe that the real origin of CO_2 is the interaction between transitional metals and source rocks. In fact, CO_2 is generated also from the thermal degradation of kerogen, but its $\delta^{13}C$ is less than $-14‰$ (PDB) (Dai et al., 2002). This paper introduces the thermal simulation to illustrate how much CO_2 that can be generated in the interaction.

2 GEOLOGICAL SETTING AND RESEARCH BACKGROUND

The basalt and source rock were discovered in the Binnan area in the northwest Dongying depression (Fig. 1 Wang and Qian, 1992), where a listric fault cut 15 km deep and served as a passage for the basaltic magma (He et al., 2001). During the source rock deposition (Late Oligocene, Member 3 of Shahejie Formation) basaltic magma erupted 5 or 6 times, but its distribution was limited just in the Binnan area (no more than 80 km^2). Each eruption resulted in a circle of rocks from the bottom to the top: black-coloured basalt, light gray-coloured tuff, dark gray-coloured muddy lacustrine carbonate, and dark gray-coloured calcareous shale (Fig. 2), in which the tuff and carbonate are in thin layers (<1.5 m), and the carbonate and shale are defined as source rocks (Table 1). Inside and outside of the rock circle, oil and gas reserve has been found, including natural gases, immature and mature oils, and high CO_2 content gas (CO_2 up to 40%~90%, its $\delta^{13}C$ in −7.5‰~−10.4‰).

Presently volcano eruption provides volumes of materials into surrounding environments, including CO_2, H_2 and volcanic minerals (Avila et al., 2005; Simoneit, 1988; Covenney et al., 1987). This kind of CO_2 is hardly to be preserved in any gas reservoirs because there are no seals or traps formed before the volcano eruption. However, it is possible that the CO_2 was dissolved into the lacustrine water when the volcano eruption took place in a lake for the deposition of Member 3 of Shahejie Formation in the Binnan area, which resulted in the formation of carbonate rocks or carbonate minerals (such as calcite and dolomite) in the lacustrine shale sequence. Meanwhile, some volcanic materials served as nutrients to hydrophytes and aquatic livings, so that the livings were in bloom of the environment (Simoneit, 1992; Jin, 1998b). Therefore,

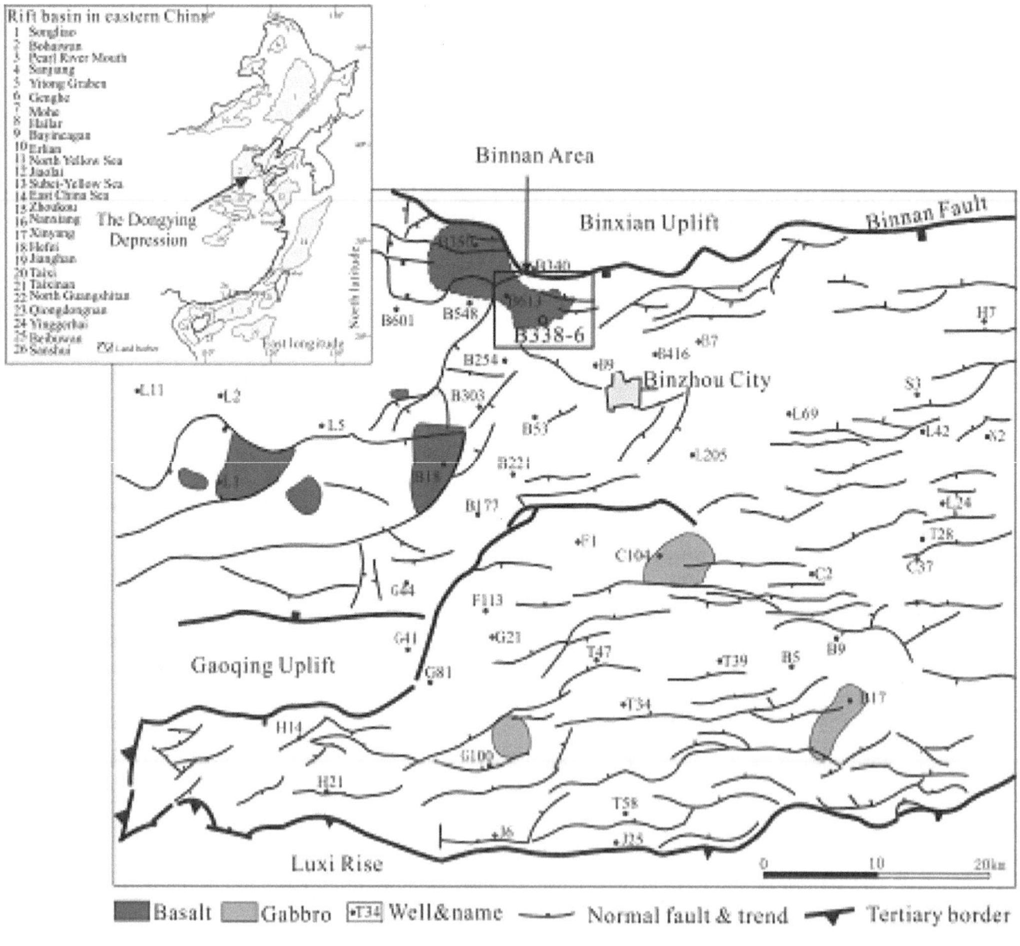

Figure 1. Map showing the Binnan area in Dongying Depression and the location in eastern China.

after an eruption of basalt, shale was either rich in organic matter or rich in carbonate minerals, in which the calcite content was often over 20%.

We conducted $\delta^{13}C$ measurements on carbonate minerals both in basalt and shale in the Binnan area. Air bubbles are a common structure of the basalt in the Binnan area, they are filled mostly by white-coloured calcite, and $\delta^{13}C$ of the calcite ranges between +5‰ and +7‰. The $\delta^{13}C$ of the carbonate minerals in the shale is in the same range as those of air-bubble calcite in the basalt. In contrast, $\delta^{13}C$ of ordinary lacustrine carbonate ranges in −2‰ ~ +3‰ (Yin et al., 1994; Cartwright, et al., 2004). Therefore, carbonate minerals in the shale are genetically related with basaltic eruption. During a long-term burial history, some calcite in the basalt air-bubbles was dissolved, what is the relation between the calcite and CO_2? This is a focus of this paper.

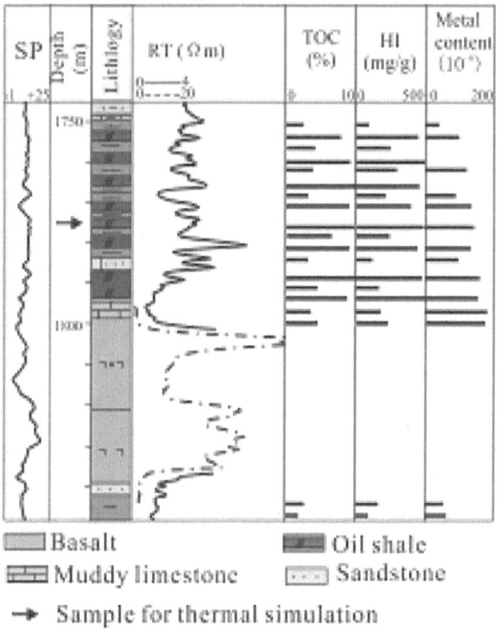

Figure 2. A column of the basalt-source rock assemblage of Well B338-6 in the Binnan area in Dongying Depression.

No evidence shows that basalts bake the shale in the Binnan area so that the interaction between volcanic materials and organic matter does not occur during deposition of the shale. But the shale contains many volcanic minerals, such as zeolite, olivine and pyrite (Coveney et al., 1987; Jin, 2001; Xiao and Hu, 2005), and the shale enriches many kinds of transition metal elements, including Co, Ni, Cr, V, Ti, etc. (Fig. 3). Volcanic minerals were distributed in the shale during volcano eruption, but the transition metals could be derived from the thermal fluid in and after the volcano eruption. The authors found that the content of transition metals in the shale near the basalt is as twice as those 15 m away from the basalt (Fig. 3), for example, Ni in the shale is 146 ppm about 0.3 m apart from the basalt, and reduces to 76 ppm at about 15 m to the basalt. In the burial history, the volcanic materials in the shale could probably interact with the organic matter to generate more hydrocarbons (Wan and Jin, 2001), and transition metals (especially Co and Ni) are of strong catalysis on hydrocarbon generation from organic matter (Mango and Hightower, 1994).

Recently, we made a series of thermal simulations of hydrocarbon generation from the shale sample with transition metals, and found that not only hydrocarbon generation increased greatly, but also the CO_2 increased obviously. As this kind of CO_2 and hydrocarbons are generated at the same time, it is possible for them to be accumulated together. Meanwhile, the carbon isotope composition of the CO_2 is similar to that of CO_2 discovered from the Binnan reservoir. Therefore, to better understand this origin of CO_2 is helpful to the explanation of CO_2 distribution in rift basins in eastern China.

3 MATERIALS & METHODS

One shale sample was selected from B338-6 Well in the Binnan area for the thermal simulation, and its basic geochemical properties listed in Table 2. The sample is rich in both organic matter and carbonate minerals, and is a potential source rock.

As Ni and Co are major transitional metals derived from basalts, we chose them as catalysts for the thermal degradation of organic matter and carbonate. The

Table 1. Basic data of the source rock in the basalt-source rock assemblage in Member 3 of Shahejie Formation in the Binnan area.

Lithology	TOC wt.%	HI mg/g	Tmax °C	Kerogen type	Carbonate content wt.%
oil shale	8.38(34)*	627(28)	331(28)	I(19)	18.6(25)
lime mudstone	1.75(26)	469(21)	334(21)	II_1(11)	26.4(18)
mud limestone	1.29(17)	417(15)	332(15)	II_1(10)	52.7(9)

*Average value or the sample number for those in the bracket.

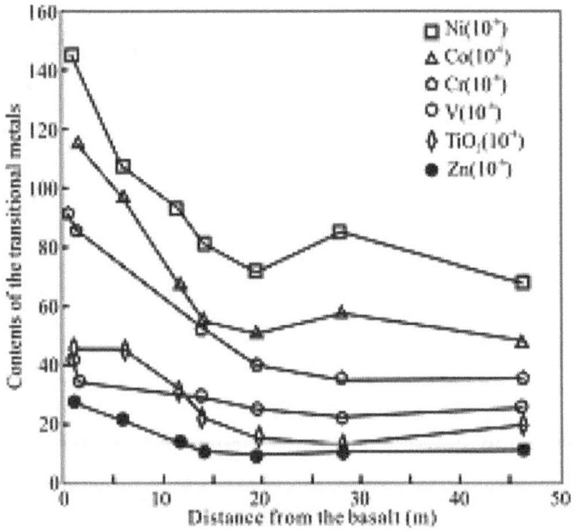

Figure 3. Plot showing migration of transitional metal elements from basalt to source rocks in the Binnan area.

Table 2. Geochemical data of the shale from Well B338-6.

Lithology	TOC wt.%	Kerogen type	Carbonate content wt.%
Oil shale	7.26	I	21.3
δ^{13} C of organic matter		δ^{13}C of carbonate	
PDB‰ −25.65		PDB‰ +4.63	

catalysts were prepared in two ways according to their natural states in the shale, i.e. absorptive state or ionic state. For the first situation, NiS and CoS were prepared, so that the solid powders are easily absorbed by organic matter in the shale. For the second situation, $NiCl_2{\cdot}6H_2O$ and $CoCl_2{\cdot}6H_2O$ are prepared, they are easy to dissolve in water, so that Ni and Co return to ions and closely contact with clay and organic matter in the shale. The detailed steps of the simulation are described as the following:

(1) To crush the shale sample to 80mush and to extract soluble organic matter with CH_3Cl_3;
(2) To put 50 g $NiCl_2{\cdot}6H_2O$ and $CoCl_2{\cdot}6H_2O$ into two beakers separately, and add 51 g $Na_2S{\cdot}9H_2O$, respectively, then agitate with water until being mixed completely;
(3) To dry the mixed materials with spin-drier for three times to eliminate NaCl from the mixed materials completely so that pasty NiS and CoS get ready. Then put the pasty NiS and CoS into desiccators below 50°C temperature, and obtain black powder of NiS and CoS;
(4) To take 0.4 g NiS and CoS, and 1.04 g $NiCl_2{\cdot}6H_2O$ and $CoCl_2{\cdot}6H_2O$, respectively (in which the transitional metal is 0.26 g) for mixing with 200 g shale powder (the mass ratio of the shale and transition metal is 1: 0.0013);
(5) To put the shale, the mixture of the shale and NiS, CoS, $NiCl_2{\cdot}6H_2O$, or $CoCl_2{\cdot}6H_2O$, respectively, into different autoclaves, and add 1 g water to each autoclave. Seal the autoclaves under inert gas and keep temperature at 50°C constantly for 24 hours to make the shale and catalyst mixing completely;
(6) To heat the autoclaves at different temperatures (such as 120°C, 160°C, 200°C and so on) for 8 hours in order to let organic matter, transitional metal, and carbonate in the autoclaves interact or react completely; then cool the autoclaves and conduct analyses on the gaseous products by gas chromatography, and on δ^{13}C of CH_4 and CO_2;
(7) To extract soluble organic matter from the solid residues with chloroform, and to measure carbonate content in the solid residues, respectively.

Composition of the gaseous products is analyzed on Agilent 6890 Gas Chromatograph, in which there are two chromatographic columns, one is purged-packed stainless steel column connected with TCD for CO_2 detection, and the other is DB-5 capillary column (30 m, 0.25 mm, 0.25 um) connected with FID for hydrocarbon detection.

The stable carbon isotope was measured on a Finnigan MAT 252 Mass Spectrometer (Finnigan MAT, Bremen, Germany) for the CO_2 and CH_4, using standard procedure (such as Zhang et al., 2003). The results are all reported in PDB standard.

4 RESULTS & DISCUSSION

Hydrocarbon generation from thermal degradation of organic matter under catalysis of volcanic minerals in simulation experiments has been reported before (Jin, 1998a; Ceron et al., 1998). This thermal simulation revealed CO_2 and hydrocarbons are co-generated from the source rock sample with transitional metals under 240°C.

In ordinary thermal simulation of source rocks, there are a few of CO_2 generated under the temperature 300°C (Lewan et al., 1995; Wan, 2002). In our simulations, if Co and Ni catalysts are not put in the samples, very few CO_2 and other hydrocarbon gases can be detected in the gaseous product under the temperature lower than 280°C. While with the presence of Co and/or Ni catalysts, the gaseous product increases greatly (Fig. 4), from 120°C the total gas productivities are at least two or three times more than those from the shale alone, which indicates that Co and Ni play a significant role of catalysis in the generation of gases. Among the gases produced from the thermal

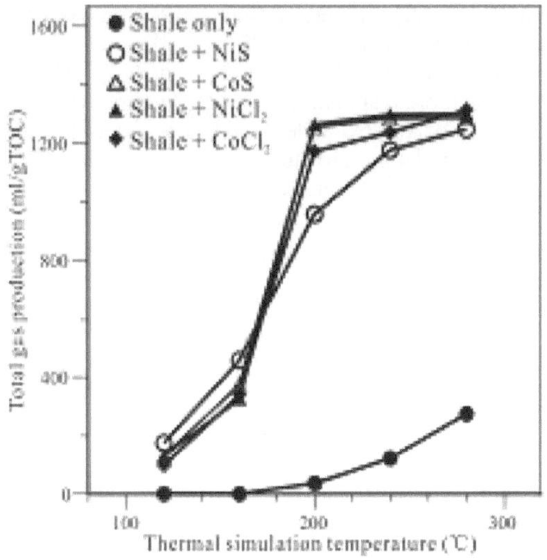

Figure 4. A plot of total gas production vs thermal simulation temperature.

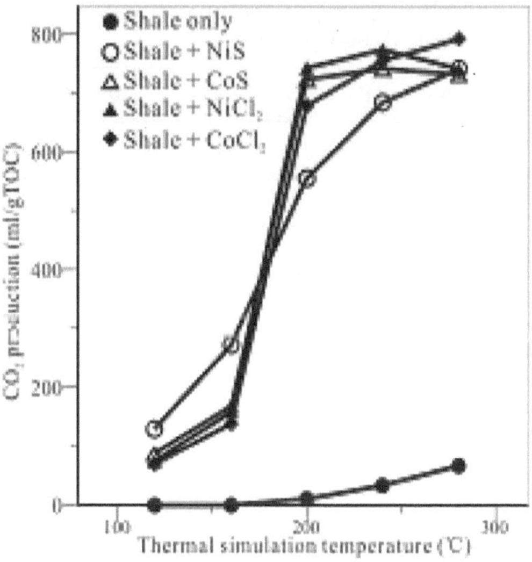

Figure 5. A plot of CO_2 production vs thermal simulation temperature.

simulations, CO_2 takes about two-thirds. When the temperature is over 120°C, the CO_2 is increasing very quickly from the mixtures of the shale and Co and/or Ni catalysts (Fig.5). At 200°C, the generated CO_2 from the mixture of the shale and catalysts is about 80 times more than that from the shale alone. At 240°C, the generated CO_2 and other gases all decrease a little (Fig. 5), which suggests that the catalysis of Co and Ni on the generation of gases including CO_2 is abated. Therefore, the efficient catalysis of Co and Ni on generation of CO_2 and hydrocarbon gases is available at temperatures between 120°C and 240°C under the simulation conditions.

From the previous discussion, we just know that CO_2 and hydrocarbon gases are generated at same time. However, we want to know whether the CO_2 is generated from thermal degradation of organic matter in the shale or from the decomposition of carbonate minerals in the shale. Before the thermal simulations, the carbonate content in the shale sample was measured. After the simulations, the carbonate content of solid residues was also measured, which showed that carbonate content in the solid residue decreased obviously during the temperatures between 120°C and 240°C (Fig. 6). The decrease of carbonate in the shale sample is just corresponding to the increase of CO_2 in the produced gases, therefore, the CO_2 is possibly generated from decomposition of the carbonate minerals in the shale.

δ^{13} C of CO_2 is a good indicator for identifying its origin (Dai, 1992). At low simulation temperature (such as 120°C), the CO_2 production from the mixture of shale and Co and/or Ni catalysts is very few, its δ^{13}C

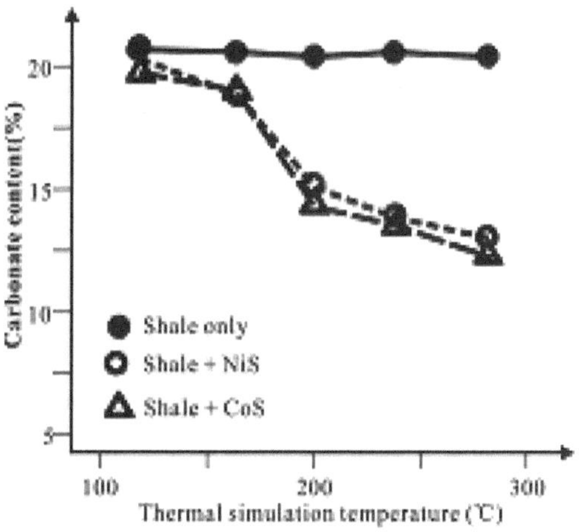

Figure 6. Relationship between the carbonate content in the solid residues and the simulation temperatures.

is about or less than −10‰ (Table 3), which indicates that the CO_2 is generated either from carbonate or from organic matter (Yoshimura et al., 2004). When simulation temperature is over 160°C, all the δ^{13}C of CO_2 is greater than −10‰ (Table 3), similar to that of CO_2 discovered from the Binnan area, and it also coincides with that of CO_2 degassed from magma (Fiebig et al. 2004). Because the carbonate content in the residue decreases dramatically in simulation temperature of 160°C~240°C and δ^{13}C of the carbonate minerals in the shale is as high as +4.63‰, we believe that the CO_2

Table 3. $\delta^{13}C$ (PDB‰) of CO_2 in the gaseous product generated in the thermal simulation

Temperature °C	120	160	200	240	280
Shale only				−15.69	−14.42
Shale and NiS		−10.46	−8.72	−7.75	−7.38
Shale and NiCl		−9.35	−8.29	−7.21	−6.89
Shale and CoS	−10.72	−9.61	−9.25	−8.27	−7.65
Shale and CoCl	−11.19	−9.73	−9.65	−8.93	−7.21

generated in the simulations is most from decomposition of carbonate in the shale, and the minor from the degradation of organic matter under catalysis of Co and/or Ni catalysts.

It is important that the CO_2 generation occurs under 300°C (actually the maximum temperature is 280°C) in the simulations, which avoids carbonate decomposition at temperature over 350°C (Jin et al., 2002; Cody et al., 2004). In our simulations, the total gases including the CO_2 generated from the shale alone are in a very small quantity compared with those generated from the mixture of the shale and Co and/or Ni catalysts, which indicates that the transitional metals play a key role in carbonate decomposition. The CO_2 co-generation with oil and gas is very useful to understanding why CO_2 and hydrocarbons are co-accumulated in traps within or around the basalt-source rock assemblage in rift basins in eastern China. If CO_2 comes directly from the magma degassing or the mantle, it is difficult to accumulate together with the oil and gas generated from organic matter in the source rocks. For example, in the studied area, volcano erupted during the deposition of source rocks of Member 3 of Shahejie Formation (Jin, 1998a), it was not possible for the CO_2 to be accumulated in the overlying reservoirs (the upper part of Shahejie Formation and younger strata). The source rock in Member 3 of Shahejie Formation did not generate oil and gas as well as CO_2 until Miocene (the deposition of Minghuazhen Formation). Meanwhile, many traps in Shahejie Formation and younger formations were formed and were charged with hydrocarbons and CO_2 together in Miocene (Duan, 1990).

It is noteworthy that the Ni and Co contents in the muddy limestone closely contacting with the basalt are in the maximum in the source rock sequence in Well B338-6 (i.e. 146×10^{-6} and 118×10^{-6}, respectively, seen in Fig. 3). But the Ni and Co contents in the mixture for the thermal simulations are 975×10^{-6}, and six or eight times more than those in the muddy limestone. Is it artificially enlarging the catalysis of Ni and Co on decomposition of carbonate minerals and on thermal degradation of organic matter? The answer is yes. Because we also conducted three thermal simulations on the mixture of Ni, Co and the shale at 160°C, 200°C and 240°C temperatures, in which the Ni and Co contents reduced to a third (i.e. 325×10^{-6}), we observed that the product of hydrocarbon gases and CO_2 is about 50~70% more than that from the shale sample alone. However, Ni and Co added into the reactants in our thermal simulations are much less than those in previous thermal experiments (e.g. Mango and Hightower, 1994). Therefore, Ni and Co transitional metals are of significant catalysis on the decomposition of carbonate minerals and on thermal degradation of organic matter in source rocks, and the more the transitional metals, the more the CO_2 and hydrocarbons.

5 CONCLUSIONS

Basaltic volcanoes and thermal fluid occur frequently in rift basins of eastern China, even in depositional environments of hydrocarbon source rocks (such as shale and muddy limestone). Within and around the basalts, hydrocarbons and CO_2 are accumulated. The authors believe that the CO_2 is not derived directly from volcanoes or thermal fluid though its $\delta^{13}C$ is heavier than −10‰. Having conducted a series of thermal simulation on the source rock sample with Co and Ni catalysts, the authors recognized that the CO_2 comes from interaction of transitional metals with carbonate in the shale during thermal degradation of organic matter to hydrocarbons, so that the CO_2 and hydrocarbons are generated together. Only this kind of CO_2 is possible to be accumulated in one trap with hydrocarbons simultaneously though it shows the same stable isotope composition as that degassed from present volcanoes (Chiodini and Frondini, 2001; Dai et al., 1994).

The case study on basalt-source rock assemblages in the Binnan area in Dongying Depression shows obviously that basaltic volcanoes provide transitional metals to the shale, and transitional metals act as catalysts on thermal degradation of organic matter into hydrocarbons and on decomposition of carbonate in the shale. At temperatures from 120°C to 240°C in the simulation, carbonate contents in the shale sample with Co or Ni as catalysts decrease obviously, the production of the hydrocarbon gases and CO_2 increases quickly, and the CO_2 is up to 2/3 in the gaseous product. Although CO_2 could be generated from degassing of basaltic magma in the area, it is impossible for the CO_2 to be accumulated in the traps with hydrocarbons at the same time because the volcanic eruption is much earlier than that of the trap formation in the Binnan area. Therefore, it is significant to better understand CO_2 generation from decomposition of carbonate minerals in source rocks under catalysis of transitional metals as well as the CO_2 and hydrocarbon co-accumulation in rift basins of eastern China, where basalts are distributed.

ACKNOWLEDGEMENT

This research was funded by National Nature Science Foundation of China (Grant No. 40272061). The authors are grateful to the Shengli Oilfield Co., Sinopec, for supplying the material and basic data, to Geochemical Lab of Geology Institute of the Shengli Oilfield Co. Ltd. for the GC and stable carbon isotope analyses, and to Ms Hu Xiaoqing for mapping for the manuscript.

REFERENCES

Avila P.F., Oliveira J.M., Silva E.F., and Fonseca E.C. (2005) Geochemical signatures and mechanisms of trace elements dispersion in the area of the Vale das Gatas mine (Northern Portugal). *Journal of Geochemical Exploration*. 85(1): 17–29.

Berner, R.A., Lasaga, A.C., and Garreles, R.M. (1983) The carbonate–silicate geochemical cycle and its effect on atmospheric carbon dioxide over the past 100 million years. *Am. J. Sci.* 283: 641–683.

Cartwright T.R., Weaverb S., Fultona C., Nichola M., and Cheng X. (2004) Hydrogeochemical and isotopic constraints on the origins of dryland salinity, Murray Basin, Victoria, Australia. *Applied Geochemistry*, 19: 1233–1254.

Ceron J.C., Bosch P., and Carlos L.S. (1998) Isotopic identification of CO_2 from a deep origin in thermaomineral waters of southeastern Spain. *Chemical Geology*, 149(3–4): 251–258.

Chiodini G., and Frondini F. (2001) Carbon dioxide degassing from the Albani Hills volcanic region, Central Italy. *Chemical Geology*, 177: 67–83.

Cody G.D., Boctor N.Z., Brandes J.A., Filley T.R., Hazen R.M., and Yoder H.S. JR. (2004) Assaying the catalytic potential of transition metal sulfides for abiotic carbon fixation. *Geochimica et Cosmochimica Acta*, 68(10): 2185–2196.

Coveney Jr.R.M., Goebel E.D., and Zeller E.J. (1987) Serpentinization and the origin of hydrogen gas in Kansas. *American Association of Petroleum Geologists Bulletin*, 71(1): 39–48.

Dai Jinxing, Dai Changshui, and Song Yan (1994) Inorganic genetic carbon dioxide gas accumulations and their characteristics in east part of China. *China Offshore Oil and Gas (Geology)*, 8(4): 215–222 (in Chinese with English abstract).

Dai Jinxing, Song Yan, and Hong Xing. (2002) The Inorganic Gas in Eastern China and the Accumulation Condition of the Gas Pool. Beijing: Science Press (in Chinese).

Du Xianyue (1998) The development of study of exploration of igneous rock oil and gas pool in Bohai basin. *Petroleum Geology and Recovery Efficiency*, 5(4): 1–4 (in Chinese with English abstract).

Duan Zhibin (1990) The discussions of the source of gas in igneous rock oil pool and the affection of igneous activity to the gas. *Petroleum Technology*, 2(3): 36–43 (in Chinese with English abstract).

Fiebig, J., Chiodini G., Caliro S., Rizzo A., Spangenber J., and Hunziker J.C. (2004) Chemical and isotopic equilibrium between CO2 and CH4 in fumarolic gas discharges: Generation of CH4 in arc magmatic-hydrothermal systems. *Geochimica et Cosmochimica Acta*, 68(10): 2321–2334.

He Yuan, Wang Dayun, and Liao Yunsheng (2001) Volcanic rocks, basin evolution and related CO_2 – Au mineralization in Shengli Oilfield, China. *Chinese Journal of Geology*, 36(4): 454–464.

Jeffery A.W., and Kaplan I.R. (1988) Hydrocarbons and ino rganic gases in the Gravberg-1 well, Siljan Ring, Sweden. *Chemical Geology*, 71 (2) : 237–255.

Jin Qiang (1998a) Volcanic activity in major source rocks in the faulted basins of China and its significance. *Geological Review*, 44(2): 136–142 (in Chinese with English abstract).

Jin Qiang (1998b) Hydrocarbon generation in rift basins, eastern China: catalysis and hydrogenation interaction between volcanic minerals and organic matter. *Advance in Earth Sciences*, 13(6): 542–546 (in Chinese with English abstract).

Jin Qiang (2001) Volcanic activities and oil/gas accumulation in the rift basins of the eastern China. *Journal of the University of Petroleum, China*, 25(1): 26–29 (in Chinese with English abstract).

Jin Qiang, Xiong Shousheng, and Lu Peide (1999a) Catalysis and hydrogenation: volcanic activity and hydrocarbon generation in rift basins, eastern China. *Applied Geochemistry*, 14(5): 547–558.

Jin Qiang, Rong Qihong, and Wan Congli (1999b) Natural gas reserves mixed from inorganic and organic sources. *Journal of Mineral petrology*, 17(4): 332–337 (in Chinese with English abstract).

Jin Zhijun, Zhang Liuping, Zeng Jianhui, and Yang Lan (2002) The multiplex-origin alkenes about the mantle fluid rich in CO_2 in Dongying depression. *Journal of Science Sinica*, 47(16): 1276–1280.

Lewan, M.D., Comer, J.B., Hamilton-Smith, T., Hansenmuller, N.R., Guthrie, J.M., Hatch, J.R., Gantier, D.L., and Frankie, W.T., (1995) Feasibility study on material-balance assessment of petroleum, from the New Albany Shale in the Illinois Basin. *U.S., Geological Survey Bulletin* 2137, pp. 1–31.

Mango F.D., and Hightower J.W. (1994) Role of transition-metal catalysis in the formation of natural gas. *Nature*, 368: 536–538.

Patrick J. M., Banner J.M., Stern L., Asmerom Y., Edwards R.L., and Emi I. (2004) Stable isotope variations in modern tropical speleothems: Evaluating equilibrium vs. kinetic isotope effects. *Geochimica et Cosmochimica Acta*, 68(21): 4381–4393.

Simoneit B.R.T. (1988) Origin of gases and condensates in the Guaymas basin hydrothermalsystem (Gulf of California). *Chemical Geology*, 71: 169–182.

Simoneit B.R.T. (1992) Aqueous organic geochemistry at high temperature/high pressure. *Origins Life Evol. Biosphere*. 22(1): 43–46.

Wan Congli, and Jin Qiang (2001) Current research situation on hydrocarbon-generating evolution of volcanic minerals upon hydrocarbon source rocks. *Petroleum Geology and Recovery Efficiency*, 8(4): 9–11 (in Chinese with English abstract).

Wan Congli, Zhai Qinglong, and Jin Qiang (2001) Interaction between source rocks and igneous rocks-a preliminary investigation of dissolution of igneous rocks by organic acids and catalysis of transition metals in the generation of hydrocarbons in the source rocks. *Geology & Geochemistry*, 29(2): 46–51 (in Chinese with English abstract).

Wan Congli (2002) Interaction between Igneous and Source Rocks and Oil Accumulation Model in Igneous-Source Rock Assembly in the Dongying Depression. The PhD Dissertation, University of Petroleum, Dongying, Shandong, China. (in Chinese)

Wang Binhai, and Qian Kai (1992) The Study and Exploration of Shengli Oilfield. The Press of Petroleum University, 354pp (in Chinese).

Xiao Jiafei, and Hu Ruizhong (2005) Sedimentary-volcanic tuffs formed during the early Middle Triassic volcanic event in Guizhou Province and their stratigraphic significance. *Chinese Journal of Geochemistry*, 24(4): 338–344.

Xiong Shousheng, and Lu Peide (1996) Volcanic eruption-effusion and their relationship with the formation and type of semi-inorganic gas. *Experimental Petroleum Geology*, 18(1): 13–26 (in Chinese with English abstract).

Yin J.R., Fursich F.T., and Werner W. (1994) Compartive study on salinity analysis between stable carbon isotopes and the benthic associations. *Geoscience — Journal of Graduate School, China University of Geosciences*, 8(4): 380–392 (in Chinese with English abstract).

Yoshimura K., Liu, Z. Cao J., Yuan D., Inokura Y., and Noto M. (2004) Deep source CO_2 in natural waters and its role in extensive tufa deposition in the Huanglong Ravines, Sichuan, China. *Chemical Geology*, 205: 141–153.

Zhai Qinglong, Jin Qiang, Zeng Yi, and Wan Congli (2004) The geological characteristics of oil and gas in the basalt-source rock combination at Binnan area in Dongying depression. *Experimental Petroleum Geology*, 26(5): 468–473 (in Chinese with English abstract).

Zhang Chanlin, Li Yongliang, Ye Qing, Fong J., Peacock A.D., Blunt E., Fang J.S., Lovley D.R., and White D.C. (2003) Carbon isotope signatures of fatty acids in Geobacter metallireducens and Shewanella algae. *Chemical Geology* 195: 17–28.

Zhang Shungan (1998) The distribution and significance of igneous rocks in the Middle of Shahejie Formation in Huanghua depression. *Experimental Petroleum Geology*, 10(3): 392–399 (in Chinese with English abstract).

3. Molecular and isotopic geochemistry

Petroleum Geochemistry and Exploration in the Afro-Asian Region – Digang Liang, Darui Wang & Zhenxi Li (eds)
© 2008 Taylor & Francis Group, London, ISBN 978-0-415-44083-7

Surface geochemical exploration practice for oils and gases in China

Chongxi Liu
Hefei Research Institute of Petroleum Geochemical Exploration, Sinopec Exploration and Production Research Institute, Anhui, China

Kebin Zhao & Chuanzhi Wu
Wuxi Research Institute of Petroleum Geology, Sinopec Exploration and Production Research Institute, Wuxi, China

ABSTRACT: The general development course, the method evolution and the main advance of surface oil and gas geochemical exploration techniques in China have been presented in the present paper. The determining method of geochemical anomalies and the geochemical significance of anomalies at various levels have been described in detail as well. The typical field practice demonstrates the important role of this technique in oil and gas exploration.

1 DEVELOPMENT COURSE AND METHOD EVOLUTION

Oil and gas geochemical exploration technique was first introduced into China from the former Soviet Union in the early 1950s just after the liberation of China. Under the advocacy and leadership of Academician Weng Wenhao and Academician Guan Shicong, the exploration technicians, through applying fluorescence spectrum method and hydrogeochemical method, set a precedent of providing geochemical services to oil and gas exploration in China. 50 years of practical application and research has made this technique an important oil and gas exploration tool in China.

From the 1950s, a variety of geochemical exploration methods, including hydro-geochemistry, soil hydrocarbon gas, microbe, soil salt, radioactivity and fluorescent bitumen, has been tested in areas such as Zhungaer Basin, Jiuquan Basin, Ordos Basin and Songliao Basin. After two decades of field tests and application research conducted under various geological and natural geophysical settings, hydrogeochemistry and hydrocarbon gas methods were identifed as the effective oil and gas geochemical methods. From the late 1970s, some geochemical exploration experiments and research have been carried out in a number of the known areas to further determine the suitable application conditions of various geochemical methods and to improve the exploration procedures (Liu and Sun, 1988) In the 1980s, some geochemical indicator combinations were proposed based on the systematic research of various geochemical methods (Cheng et al., 1999; Liu et al., 2001). These geochemical indicator combinations are: major indicators: integrated hydrocarbon gases including water dissolved gas, free gas, adsorbed gas and occluded gas;
minor indicators: altered products due to the intrusion of migrated hydrocarbons; discriminating indicators: synchronous fluorescence spectrum, and 3D fluorescence spectrum; and source indicator: carbon isotope of methane dissolved in underground water (i.e. $\delta^{13}C_1$).

All the researches of these geochemical indicators (Li and Yong, 1996; Jia, 1997; Tang et al., 2000; Zhao and Xu, 2000; Li et al., 2002; Zhang and Wang, 2004) have paved the way for the quick development, the wide application and the upsurge of surface oil and gas geochemical exploration in China.

2 ADVANCES OF SURFACE OIL AND GAS GEOCHEMICAL EXPLORATION TECHNIQUE

2.1 *The understanding of the theoretical basis has been deepened*

The geochemical evidence of vertical hydrocarbon migration has been systematically studied based on the abundant near-surface and drilling geochemical data (Liu and Xu, 1992; Cheng et al., 2000; Zhao et al., 2000; Tang and Liu, 2002). The near-surface geochemical expressions and the variation regularities of migrated hydrocarbons have been traced vertically

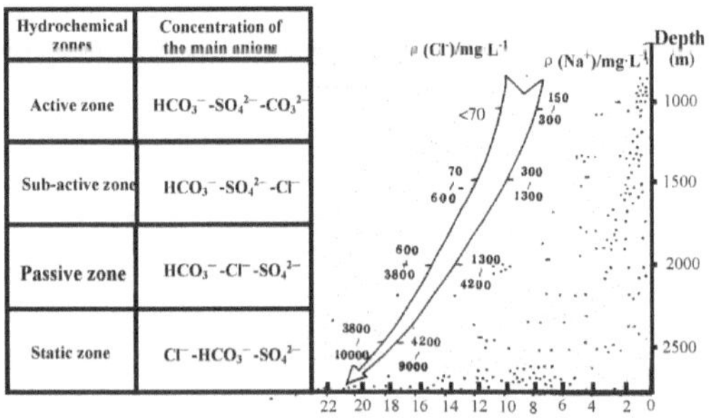

Figure 1. Vertical composition variation features of the underground water, Biyang Depression.

from the view of the development stage of the geological history (Liu et al., 1999; Cheng et al., 2000) (Fig. 1). Meanwhile, theoretical studies have also been conducted from the angles of the water-rock equilibrium, the composition transformation with surrounding rocks, simulation experiments of the relationship between the geophysical field and geochemical field, and the multi-dimensional dynamic analysis(Qi et al., 1994; Bao et al., 1997; Cai et al., 1997; Tang et al., 2001).

2.2 *The application conditions have been strengthened*

Surface oil and gas geochemical exploration technique is used to predict oil and gas potentials based on the subtle information resulting from the dispersion of oils and gases underground, and has its own application prerequisites. Different levels of anomalies reflect their corresponding geological regularities. Only when the regional, zonal and local background values are determined based on the variation regularities of the geochemical field and the concentration variations of the geochemical indicators, can the corresponding anomalies be properly delineated, and the oil and gas reservoirs be possibly traced.

The application conditions in different sedimentary basins, different oil formation systems, and different surface landscapes may require different geochemical methods. So selection of suitable geochemical methods for a given survey area is an important aspect to improve the prediction results of the oil and gas geochemical exploration technique.

2.3 *Great achievements have been made*

Supported by long period of field practice and scientific research, the surface oil and gas geochemical exploration technique has evolved gradually toward standardization in China (Liu et al. 1997). In 1987, this technique was formally accepted into the normal oil and gas exploration technique series in China. The Technique Norm for Oil and Gas Geochemical Exploration and the Analyzing and Testing Regulation was officially issued in 1997 and has taken effect.

The standard samples for analyzing and testing are well developed, the geochemical exploration procedures are summarized, and the methods for diminishing the interference factors are further perfected.

Recently, the China Oil and Gas Geochemical Database and the Geochemical Data Processing System were established, which has greatly improved the data interpretation and evaluation level in oil and gas geochemical exploration. All these achievements have made the oil and gas geochemical exploration technique a modern exploration technique that is able to match with geology, geophysics and drilling in oil and gas exploration.

In addition, information exchanges and academic activities also contribute a great deal to the development of the surface oil and gas geochemical exploration technique in China. A number of research papers and monograph books related to the surface oil and gas geochemical exploration have been published (Ruan and Fei, 1992; Yang et al., 1995; Wu et al., 1996; Cheng et al., 1999; Liu et al., 2001), and many young geochemical exploratonists and experts have joined into this field.

3 ANOMALY DETERMINING METHODS AND GEOCHEMICAL SIGNIFICANCE OF VARIOUS ANOMALIES

An anomaly is an important parameter in the surface and near-surface oil and gas geochemical exploration. It has expanded its concept besides the conventional concentration anomaly. Anomalies today, besides concentration anomaly, may refer to contrast anomaly, anomaly of the ratios between concentration values and background values, fabric anomaly, anomaly determined by the ranking order of the indicator concentration, structural anomaly, transformed anomaly based on statistical processing of multi-indicators; gradient anomaly, variation trend anomaly of an indicator within a geological unit; trend anomaly, anomaly determined by trend surface analyses (including residual anomaly), analogy anomaly, anomaly acquired by comparing with the shallow geochemical effect in the known areas, and discrimination anomaly acquired by discrimination analyses of an indicator.

Anomalies determined in different ways may have some variations in their space distribution. When these geochemical anomalies are combined with geological and geophysical data as well as the exploration experience of the users, the final integrated anomaly can be delineated, which may play an important role in improving the accuracy rate of predicting the underground oil and gas potentials.

Anomalies can also be classified into the following categories according to the distribution coincidence

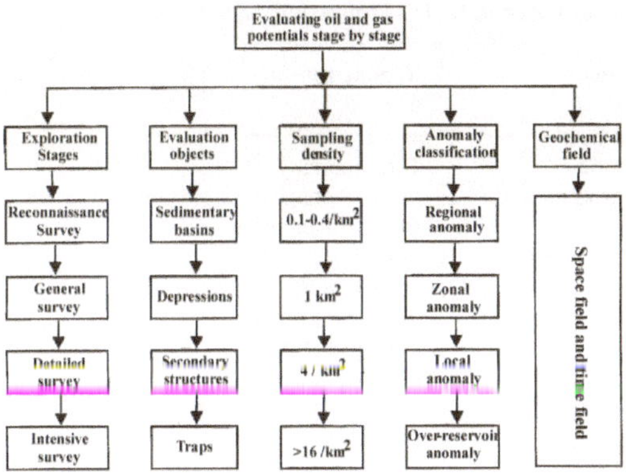

Figure 2. Proper exploration procedures in oil and gas geochemical exploration.

of the indicators, such as the overlapped anomaly, the most reliable anomaly type with the anomaly areas of the main indicators coinciding well to each other; the combined anomaly, the more reliable anomaly type with the main anomaly areas being separated but still having some trend to form a completely integrated anomaly, and the united anomaly, the less reliable anomaly type with each indicator having its own isolated anomaly area and little coincidence degree to each other, which is only able to form a unitedly integrated anomaly.

The practice in using oil and gas geochemical exploration technique is to follow the proper exploration procedures (Fig. 2), which starts from larger exploration areas, and then narrows down the exploration areas step by step, and finally evaluates oil and gas prospects.

Anomalies can be also classified in a third way based on the exploration stages into the regional anomaly, zonal anomaly, local anomaly and over-reservoir anomaly. These different levels of geochemical anomalies have different geochemical significance in oil and gas exploration as shown in Table 1.

Table 1. Geochemical significance of various levels of anomalies determined in different exploration stages.

Exploration stage	Appellation of anomaly levels	Anomaly significance	Trend
Reconnaissance survey	Regional anomaly	Predicting oil and gas prospects of a whole basin	Going toward the smaller targets step by step ↓
General survey	Zonal anomaly	Locating areas/zones favorable for oil and gas accumulations	
Detail survey	Local anomaly	Evaluating oil and gas potentials of a trap	
Intensive survey	Over-reservoir anomaly	Delineating the outline of an oil/gas field	
Down-hole survey	Layer anomaly	Determining certain layers/intervals for oil/gas testing	

4 TYPICAL CASE HISTORIES

4.1 *Baiyindulan (BYDL) Depression – A typical case of reconnaissance survey*

A reconnaissance geochemical survey was conducted in the late 1970s in 1450 km² of BYDL Depression. The survey design and geochemical methods used are listed in Table 2.

The results of this reconnaissance geochemical survey show that the geochemical indicators have certain characteristics in BYDL Depression, e.g. the composition of the dissolvable hydrocarbon gases in underground water covers large ranges, similar to those in the known oilfield water. Table 3 lists the typical hydrocarbon compositions and their concentrations in the underground water in this area. These geochemical data demonstrate that 1) the concentration of the total hydrocarbons in a few points is higher than that in the phreatic water above the Mid-Cenozoic oil/gas reservoirs in Eastern China; 2) heavier hydrocarbon gases such as ethane, propane, and butane have relatively high contents; 3) the contrast between anomaly and background is quite obvious, with anomaly values being normally 2–3 times of background values; 4) the anomaly points are fairly focused and have the distribution trends of clustering in the center of the depression and along the NE fracture; 5) the concentration of alkene is quite high in a few samples, indicating that recent organic geochemical processes are not the only contribution to the presence of alkene, the abundance of coal-measure strata in the depth are also related to the increase of alkene. Thus, the distribution characteristics of dissolved hydrocarbon gases in the underground water can provide the information about hydrocarbon migration and accumulation at depth.

Like dissolved hydrocarbons, the major elements in the underground water are anomalous in the studied area as well. The total dissolved solids increase up to 2.15 g/l in contrast with the background value of <0.5 g/l; and the content of HCO_3^- increases, too. The minor elements and the organic concentrations are also anomalously high in many underground water samples. For example, the anomalous concentration of phenol in the studied area is 21 ~ 43 μg/l in contrast

Table 2. Geochemical survey design and methods used in BYDL Depression.

Method	Hydrochemical methods	Soil-based methods	Remote sensing-methods
Indicator	benzene, phenol, water-dissolved hydrocarbons, U^{6+}, minor elements (Ni, V, Mn, Ca, Fe)	acid-extracted hydrocarbon, carbonate, organic carbon, high-/low-valence iron	Tone anomaly of MSS images, spectrum anomaly of plants and elemens (Cu, Zn, Mn, Ni) in plant ashes
Sampling style	sampling randomly in the entire studied area	17 survey lines with 2–3.5 km line distance and 1–2 km point distance	5 survey lines
Sample number	69	282, and 58 from drilled holes	40

Table 3. Composition of dissolvable gas hydrocarbons in the underground water.

Water type	Sample number	Depth of water table/m	Contents of hydrocarbons/10^{-6}			
			Methane	Ethane	Protane	Alkene
Phreatic water	D-5	1.5	394.96	4016	3.25	49.60
	D-43	2.0	137.97	3.43	2.30	78.65
	D-20	2.0	126.36	5.66	4.15	28.67
	D-39	2.0	808.82	3.33	3.00	20.35
	D-27	1.0	338.77	4.18	2.49	27.93
Pressure water	Q-9	46	253.47	153.72	11.50	123.69
	ZK7-1	72	693.80	0.74	0.50	142.67
	Fu-5	78	201.67	5.03	4.01	32.21
	38	80	291.87	5.17	4.87	39.14
	28	140	137.60	6.24	5.24	124.49

with the background value of 17 μg/l; and the anomaly value of Ni in the studied area is 2∼10 times of its background value, for example, the anomaly value of U^{6+} is 1.4 ∼ 3 mg/l while the background value is 0.9 mg/l.

The studies of the relationship between the geochemical anomalies determined by different methods demonstrate that the anomaly distribution of major and minor elements and organic compositions coincide with that of dissolved hydrocarbon gases, as well as tone anomaly of the satellite remote sensing images (Fig. 3).

As for soil-based adsorbed hydrocarbon gases, it is shown that methane is generally 4∼40 μg/kg, and may be as high as 1817 μl/kg; ethane is normally 0.3∼28 μl/kg, with the highest value of 209 μl/kg,; propane is normally 0.3∼8 μl/kg, and may be up to 100 μl/kg. The study also shows that iso-butane (up to 19 μl/kg) and normal butane (up to 23 μl/kg) exist in some samples. All the adsorbed hydrocarbon indicators show an evident contrast between anomaly values and background values (Table 4). As the main anomaly, methane anomaly focuses on the eastern part of the depression but coincides poorly with the anomalies of the other adsorbed hydrocarbon indicators.

The study on the contents of carbonates and organic carbons in the soil samples reflects that carbonate content is rather high and the value of organic carbon is 0.1% ∼ 0.4% in this area. Overall, the reconnaissance geochemical survey in BYDL Depression indicates, from a unique geochemical aspect, that this depression has certain oil and gas prospects. And this surface geochemical survey provides additional geochemical support to the above conclusion.

4.2 *Jiangling (JL) Depression – A typical case of general survey*

JL Depression is a sub-structural unit of 6900 km^2 in Jianghan Basin. It is controlled by fractures to be a symmetrical depression with a NW central anticline zone (Jingsha Anticline) as its center and synclines

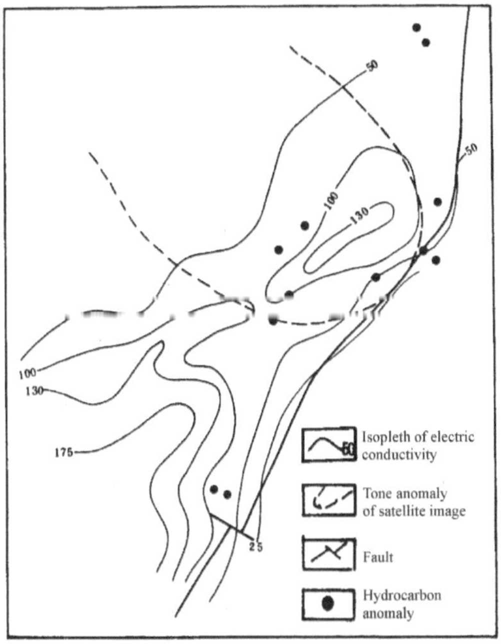

Figure 3. Water-dissolved hydrocarbon anomalies in the underground water in BYDL Depression.

Table 4. Background values and anomaly values of acid-extracted hydrocarbons in BYDL Depression.

Hydrocarbons	C_1	C_2	C_3	i-C_4	n-C_4
Background/ $\mu l \cdot kg^{-1}$	14.13	1.61	1.23	0.55	1.28
Anomaly/ $\mu l \cdot kg^{-1}$	189.70	18.79	11.22	1.45	7.61

(Meihuaihe-Huduhe-Zifusi Syncline and Guanjukou Syncline) and slopes (Lingnan Slope and Lingbei Slope) as its south and north flanks, respectively. The deep structure of this depression consists of two rock bodies, which cause the south-north symmetry of this depression, the thick deposit sequences in the center and the thin deposits in the edge.

A general geochemical survey was carried out in 1990 in JL Depression with the sampling grid of 1 km × 1 km and the sampling depth of less than 1.5 m. The soil samples with the same lithological characteristics were collected and analysed. The geochemical methods used were adsorbed hydrocarbon gas, ΔC, heat-released mercury, ultra-violet adsorption spectrum, fluorescence spectrum, carbonate and stable carbon isotope of methane. The geochemical survey results of these indicators were expressed through the isograms of methane (C_1), heavier hydrocarbons (C_{2+}), propane (C_3), C_{2+}/C_1, ΔC, heat-released mercury (Hg), fluorescence spectrum at 320 nm (F320 nm), fluorescence spectrum at 320 nm/360 nm (F320 nm/F360 nm), and ultra-violet adsorption spectrum at 360 nm (UV360 nm). Through the isograms, the relationship between the concentration variation of these geochemical indicators and the regional geology settings can be analyzed, the anomaly thresholds of various indicators can be determined zonally, and the anomaly distribution of each individual indicator can be defined (Table 5). Based on anomaly types and patterns of these anomaly zones of each indicator, an integrated anomaly zone can be delineated finally, which is normally interpreted as the favorable zone for oil and gas accumulations. The reliability of the integrated anomaly can be further evaluated by the stable carbon isotope data of methane and the Integrated Evaluation Index (IEI).

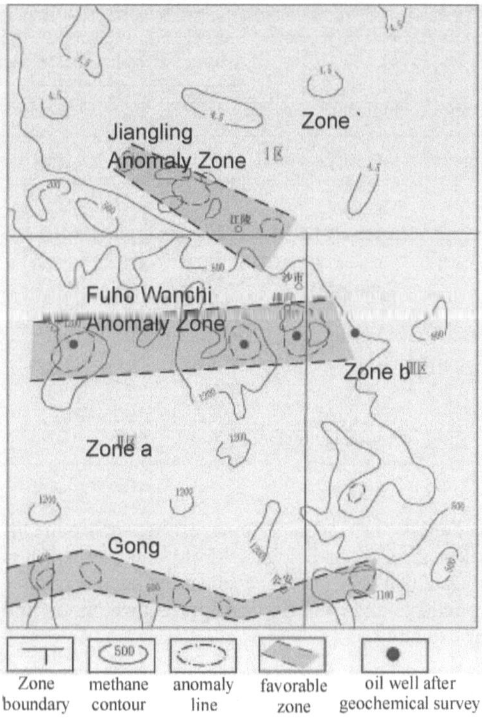

Figure 4. Distribution of geochemical anomaly zones in JL Depression.

As a result, the general geochemical survey in JL Depression delineated 3 favorable zones for further oil and gas exploration (Fig. 4), including the Jiangling anomaly zone, Fuhe-Wanshi anomaly zone and Gong'an anomaly zone. The consequent drillings within these anomaly zones have attained quite good results. More than 10 wells have been drilled in the Jiangling anomaly zone, leading to the discovery of Jiangling Oilfield; and over 20 wells drilled within

Table 5. Geochemical anomaly characteristics of different zones in JL Depression.

Zones	Indicators	Acid-extracted hydrocarbons/$\mu l \cdot kg^{-1}$			Spectrum			
		C_1	C_3	C_{2+}	U222 nm	F320 nm	ΔC, %	Heated-released mercury/10^{-9}
Zone I	Intensity	7.20		1.55	113.19	253.84	0.3	36.71
	Contrast	2.56		6.21	4.13	1.59	2.83	3.55
	Threshold	4.5		0.47	30.0	205.0	0.17	17.00
	Anomaly point	125		151	173	146	192	92
Zone II	Intensity	1437.68	8.28	42.97	596.57t	715.38	2.62	45.06
	Contrast	1.75	1.40	1.35	18.23	4.99	1.20	1.63
	Threshold	1200	7.5	39	50	192	2.4	39
	Anomaly point	167	230	194	286	257	256	136
Zone III	Intensity	1059.65	6.18	34.38	144.94	193.2	2.53	55.00
	Contrast	2.04	1.35	1.70	4.45	1.84	1.32	277
	Threshold	860.00	5.50	30.52	50.00	139.00	2.30	41.00
	Anomaly point	110	162	95	112	147	78	139

the Fuhe-Wanshi anomaly zone resulted in Shashi Oilfield and Xiefeng Bridge Oilfield. All these results indicate that the surface geochemical anomaly zones are reliable and can play an important role in oil and gas exploration.

4.3 *Zhangtianqu (ZTQ) Oilfield – A typical case of detailed survey*

4.3.1 *Geology features*

ZTQ Oilfield is 4 km north of the existing Huabaogou oilfield. Oil is produced mainly form Chang 2 Formation with a buried depth of less than 1900 m. The reservoir is dominated by fine-grained sand with the porosity of 15.47%, the rock density of 2.26 g/cm^3, and the gas permeability of 0.63×10^{-3} μm^2. It is known that the reservoir belongs to a low-resistance, low-pressure oil-bearing layer with good oil quality and a certain amount of dissolved gases. The reservoir is sourced mainly from fluvial delta deposits and the trap constituted mainly by sand bodies.

4.3.2 *Geochemical anomaly features*

A detailed surface geochemical survey was once carried out in ZTQ oilfield and its neighboring area. Before this detailed geochemical exploration survey was conducted, there had been only four oil producing wells to the east of ZTQ Oilfield, covering an area of only 0.12 km^2. The main indicator combination used in this detailed geochemical survey was the combination of adsorbed methane, adsorbed C_{2+}, and heat-released methane with some secondary indicators, such as ΔC, fluorescence spectrum and adsorbed propane. Table 6 lists the survey results of these geochemical indicators.

The distribution characteristics of the geochemical indicators are summarized as follows: 1) Heat-released methane forms a bulk top anomaly, adsorbed hydrocarbon series forms a disrupted ring around the bulk anomaly of heat-released methane, and the other indicators form anomalies coinciding quite well with the above two kinds of anomalies. On the whole, the general anomaly pattern of all the indicators presents a complex combined ring-bulk anomaly type. 2) The anomalies of the main indicators all have fairly high intensity, and coincide well with each other. For example, adsorbed hydrocarbons, not only cover all the light hydrocarbon compositions ranging from C_1 to C_4, but also have a high total hydrocarbon content, higher than the total hydrocarbon content in the surface samples of the existing Huabaogou oilfield. Anomaly contrasts are generally higher than 1.3, and may be up to 1.9 (for heat-released C_1); the Integrated Evaluation Index (IEI) reaches 3.27(Zhao, 1993); and the Space Variation Index (SVI)(the raito of standard deviation to the mean concentration) is close to that in Huabaogou Oilfield. 3) The peaks of synchronous fluorescence spectra occur mainly at 300 nm and 330 nm, and the highest intensity of 3D fluorescence spectra occurs at Ex230 nm/Em320 nm. These features of synchronous fluorescence spectra and 3D fluorescence spectra reflect that light oil dominates the studied area. 4) The concentration of the indicators (each single indicator and the population) within the complex combined anomaly has a relatively stable space structure with little concentration variation, reflecting that there exist abundant and homogenous source rocks or reservoir rocks in this area. What is especially worth mentioning is the fact that the area of the complex combined anomaly greatly exceeds the known oil producing area, and extends towards the west or the southwest direction of the existing producing area. Not only the concentration anomaly, but also the contrast anomaly, the structure anomaly, the fabric anomaly, the trend-residual anomaly and the discrimination anomaly all have high intensities, and coincide well to one another. The integrated anomaly delineated

Table 6. Characteristics of the main geochemical indicators over ZTQ Oilfield.

Indicators	Acid-extracted C_1/μl/kg	Acid-extracted C_{2+}/μl/kg	Acid-extracted C_3/μl/kg	Heat-released C_1/μl/kg	ΔC, %	Soil Susceptibility/ 10^{-5}	Integrated evaluation index
Intensity	518.18	36.09	7.98	7.78	1.09	94.30	3.27
Contrast	1.33	1.34	1.59	1.95	1.11	1.35	

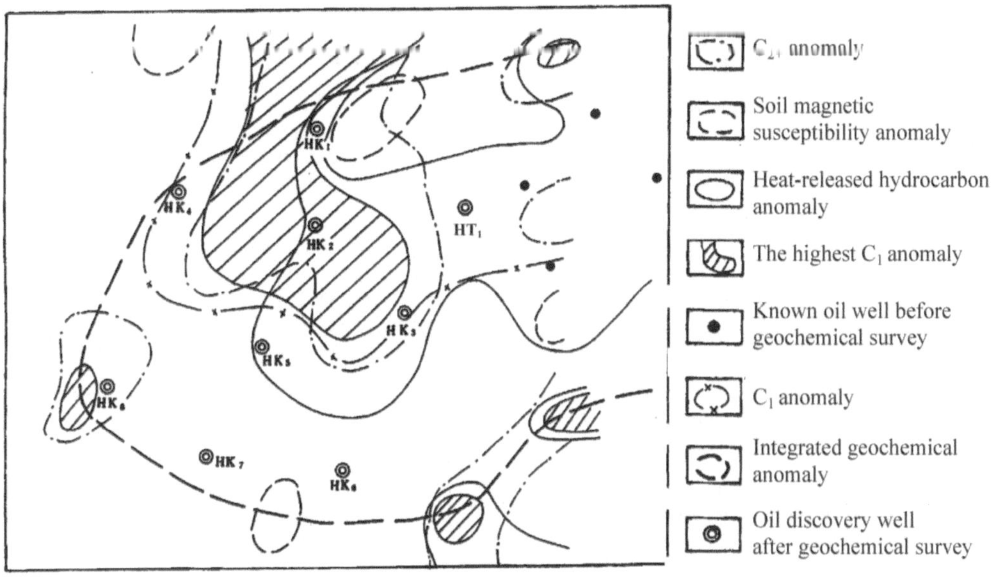

Figure 5. Distribution of the integrated anomaly in ZTQ Oilfield.

at 80% reliability level (Fig. 5) is so valuable in oil and gas exploration that it can provide reliable geochemical support for making drilling decision and deploying well sites. The outline of the integrated anomaly coincides approximately with the boundary of the oil-bearing Yanchang formation. Up to now, more than 30 development wells have been drilled within this integrated anomaly, covering an area of 1.3 km^2 and being of high economic benefits.

5 CONCLUSION

Although the surface oil and gas geochemical exploration technique has experienced several ups and downs in China, under the continuous efforts of generations of geochemical explorationists, a series of geochemical methods suitable for China's petroleum geology settings have been established, and great achievements have been made in the application of the surface geochemical exploration technique, the effective methods and indicator combinations, and the integrated anomaly determination and evaluation. The significant exploration roles in oil and gas exploration practices demonstrate that this technique has bright application prospects.

REFERENCES

Bao Z.Y., Li J.Y., Su J.Y., Li C.W., Pei T., Li F.L., 1997. Geochemical field analysis and oil/gas geochemical exploration. Petroleum Exploration and Development 6, 39–41 (in Chinese).

Cai C.F., Mei B.W., Ma T., Chen C.P., Li W., Liu C.Q., 1997. Approach to fluid-rock interaction in Tarim Basin, 1st Edition. Geology Press, Beijing (in Chinese).

Cheng J., Zhao K.B., Liu C.X., 1999. The theory, method and application of geochemical exploration of petroleum. Geology of Anhui 3, 204–208 (in Chinese).

Chen J., Liu C.X., Zhao K.B., Guo X.S., Yang Z.B., 2000, Evidence and characteristics of oil and gas vertical migration. Oil & Gas Geology 3, 236–240 (in Chinese).

Cheng T.J., Wang Z.S., Wu X.M., Yao J.M., Wu C.Z., Zhao K.B. 1999. Near-Surface Traces of Hydrocarbon Migration and Petroleum Geochemical Exploration, 1st Edtion. Petroleum Industry Press, Beijing (in Chinese).

Jia G.X., 1997. Origin of the absorbed hydrocarbon anomalies in soil in the BS basin. Mineral Sources and Geology, 58, 141–144 (in Chinese).
Li W., Yong K.L., 1996. An applied study on the fingerprinting technology of 3D fluorescence spectrum. Petroleum Exploration and Development, 4, 32–34 (in Chinese).
Li Z.M., Song X.L., Zhang C.J., 2002. The application of the free hydrocarbon technique to the evaluation of oil and gas potential of structures in desert areas. Geophysical & Geochemical Exploration, 5, 344–346 (in Chinese).
Liu C.X., Sun S.X.,1988. Theory and Methods of Hydrogeochemistry for Oil and Gas Exploration, 1st Edition. Geology Press, Beijing (in Chinese).
Liu C.X., Xu S.R., 1992. Oil and Gas Geochemical Exploration: Methods and Applications, 1st Edtion. China Scientific and Technological University Press, Hefei (in Chinese).
Lui C.X., Zhao K.B., Cheng T.J., 1997. The Technique Norm for Oil and Gas Geochemical Exploration, 1st Edition. China Standardization Press, Beijing (in Chinese).
Liu C.X., Cheng J., Zhao K.B., 1999. Geologic significance of oil and gas geochemical prospecting anomalies in different phases of exploration. Petroleum Exploration and Development 2, 42–45 (in Chinese).
Liu C.X., Zhao K.B., Yu L.Y., Cheng J., Zhang Y.M., 2001. *Application of Geochemical Prospecting for Oil and Gas*, 1st Edition. Geology Press, Beijing (in Chinese).
Qi S.H., 1994. A preliminary numerical modeling of anomaly of geochemical exploration in oil and gas. Geological Science and Technology Information 1, 76–80 (in Chinese).
Ruan T.J., Fei Q., 1992. Oil and Gas Petroleum Exploration, 1st Edition. China University of Geosciences Press, Wuhan (in Chinese).
Tang Y.P., Tang Y.L., Hu B., Wei W., Wang G.J., Yue X.Y., 2000. Application of fluorescence spectrum to studies of hydrocarbon micro-migration in vertical direction. Oil & Gas Geology, 4, 370–371 (in Chinese).
Tang Y.P., Ding X.Y., Wu X.H., Wei W., Li S.G., 2001. Feature and genesis of regional geochemical fields of the main hydrocarbon-bearing basins in China. Petroleum Exploration and Development 3, 1–4 (in Chinese).
Tang Y.P., Liu Y.L., 2002, Study of the geochemical effects of vertical hydrocarbon micromigration and their mechanism. Petroleum Geology & Experiment 5, 231–236 (in Chinese).
Yang Y.B., Zhang J.L., Wu X.M., Shao Z., Yao Z.W., 1995. Geochemical Prospecting for Oil and Gas, 1st Edition. China University of Geosciences Press, Wuhan (in Chinese).
Wu C.B., Qiu Y.W., Chen Y.M., Shi J.F., 1996. Development Course of and Ponderation on Surface Oil and Gas Geochemical Exploration, 1st Edition. Geology Press, Beijing (in Chinese).
Zhao K.B., 1993. The evaluation methods and effects of the integrated oil and gas geochemical anomaly. Geophysical & Geochemical Exploraion, 2, 89–97 (in Chinese).
Zhao K.B., Sun C.Q., Chen X.H., 2000. The vertical micromigration of hydrocarbon and its near surface geochemical effect. Petroleum Exploration and Development 4, 99–101 (in Chinese).
Zhao X.H., Xu G.H., 2000. The anti-interference capacity of adsorbed filaments and some other geochemical techniques. Geophysical & Geochemical Exploration, 3, 212–214 (in Chinese).
Zhang Z.Y., Wang Guojian, 2004. Petroleum geochemical exploration significance of soil free hydrocarbon technique. Natural Gas Industry, 6, 30–32 (in Chinese).

Petroleum Geochemistry and Exploration in the Afro-Asian Region – Digang Liang, Darui Wang & Zhenxi Li (eds)
© 2008 Taylor & Francis Group, London, ISBN 978-0-415-44083-7

Maximising the yield of bound aliphatic biomarkers via a convenient two-stage hydropyrolysis procedure

W. Meredith, C.E. Snape, & C. Uguna
Nottingham Fuel & Energy Centre, School of Chemical, Environmental and Mining Engineering, University of Nottingham, UK

G.D. Love
Department of Earth, Atmospheric and Planetary Sciences, Massachusettes Institute of Technology, USA

ABSTRACT: Analytical hydropyrolysis performed under high hydrogen gas pressure (15 MPa) in the presence of a dispersed sulphided molybdenum catalyst has been demonstrated to possess the unique ability to release high yields of biomarker hydrocarbons covalently bound within the macromolecular fraction of source rocks and heavy crude oil fractions. This study describes the further development of the experimental procedure by means of introducing a bed of the same Mo catalyst into the reactor immediately directly below the catalyst-impregnated sample. For a sample of Göynük oil shale (immature Type I kerogen), the presence of this hydrotreating catalyst results in the defunctionalisation of the primary oil vapours, so doubling of the yield of GC amenable aliphatic hydrocarbons with a similar increase in the yield of hopane and sterane biomarkers. There is no evidence that the second hydrotreating stage leads to the thermal cracking or structural rearrangement of the generated hydrocarbons, resulting in a biomarker profile that is fully representative of the original kerogen. A similar approach to the hydropyrolysis of oil asphaltenes has also investigated, with a bed of catalyst replacing the catalyst-impregnated silica as the support for the asphaltenes. Due to their greater maturity the second stage of hydropyrolysis does not lead to an increase in the overall aliphatic yield from the asphaltenes, although the yield of polar material approximately doubles, probably as a result of weaker interactions between the functional groups within the asphaltenes and the catalyst than was the case the with the silica.

1 INTRODUCTION

The hydropyrolysis of coals and oil shales at high hydrogen pressures (>10 Mpa), together with the use of a dispersed sulphided molybdenum catalyst has been demonstrated to give rise to high yields (>65%) of dichloromethane (DCM) soluble oil, with conversions of organic matter >85% (Lafferty et al. 1993; Snape et al. 1994; Roberts et al. 1995). Further studies have shown that the use of a slow heating rate (8°C min.$^{-1}$) for the hydropyrolysis of organic matter rich samples generates high yields of hydrocarbon biomarkers such as hopanes and steranes, whilst minimising the alteration to their isomeric distributions (Love et al. 1995; 1996; 1997; Bishop et al. 1998). These characteristics make the bound biomarkers released by hydropyrolysis a powerful tool in tackling key problems in oil exploration, such as for severely biodegraded crude oils, oil field solids and oil-contaminated drill cuttings, where the conventional approach using the free hydrocarbon biomarkers fails (Murray et al. 1998; 1999; Meredith et al. 2003). In addition to petroleum geochemistry, hydropyrolysis has recently been applied to the characterisation of organic matter in meteorites (Sephton et al. 2004), archaeological samples (Craig et al. 2004) and Archaen bitumens (Brooks et al. 2003), and it is currently being employed as a preparative method for the compound-specific carbon isotope analysis of fatty acids (Sephton et al. 2005).

As it is generally more rapid than the closed system pyrolysis methods utilised for generating bound biomarker profiles from geomacromolecules in correlation studies (Rubinstein et al. 1979; Curiale et al. 1983; Cassani & Eglinton 1986; Jones et al. 1987; 1988; Connan 1993), analytical pyrolysis has been widely used for the structural characterisation of asphaltenes, with the pyrolyser coup led directly to a gas chromatograph (py-GC) (e.g. Behar et al. 1984; Behar & Pelet 1984; 1985) a mass spectrometer (py-MS) (e.g. Eglinton et al. 1991) or a py-GC-MS system (e.g. Philip & Gilbert 1985; van Graas 1986). However, due to the low yields and the high degree of cracking undergone, the pyrolysates generated by such methods are, at best, only partially representative of the bulk asphaltene structure (Love et al. 1995). Hydropyrolysis, however provides a methodology for the more comprehensive characterisation of oil asphaltenes (Russell et al. 2004).

Developments to the hydropyrolysis procedure have included the introduction of a second stage in the reactor, composed of a bed of hydrotreating (Ni/Mo or Co/Mo) catalyst below the sample which results in the defunctionalisation of the primary oil vapours (Love et al. 1999). In order to prevent the thermal cracking of these vapours, this lower zone of the reactor tube was held isothermally at a temperature (320°C) below that of the maximum temperature of the pyrolysis zone (520°C). This technique results in the doubling of the yield of GC amenable aliphatic hydrocarbons, with a similar increase in the yield of hopane and sterane biomarkers. However, the requirement for two different heating zones, introduced too many complications into the hydropyrolysis procedure for routine operation (Love et al. 1999). This study describes attempts to replicate these results using a much simpler reactor configuration allowing for much faster, more reproducible analyses.

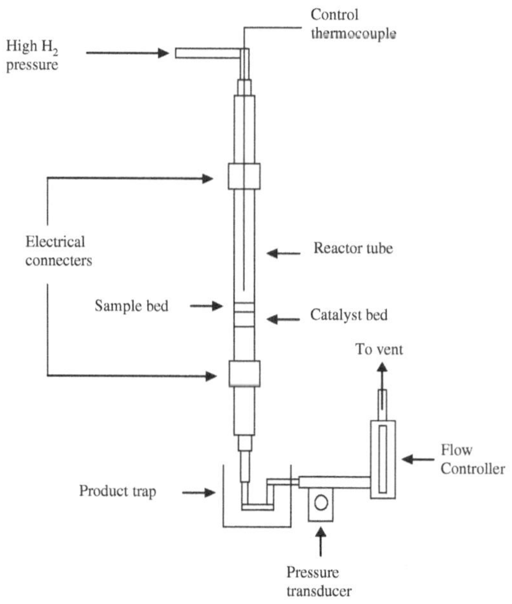

Figure 1. Schematic of the hydropyrolysis apparatus.

2 METHODS

Fixed bed hydropyrolysis of the samples were conducted using the apparatus and procedure that have been described in detail elsewhere (Love et al. 1995). Briefly, the samples were pyrolysed with resistive heating from 50°C to 250°C at 300°C $min.^{-1}$, and then from 250°C to 520°C at 8°C $min.^{-1}$, under a hydrogen pressure of 15 MPa. A hydrogen sweep gas flow of 5 l $min.^{-1}$, measured at ambient temperature and pressure, ensured that the products were quickly removed from the reactor vessel. The products were trapped on dry ice cooled silica as previously described (Meredith et al. 2004).

The samples used to assess potential improvements to the hydropyrolysis procedure were the Göynük oil shale (immature lacustrine Type I kerogen), and the asphaltene fraction isolated from a biodegraded crude oil (Soldado Field, Trinidad). The oil shale was pre-extracted with DCM/methanol (93:7 v/v), washed with dilute HCl (2M) at 50°C for 3 hours and re-extracted with DCM/methanol (93:7 v/v), so any products generated were predominantly released from the kerogen phase. The shale was loaded with Mo catalyst as previously described (Roberts et al. 1995), with an aqueous/methanol solution of ammonium dioxydithiomolybdate $[(NH_4)_2MoO_2S_2]$ to give a nominal molybdenum loading of 1 wt. % sample. In a simplified version of the two-stage procedure described by Love et al. (1999), tests were also performed with the catalyst loaded oil shale resting on a 250 mg bed of the same Mo catalyst, as shown in Figure 1.

The asphaltenes were isolated by four sequential dissolutions of the maltene fraction in an excess of *n*-heptane, with the insoluble asphaltenes removed from the suspension by centrifugation (2500 r.p.m. for 5 min.). The asphaltenes (~50 mg) were then adsorbed to a silica substrate (30 mg/g) and loaded with catalyst as for the oil shale. As an alternative method the solid asphaltenes (~50 mg) were placed directly onto a 250 mg bed of Mo catalyst.

The aliphatic, aromatic and polar (NSO) fractions of the oil shale and asphaltene hydropyrolysates were separated by silica gel adsorption chromatography with successive elutions of *n*-hexane, *n*−hexane/DCM (4:1 v/v) and DCM/methanol (1:1 v/v). The yields of each fraction were determined by evaporation under a stream of dry nitrogen in pre-weighed vials.

Analysis of the recovered hydrocarbons was by gas chromatography-flame ionisation detection (GC-FID), performed on a Carlo Erba HRGC instrument, with separation achieved on a fused silica capillary column (50 m × 0.32 mm i.d.) coated with BPX5 phase (0.25 μm thickness). Helium was employed as the carrier gas, with a temperature programme of 50°C (2 min.) to 300°C (28 min.) at 5°C $min.^{-1}$. Gas chromatography-mass spectrometry (GC-MS) analyses of the fractionated samples (1 μl in DCM) were performed on a Fisons Instruments 8000 gas chromatograph interfaced to a MD 800 mass spectrometer (ionising energy 70 eV, source temperature 280°C). Separation was performed on a column and under conditions similar to that used for the GC analyses.

The selected ions monitored included *m/z* 69 (*n*-alk-1-enes), *m/z* 71 (*n*-alkanes), *m/z* 191 (hopanes) and *m/z* 217 (steranes). The addition of the standards 5β(H)-cholane (Chiron) and squalane (Sigma-Aldrich) to the aliphatic fraction prior to analysis, allowed for the quantification of individual compounds, assuming a response factor of one (Love et al. 1995).

3 RESULTS AND DISCUSSION

3.1 *Göynük oil shale hydropyrolysis*

The introduction of the Mo catalyst bed in the simplified two-stage hydropyrolysis of the Göynük oil shale resulted in a significant increase (179 to 347 mg/g TOC) in the yield of aliphatic components generated, as shown in Figure 2. The amount of aromatic components was also seen to increase (126 to 171 mg/g TOC) while the yield of polar compounds was significantly reduced (502 to 276 mg/g TOC). Therefore, while the total pyrolysis yield did not change, the yield of GC amenable aliphatic and aromatic compounds approximately doubled, with a concurrent decrease in the yield of the polar fraction. Clearly the passage of the primary oil vapours through the catalyst bed resulted in a more complete defunctionalisation of the polar fraction, so releasing more aliphatic and aromatic components than was possible with just the dispersed catalyst. These results mirror those of Love et al. (1999), but by replacing the highly active Ni/Mo hydrotreating catalyst used in that study with a bed of the same sulphided Mo catalyst as that dispersed within the sample allows for it to be placed directly beneath the sample bed. The whole reactor can then be heated under the same temperature regime, thus obviating the need for a separate furnace to be incorporated into the set-up.

While the presence of the additional catalyst bed can be seen to have a major impact on the amount of aliphatic hydrocarbons generated, it appeared to have a relatively minor influence on their distribution. Figure 3 shows the gas chromatographs of the aliphatic hydrocarbons from the conventional single-stage hydropyrolysis (using just the dispersed sulphided Mo catalyst) of Göynük oil shale, together with the product after it had passed through the hydrotreating bed of the Mo catalyst. The overall distributions of *n*-alkanes appear fairly similar, although the additional Mo catalyst bed resulted in the virtual complete hydrogenation of the *n*-alk-1-enes generated during the hydropyrolysis of the oil shale.

Doublets of *n*-alk-1-enes and *n*-alkanes are ubiquitous components of kerogen pyrolysates (Larter & Horsfield 1993), and their hydrogenation results in a more accurate characterisation of the kerogen, and will be beneficial in improving the chromatographic resolution if the samples are to be subject to compound-specific carbon isotope analysis. There is no evidence of thermal cracking during the second hydrotreating stage, as there is no reduction in the abundance or even-over-odd predominance of the longer chained *n*-alkane homologues. Indeed the slightly higher abundance of longer chained *n*-alkanes in the two-stage hydropyrolysates points to more efficient preservation of the generated hydrocarbons, suggesting that the conventional one-stage procedure may result in minor cracking of the longer chained components.

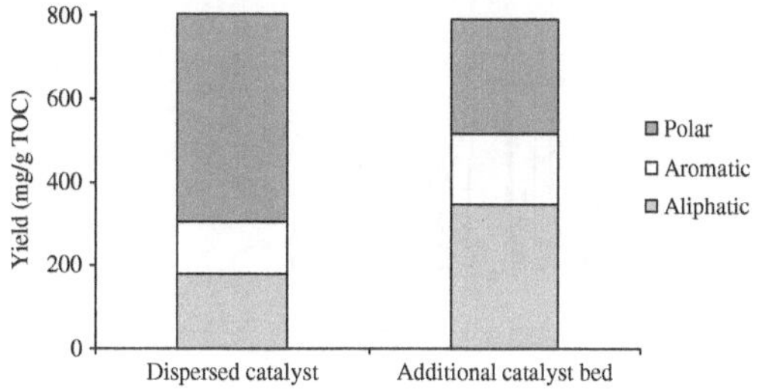

Figure 2. Yield (mg/g shale TOC) of the aliphatic, aromatic and polar fractions from the hydropyrolysis of the Göynük oil shale using just a dispersed Mo catalyst (single-stage), and with an additional Mo catalyst bed (two-stage).

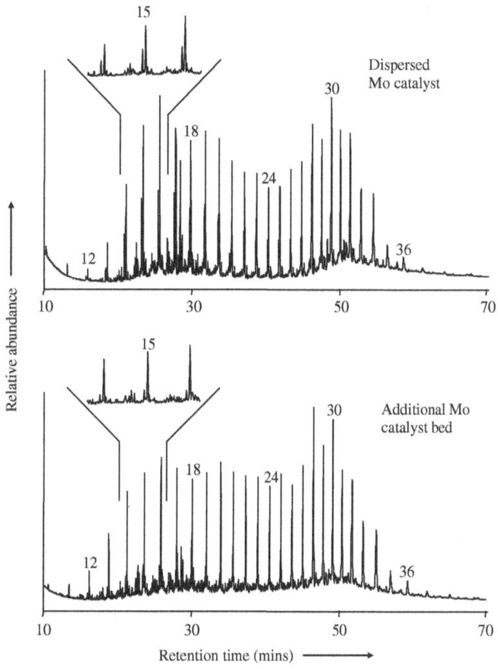

Figure 3. Saturated hydrocarbon gas chromatograms from the single and two-stage hydropyrolysis of the Göynük oil shale.

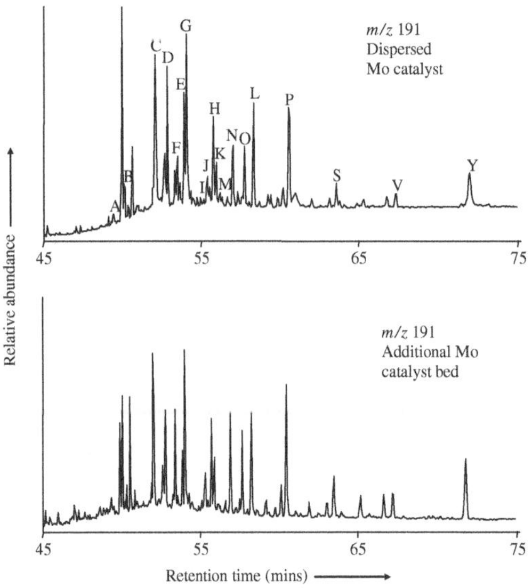

Figure 4. Partial mass chromatograms (m/z 191 – hopanes) from the single and two-stage hydropyrolysis of the Göynük oil shale (for peak assignments see Table 1).

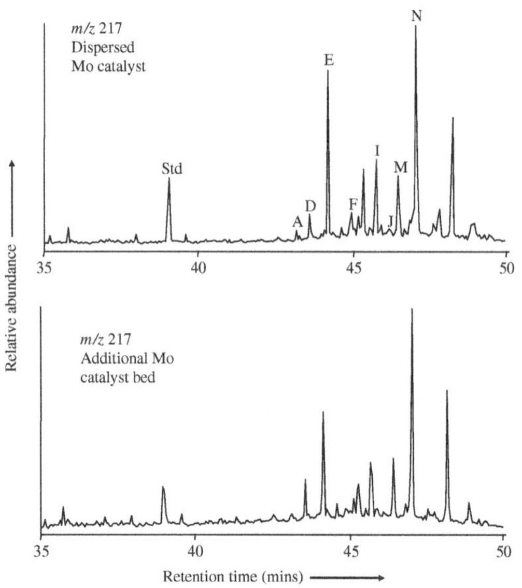

Figure 5. Partial mass chromatograms (m/z 217 – steranes) from the single and two-stage hydropyrolysis of the Göynük oil shale (for peak assignments see Table 1).

The distribution of the hopanes and sterane biomarkers produced from the Göynük oil shale by the two reactor configurations are shown in Figures 4, and 5 respectively, and can be seen to be broadly similar. However, the concentration of individual compounds shown in Table 1, illustrate important differences in the product distribution generated by the two methods. Generally, the concentration of both the hopanes and steranes can be seen to have approximately doubled with the additional catalyst bed present in the reactor (total hopanes from 724 to1430 μg/g TOC; total steranes from 90 to 189 μg/g TOC), which is consistent with the increase in abundance of aliphatic compounds as a whole. However, individual compounds, especially amongst the hopanes are seen to increase in abundance to varying degrees. This may reflect the initial bonding distributions (Love et al. 1999), but may also be due to better preservation of extended hopanes (C_{31}-C_{35}) with the thermodynamically unstable $17\beta(H),21\beta(H)$ stereochemistry, which are most susceptible to isomerisation (Love et al. 1995). Evidence of the enhanced preservation of these most vulnerable compounds can be detected in the much greater than average increase in their abundance in the two-stage product compared to that from the one-stage method, e.g. the $C_{34}\beta\beta$ hopane increases from 13.7 μg/g TOC to 40.0 μg/g, while the $C_{35}\beta\beta$ hopane increased from 27.9 μg/g to 117.1 μg/g. Furthermore, the $C_{29}\beta\beta$ hopane, which might be formed from the cracking of the more extended homologs is seen to decrease in abundance in the two-stage hydropyrolysates (28.4 μg/g TOC) compared to the one-stage (58.5 μg/g TOC). It should be noted that this variation in hopane preservation does not appear to influence maturity dependant biomarker ratios such as those listed in Table 2, which show good consistency for the two methods, indicating that the one-stage method used in previous studies provides an accurate characterisation of such kerogens.

3.2 *Oil asphaltene hydropyrolysis*

Following the successful trial with the oil shale, a similar approach to the hydropyrolysis of oil asphaltenes was attempted. When introduced into the reactor as solid flakes directly on top of a bed of catalyst, it was assumed that the asphaltenes would soften on heating and the flow of gas would then encourage the asphaltenes into intimate contact with the catalyst at the top of the bed. Therefore, the need for a time consuming process of adsorbing each sample onto catalyst-impregnated silica is removed. The remainder of the bed would then act as the second stage in the procedure to fully defunctionalise the primary oil vapours.

The utilisation of a 250 mg catalyst bed as the support for the asphaltenes had a beneficial effect on the total product yield, as shown in Figure 6. Whereas for the Göynük oil shale, hydrotreating of the primary oil vapours resulted in an increase in the aliphatic and aromatic yields, and a concurrent decrease in the polar yield, this approach for the oil asphaltenes resulted in a significant increase in the total product yield (534 to 694 mg/g TOC). However, this was

Table 1. Yield (μg/g initial TOC) of selected hopanes and steranes from the Göynük oil shale and Soldado oil asphaltene hydropyrolysates (n.d. – not detected).

		Göynük oil shale		Soldado oil asphaltene	
	Peak	Dispersed catalyst bed	Additional catalyst bed	Silica supported	Catalyst bed
Ts	A	5.2	15.5	n.d.	n.d.
Tm	B	17.9	57	12.6	25.5
C_{29} $\alpha\beta$ hopane	C	68.8	149.5	12	27.1
C_{29} $\beta\alpha$ hopane	D	59.1	131.8	2.4	5.5
C_{29} $\beta\beta$ hopane	E	58.5	28.4	n.d.	n.d.
C_{30} $\alpha\beta$ hopane	F	34.6	55.8	8	24.3
C_{30} $\beta\alpha$ hopane	G	72.1	194	2.8	7.8
C_{30} $\beta\beta$ hopane	H	52.8	109	n.d.	n.d.
C_{31} $\alpha\beta$ hopane (22S)	I	4.1	9.2	3.5	6.5
C_{31} $\alpha\beta$ hopane (22R)	J	17.4	32.3	2.9	5.2
C_{31} $\beta\alpha$ hopane	K	16	5.8	n.d.	n.d.
C_{31} $\beta\beta$ hopane	L	67.2	112.3	n.d.	n.d.
C_{32} $\alpha\beta$ hopane (22S)	M	5.5	11.8	3.8	6.2
C_{32} $\alpha\beta$ hopane (22R)	N	52.4	89.2	3	4.4
C_{32} $\beta\alpha$ hopane	O	36.1	73.8	n.d.	n.d.
C_{32} $\beta\beta$ hopane	P	96.8	146	n.d.	n.d.
C_{33} $\alpha\beta$ hopane (22S)	Q	n.d.	n.d.	3.8	5.9
C_{33} $\alpha\beta$ hopane (22R)	R	n.d.	n.d.	2.2	3.5
C_{33} $\beta\beta$ hopane	S	17.9	51.4	n.d.	n.d.
C_{34} $\alpha\beta$ hopane (22S)	T	n.d.	n.d.	1.7	2.7
C_{34} $\alpha\beta$ hopane (22R)	U	n.d.	n.d.	1.6	2
C_{34} $\beta\beta$ hopane	V	13.7	40	n.d.	n.d.
C_{35} $\alpha\beta$ hopane (22S)	W	n.d.	n.d.	1.8	2
C_{35} $\alpha\beta$ hopane (22R)	X	n.d.	n.d.	1.7	1.8
C_{35} $\beta\beta$ hopane	Y	27.9	117.1	n.d.	n.d.
C_{27} $\alpha\alpha\alpha$ sterane (20S)	A	2.3	3.8	6.2	11
C_{27} $\alpha\beta\beta$ sterane (20R)	B	n.d.	n.d.	3.3	5.9
C_{27} $\alpha\beta\beta$ sterane (20S)	C	n.d.	n.d.	2.7	4.8
C_{27} $\beta\alpha\alpha$ sterane (20R)	D	5.8	11.6	n.d.	n.d.
C_{27} $\alpha\alpha\alpha$ sterane (20R)	E	20.8	42.9	8.8	14.9
C_{28} $\alpha\alpha\alpha$ sterane (20S)	F	1.4	5.4	1.7	2.9
C_{28} $\alpha\beta\beta$ sterane (20R)	G	n.d.	n.d.	3.8	6.6
C_{28} $\alpha\beta\beta$ sterane (20S)	H	n.d.	n.d.	2.5	4.5
C_{28} $\alpha\alpha\alpha$ sterane (20R)	I	11.3	27.8	7.2	12.7
C_{29} $\alpha\alpha\alpha$ sterane (20S)	J	1.3	2.9	3.6	6.4
C_{29} $\alpha\beta\beta$ sterane (20R)	K	n.d.	n.d.	3.6	6.6
C_{29} $\alpha\beta\beta$ sterane (20S)	L	n.d.	n.d.	1.5	3.1
C_{29} $\beta\alpha\alpha$ sterane (20R)	M	10.7	22.6	n.d.	n.d.
C_{29} $\alpha\alpha\alpha$ sterane (20R)	N	35.9	72.4	7.1	12.9

composed almost entirely of an increase in the abundance of the polar components (157 to 299 mg/g TOC), while there was no significant change in the yield of aliphatic and aromatic constituents of the product. The observed increase in the polar yield may well be due to decreased interaction between the functionalised polar constituents of the asphaltene pyrolysates and the catalyst bed, than was the case with the catalyst-impregnated silica substrate. A suppressed hydropyrolysis yield due to the use of a silica substrate has previously been observed for the hydropyrolytic decarboxylation of fatty acids (Snape et al. 2003), where it was suggested that interactions between the acids and substrate leading to the formation of stable entities (Si-O-C linkages) retarded volatilisation of the acids.

The reduced total product yield for the hydropyrolysis of oil asphaltenes in relation to the Type I kerogen of the Göynük oil shale is to be expected given the much higher degree of maturity of the asphaltenes. This is illustrated by both the specific maturity dependant biomarker ratios listed in Table 2, and by the overall biomarker assemblages which for the oil shale contain immature species such as 17β(H),21β(H) hopanes,

Table 2. Selected aliphatic biomarker maturity dependant parameters for the Göynük oil shale and Soldado oil asphaltene.

	Göynük oil shale		Soldado oil asphaltene	
	Dispersed catalyst bed	Additional catalyst bed	Silica supported	Catalyst bed
C_{31} hopane $\alpha\beta$ S/S + R	0.19	0.22	0.55	0.56
C_{32} hopane $\alpha\beta$ S/S + R	0.10	0.12	0.56	0.58
C_{29} hopane $\beta\alpha/\alpha\beta$	0.86	0.88	0.20	0.20
C_{29} sterane $\alpha\alpha\alpha$ S/S + R	0.03	0.04	0.33	0.33
C_{29} sterane $\alpha\beta\beta/\alpha\alpha\alpha + \alpha\beta\beta$	n.d.	n.d.	0.32	0.34

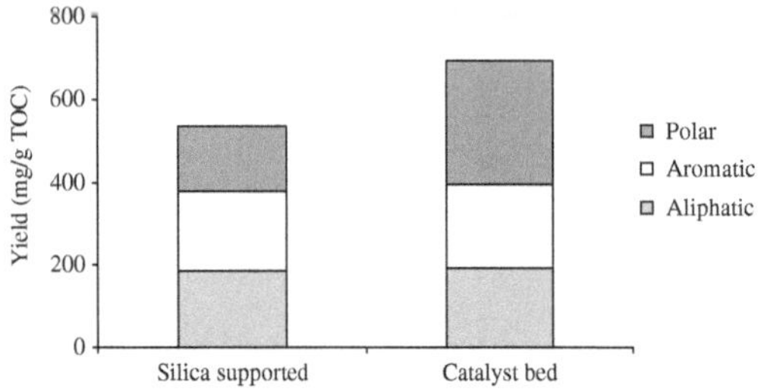

Figure 6. Yield (mg/g shale TOC) of the aliphatic, aromatic and polar fractions from the hydropyrolysis of the Soldado oil asphaltene using the silica supported and catalyst bed procedures.

and 5β(H),14α(H),21α(H) steranes which do not occur in produced oils. Such a decrease in hydropyrolysis yield with an increasing degree of maturity has previously been reported for a rank series of coals (Love et al. 1996), and is due to the lower abundance of labile C-O and C-S bonds in samples of increasing maturity.

The distributions of aliphatic compounds in the hydropyrolysates generated by the two methods are very similar. The shape of the *n*-alkane envelope, shown in Figure 7 for the two samples are virtually identical, and the distribution of hopane and sterane biomarkers are also very similar in terms of both their appearance (Figure 8 – hopanes, and Figure 9 – steranes), and the maturity dependant biomarkers listed in Table 2. Therefore, as with the Göynük oil shale it appears that the presence of the catalyst bed in the reactor does not result in cracking or isomerisation of the product, although as the asphaltenes are so much more mature, the hydropyrolysates will not be as susceptible to such rearrangements as the oil shale.

While the distribution of the hopane and sterane biomarkers are the same for the two methods, it can be seen in Table 1 that the concentration of individual compounds generated by the two-stage procedure is approximately double that for the one-stage. Surprisingly this increase in yield is not reflected by the total *n*-alkane (*n*-C_{12} to *n*-C_{36}) concentration in the two products (20.8 mg/g TOC for the one-stage and

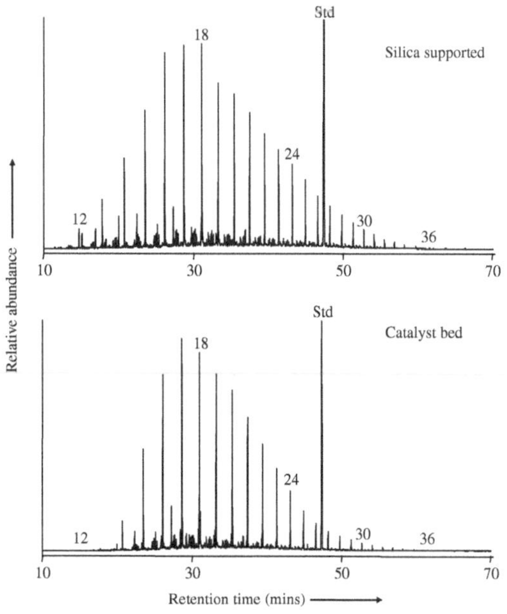

Figure 7. Saturated hydrocarbon gas chromatograms from the silica supported and catalyst bed hydropyrolysis of the Soldado oil asphaltene.

21.9 mg/g TOC for the two-stage procedure), or by the total abundance of aliphatic components as a whole. The reason for the apparently contradictory increase in the hopane and sterane concentration with respect to that of the *n*-alkanes is not presently understood but implies that they are bound into the asphaltene matrix through weaker bonds than the *n*-alkyl moieties.

4 CONCLUSIONS

For the hydropyrolysis of an immature oil shale the addition of a hydrotreating Mo catalyst bed operating under the same temperature regime as the catalyst-impregnated sample offered a practical method to

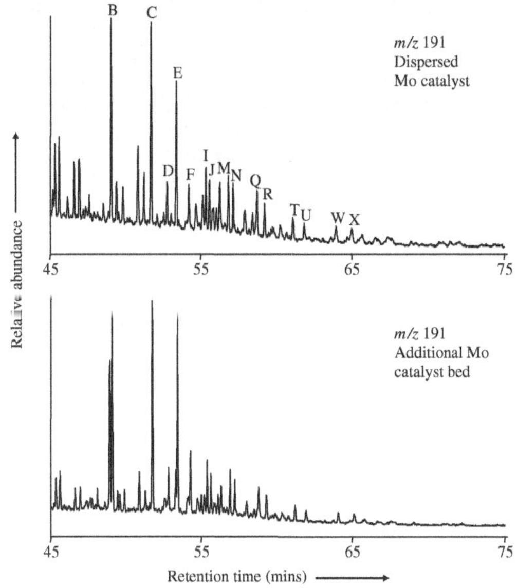

Figure 8. Saturated hydrocarbon gas chromatograms from the silica supported and catalyst bed hydropyrolysis of the Soldado oil asphaltene.

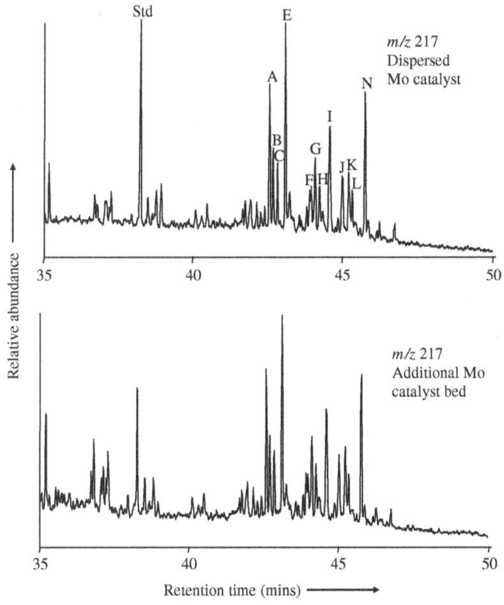

Figure 9. Partial mass chromatograms (*m/z* 217 – steranes) from the silica supported and catalyst bed hydropyrolysis of the Soldado oil asphaltene (for peak assignments see Table 1).

maximise the yield of aliphatic biomarkers without the need for a relatively complicated reconfiguration of the system. Overall the yield of GC amenable aliphatic and aromatic compounds approximately doubled, with a concurrent decrease in the yield of the polar fraction. There was no evidence of thermal cracking during the second hydrotreating stage, with increased preservation of extended hopanes with the thermodynamically unstable 17β(H),21β(H) stereochemistry.

A similar approach for the hydropyrolysis of oil asphaltenes resulted in an increased total product yield, although this is made up almost entirely of polar compounds. However, while the *n*-alkane yield is similar to that obtained with the conventional silica supported approach the biomarker yields are notably higher, and there are also significant advantages in terms of a simplified sample preparation procedure.

REFERENCES

Behar, F. & Pelet, R. 1984. Characterisation of asphaltenes by pyrolysis and chromatography. Journal *of Analytical and Applied Pyrolysis* 7: 121–136.

Behar, F. & Pelet, R. 1985. Pyrolysis-gas chromatography applied to organic geochemistry. Structural similarities between kerogens and asphaltenes from related extracts and oils. *Journal of Analytical and Applied Pyrolysis* 8: 173–187.

Behar, F., Pelet, R. & Roucache, J. 1984. Geochemistry of asphaltenes. *Organic Geochemistry* 6: 587–595.

Bishop, A.N., Love, G.D., Snape, C.E. & Farrimond, P. 1998. Release of kerogen-bound hopanoids by hydropyrolysis. *Organic Geochemistry* 29: 989–1001.

Brooks, J.J., Love, G.D., Snape, C.E., Logan, G.A., Summons, R.E. & Buick, R. 2003. Release of bound aromatic hydrocarbons from late Archean and Mesoproterozoic kerogens via hydropyrolysis. *Geochimica et Cosmochimica Acta* 67: 1521–1530.

Cassani, F. & Eglinton, G. 1986. Organic geochemistry of Venezuelan extra heavy oils, 1. Pyrolysis of asphaltenes: a technique for the correlation and maturity evaluation of crude oils. *Chemical Geology* 56: 167–183.

Connan, J. 1993. Origin of severely biodegraded oils: a new approach using biomarker pattern of asphaltene pyrolysates. In M.L. Bordenave (ed), *Applied Petroleum Geochemistry*: 456–463. Paris: Editons Technip.

Craig, O.E., Love, G.D., Isaksson, S., Taylor, G. & Snape, C.E. 2004. Stable carbon isotopic characterisation of free and bound lipid constituents of archaeological ceramic vessels released by solvent extraction, alkaline hydrolysis and catalytic hydropyrolysis. *Journal of Analytical and Applied Pyrolysis* 71: 613–634.

Curiale, J.A., Harrison, W.E. & Smith G. 1983. Sterane distributions of solid bitumen pyrolysates. Changes with biodegradation of crude oils in the Ouachita, Oklahoma. *Geochimica et Cosmochimica Acta* 47: 517–523.

Eglinton, T.I., Larter, S.R. & Boon, J.J. 1991. Characterisation of kerogens, coals and asphaltenes by quantitative pyrolysis-mass spectrometry. *Journal of Analytical and Applied Pyrolysis* 20: 25–45.

Jones, D.M., Douglas, A.G. & Connan, J. 1987. Hydrocarbon distributions in crude oil asphaltene pyrolysates. 1. Aliphatic compounds. *Energy & Fuels* 1: 468–476.

Jones, D.M., Douglas, A.G. & Connan, J. 1988. Hydrous pyrolysis of asphaltenes and polar fractions of biodegraded oils. *Organic Geochemistry* 13: 981–993.

Lafferty, C.J., Mitchell, S.C., Garcia, R. & Snape, C.E. 1993. Investigation of organic sulphur forms in coals by high temperature-programmed reduction. *Fuel* 72: 367–371.

Larter, S.R. & Horsfield, B. 1993. Determination of structural components of kerogen by the use of analytical pyrolysis methods. In: M.H. Engel & S.A. Macko (eds), *Organic Geochemistry: 271–287.* New York: Plenum Press.

Love, G.D., Snape, C.E., Carr, A.D. & Houghton, R.C. 1995. Release of covalently-bound alkane biomarkers in high yields from kerogen via catalytic hydropyrolysis. *Organic Geochemistry* 23: 981–986.

Love, G.D., Snape, C.E., Carr, A.D. & Houghton, R.C. 1996. Changes in molecular biomarker and bulk carbon skeletal parameters of vitrinite concentrates as a function of rank. *Energy & Fuels* 10: 149–157.

Love, G.D., McAulay, A., Snape, C.E. & Bishop, A.N. 1997. Effect of process variables in catalytic hydropyrolysis on the release of covalently-bound aliphatic hydrocarbons from sedimentary organic matter. *Energy & Fuels* 11: 522–531.

Love, G.D., Murray, I.P. & Snape, C.E. 1999. Maximising the yields of covalently-bound biomarkers from sedimentary organic matter. In *Abstracts of the 19th International Meeting on Organic Geochemistry, Istanbul*: 135–136.

Meredith, W., Russell, C.A., Snape, C.E., Love, G.D., Clarke, E., Moffatt, B., Carr, A.D. & Scotchman, I.C. 2003. Potential of bound biomarkers released via hydropyrolysis for the characterisation of pyrobitumens and tar mats. In *Abstracts of the 21st International Meeting on Organic Geochemistry, Krakow*: 305–306.

Meredith, W., Russell, C.A., Snape, C.E., Fabbri, D., Vane, C.H. & Love, G.D. 2004. The thermal desorption of hydropyrolysis oils from silica to facilitate rapid screening by GC-MS. *Organic Geochemistry* 35: 73–89.

Murray, I.P., Love, G.D., Snape, C.E. & Bailey, N.J.L. 1998. Comparison of covalently-bound aliphatic biomarkers released via hydropyrolysis with their solvent-extractable counterparts for a suite of Kimmeridge clays. *Organic Geochemistry* 29: 1487–1505.

Murray, I.P., Snape, C.E., Love, G.D. & Bailey, N.J.L. 1999. Hydropyrolysis of heavy oils for source correlation studies. In *Abstracts of the 19th International Meeting on Organic Geochemistry, Istanbul*: 341–342.

Philip, R.P. & Gilbert, T.D. 1985. Source rock and asphaltene biomarker characterisation by pyrolysis-gas chromatography-mass spectrometry-multiple ion detection. *Geochimica et Cosmochimica Acta* 49: 1421–1432.

Roberts, M.J., Snape, C.E. & Mitchell, S.C. 1995. Hydropyrolysis: fundamentals. Two-stage processing and PDU operation. In C.E. Snape (ed.), *Geochemistry, Characterisation and Conversion of Oil Shales*: 277–295. Dordrecht: Kluwer.

Rubinstein, I., Spyckerelle, C. & Strausz, O.P. 1979. Pyrolysis of asphaltenes: a source of geochemical information. *Geochimica et Cosmochimica Acta* 43: 1–6.

Russell, C.A., Snape, C.E., Meredith, W., Love, G.D., Clarke E. & Moffatt, B. 2004. The potential of bound biomarker profiles released via catalytic hydropyrolysis to reconstruct basin charging history for oils. *Organic Geochemistry* 35: 1441–1459.

Sephton, M.A., Love, G.D., Watson, J.S., Verchovsky, A.B., Wright, I.P. Snape, C.E. & Gilmour, I. 2004. Hydropyrolysis of insoluble carbonaceous matter in the Murchison meteorite: New insights into its macromolecular structure. *Geochimica et Cosmochimica Acta* 68: 1385–1393.

Sephton, M.A., Meredith, W., Sun, C. & Snape, C.E. 2005. Hydropyrolysis as a preparative method for the compound-specific carbon isotope analysis of fatty acids. *Rapid Communication in Mass Spectrometry* 19: 323–325.

Snape, C.E., Lafferty, C.J., Eglinton, G., Robinson, N. & Collier, R. 1994. The potential for hydropyrolysis as a route for coal liquefaction. *International Journal of Energy Research* 18: 233–242.

Snape, C.E., Love, G.D. & Meredith, W. 2003. Behaviour of model substrates in catalytic hydropyrolysis to investigate preservation of biomarkers released from kerogens and asphaltenes. *Abstracts of papers of the American Chemical Society* 226.

van Graas, G. 1986. Biomarker distributions in asphaltenes and kerogens analysed by flash pyrolysis gas chromatography-mass spectrometry. *Organic Geochemistry* 10: 1127–1135.

Petroleum Geochemistry and Exploration in the Afro-Asian Region – Digang Liang, Darui Wang & Zhenxi Li (eds)
© 2008 Taylor & Francis Group, London, ISBN 978-0-415-44083-7

Comparison of hydrocarbon generation simulation of source rocks in close and open system: A case of oil shale from Lower Jurassic in Qaidam Basin, China

Jianzhong Qin & Zhiming Li
Wuxi Research Institute of Experimental Geology, RIPED, SINOPEC, Wuxi, Jiangsu China

Baoquan Liu
Research Institute of Petroleum Exploration and Development, Huabei Oil-field, CNPC, Renqiu, Hebei, China

ABSTRACT: Hydrocarbon generation simulations of an oil-shale sample from Lower Jurassic in Qaidam Basin, China were carried out in both close and open systems in order to compare the results of two different simulating modes which may make us better understand the process of hydrocarbon generation of source rocks in natural geological conditions. The results in close and open systems show that (1) the yields of hydrocarbon gases, CO_2 and H_2 in both close and open system are very close when the simulation temperature is lower than 350°C, but yields of the total gases and hydrocarbon gases in open system are lower than those in close system when simulation temperature is equal to or higher than 350°C, which indicates liquid hydrocarbons begin to crack into gases when the simulation temperature higher than 350°C, whereas the yield H_2 in open system is higher than that in close system, which may be due to the participation of hydrogen in second cracking of liquid hydrocarbons in close system; (2) the yields of expulsion oil and the total generation oil in open system are both higher than those of close system, whereas the yields of residual oil in open and close systems are very close, but the simulation temperature of the highest yield is 300°C and 350°C, respectively; (3) the total yields of hydrocarbon generation in close and open system are 473.6 kg/t_c and 518.0 kg/t_c, respectively, which suggests the oil-shale unit of the Lower Jurassic in Qaidam Basin of China has great hydrocarbon generation potentials; (4) therefore, it may be more reasonable to evaluate hydrocarbon generation potentials of a basin by using the average value of hydrocarbon generation simulation results of source rocks in both close and open systems.

1 INTRODUCTION

Thermal simulation experiment of source rocks is based on kerogen thermal degradation theory on hydrocarbon generation (Zeng et al.,1990). Many thermal simulation experiments indicate that the results of thermal simulation experiments for source rocks can be consistent with the evolutional results of source rocks in natural geological conditions (e.g. Jia and Zhou, 1981; Jia et al., 1987; Liu et al., 1990; Guo et al., 1991) and the thermal simulation experiment of hydropyrolysis using high-pressure autoclave is a relatively effective way nowadays (Weres et al., 1988; Stalker et al., 1998; Barth et al., 1989; Sofer, 1987). Thermal simulation experiments can be divided into close and open system modes according to different heating ways to a simulation sample. The original sample is used at every simulation temperature stage and the generated oil and gas are not separated from the reaction system in close system, whereas the same original sample is used all along and heated continuously at different simulation temperature stage and the pyrolysis gases and condensed oil are removed off at every simulation temperature stage in open system. Compared with open system simulation experiment, the close system simulation experiment is relatively simple and convenient, rapid and easy, and the final simulation results are consistent with the thermal evolutional trend on the whole (Zhao et al., 2000; Wang et al., 2004). Therefore, Hydropyrolysis hydrocarbon generation simulation carried out in close system using high-pressure autoclave is one of better methods to model the process of oil and gas formation and to study the quantity of oil and gas generated and expelled from source rocks in short time. However, the generated hydrocarbons in close simulation experiment system are not separated from the reaction system, which is different from the conditions of hydrocarbon generation in natural geological setting because the most hydrocarbons derived from source rocks in hydrocarbon-bearing basins are far away from the original thermal evolution system due to migration and they do not take part in further thermal evolution process of source rocks, but some of hydrocarbons remaining in source rock or not far away from the original thermal evolution system when they are expelled from source rocks may continue to take part in the further thermal evolution together with source rocks. This means the state of hydrocarbon generation for source rocks under natural geological conditions is more similar to open system or between close and open system. Therefore, we choose an oil-shale sample from Lower Jurassic in Qaidam Basin of China as a simulation sample for hydrocarbons generation simulation experiments in close and open systems in order to compare the results of two different simulating modes which may make us better understand the process of hydrocarbons generation of source rocks in natural geological conditions.

2 CHARACTERISTICS OF SIMULATION SAMPLE

The simulated oil shale is sampled from the upper part of the Lower Jurassic of Dameigou profile in north margin of Qaidam Basin, China. The results of Rock-Eval for simulation sample are as follows: $T_{max} = 433°C$, $S_1 + S_2 = 227.9$ mg/g, IH = 578 mg/g, Cp/TOC = 48.88, TOC = 38.7%, II_1 type of organic matter. This suggests the simulation sample belongs to the low mature and hydrocarbon-rich source rock.

3 EXPERIMENTAL EQUIPMENT, CONDITION AND METHOD

3.1 *Experimental equipments*

The equipments of simulation experiment include autoclave system, temperature-controlling system and collection and separation system of pyrolysis gas and condensed oil (or light hydrocarbons). The GCF-0.25L reaction autoclave is made by the Dalian automation equipment factory. The designed pressure of autoclave is 19.6 Mpa. The temperature-controlling system is XMT-131 digital modulation instrument. The collection and separation system of pyrolysis gas and condensed oil (or light hydrocarbons) includes the acceptor for liquids cooled by liquid nitrogen, the helical condensation tube cooled by ice water, and gas collection measure tube with scale. A sketch of experimental equipments for hydrocarbon generation simulation is shown in Fig. 1.

3.2 *Experimental conditions*

The experimental conditions of hydrocarbon generation simulation such as the weight of the simulated sample, the grain size of the sample, the weight of added water and heating time had been systemically

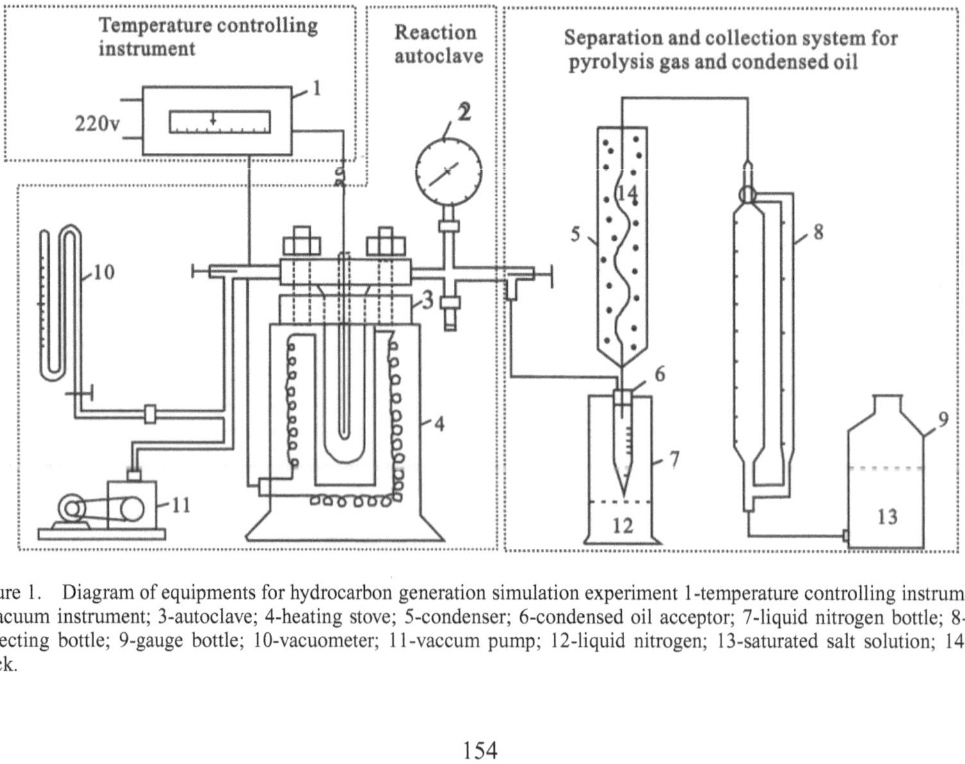

Figure 1. Diagram of equipments for hydrocarbon generation simulation experiment 1-temperature controlling instrument; 2-vacuum instrument; 3-autoclave; 4-heating stove; 5-condenser; 6-condensed oil acceptor; 7-liquid nitrogen bottle; 8-gas collecting bottle; 9-gauge bottle; 10-vacuometer; 11-vaccum pump; 12-liquid nitrogen; 13-saturated salt solution; 14-ice block.

studied (Zhao et al., 2000). The experimental conditions used in this study are as follows. The weight of the simulated sample is 15~30 g for close system and 150 g for open system. The grain size of sample is 2.5~10 mm. The weight of added water is 5~10% of the sample weight. The simulation temperature for close system includes nine stages, they are 200°C, 250°C, 275°C, 300°C, 325°C, 350°C, 400°C, 450°C and 500°C, and for open system includes seven stages, they are 200°C, 250°C, 300°C, 350°C, 400°C, 450°C and 500°C. The simulation time is 24 hours at every simulation temperature stage for both close and open system.

3.3 *Experimental method*

The simulation sample and de-ion water were put into the reaction autoclave. After the autoclave was airproofed, nitrogen with 4~6 Mpa was filled up. In seepage test, nitrogen was pumped out into vacuum and refilled again. As the above step was repeated 3~5 times and the reaction autoclave was finally pumped into vacuum. Then the heating experiment started until the set temperature was attained and kept for 24 hours. The gases were released when the temperature of the autoclave decreased to 200°C. The pyrolysis gases passed through the acceptor cooled by liquid nitrogen for liquid collection and then through the helical condensation tube cooled by ice water, and finally entered into the gas collection tube and the volumes were determined. The collected pyrolysis gases were prepared for analyses of compositions and other items. There were water and condensed oils in the liquid acceptor. Dichloromethane was added to the liquid acceptor and then separated, and the water was extracted three times with dichloromethane in order to avoid the loss of light hydrocarbons in condensation. The condensate in dichloromethane was quantified by chromatography and gravimetric analysis. The oil substances accreting on the surface of simulation sample, cover and inter-wall of the autoclave were washed with dichloromethane and the expulsion light hydrocarbons were gained after volatilization of the dichloromethane. The TOC, Rock-Eval and reflectivity of the residual simulated sample were determined after being weighed. The bituminous A or named residual oil was finally extracted with chloroform.

4 SIMULATION RESULTS

4.1 *Results of close system*

The original sample was heated individually at every simulation temperature stage and the generated oils and gases were not separated from the reaction system in the process of close hydrocarbon generation simulation experiment. The oil and gas yields of oil shale hydrocarbon generation simulation experiment in close system are shown in Table 1 and Fig. 2.

Obviously, the total-gas yield and gaseous hydrocarbon yield all increased with the simulation temperature increasing, and the carbon dioxide (CO_2) yield, hydrogen gas (H_2) yield increased with the simulation temperature increasing as well. The total gas yield was less than 126.8 m^3/t_c, the hydrocarbon gases yield less than 48.4 m^3/t_c, the CO_2 yield less than 31.9 m^3/t_c and the H_2 yield less than 39.1 m^3/t_c when simulation temperature was lower than 350°C. The total gas yield and hydrocarbon gas increased sharply after the simulation temperature was higher than 350°C. The highest yields of total gases and hydrocarbon gases were 617.2 m^3/t_c and 474.3 m^3/t_c, respectively at 500°C. This suggests the simulation sample has entered into a stage that the liquid hydrocarbon begins to crack and generate gases after the simulation temperature is higher than 350°C.

As shown in Table 1 and Fig. 2, the yield of oil generation for the simulated sample is very high. The oil generation yield increases slowly before the simulation temperature is lower than 275°C, whereas it increases sharply when the simulation temperature is higher than 275°C. The total-oil yield, expulsion oil yield, condensate and residual oil yields all attain the maximum at 350°C, the total oil generation yield is 498.4 kg/t_c, among which the expulsion oil yield and the residual oil yield are 227.0 kg/t_c and 271.4 kg/t_c, respectively. The ratio of oil expulsion is 45.5%. Although the efficiency of oil expulsion is relatively low, this indicates the oil shale unit of Lower Jurassic in Qaidam Basin has great hydrocarbon generation potential. The oil generation yield decreases sharply after simulation temperature is higher than 350°C and the residual oil yield sharply decreases to 3.7 kg/t_c at 400°C. This suggests the sample has entered into wet gas formation stage at 400°C, which corresponds to the sharp increases of the hydrocarbon-gas yield of after simulation temperature is higher than 350°C.

The peak of hydrocarbon generation is also at 350°C and the yield of total hydrocarbon generation is 566.3 kg/t_c, but it decreases obviously after 350°C and to 473.6 kg/t_c at 500°C. This may be related to the reaction products (pyrolysis oils), which are not separated from reaction system in the process of close simulation experiment. When the simulation temperature is higher than 350°C, the polar big molecule compounds such as non-hydrocarbons and bitumen except for those cracked into some light hydrocarbons, they can also further polymerise to form more bigger molecules which are insoluble in solvent, and become into kerogen with much higher thermal maturity. Thus the total hydrocarbon generation yield decreases, whereas the content of TOC of the reaction residue increases. This is the shortcoming of hydrocarbon generation simulation experiment in close system.

Table 1. Oil and gas yields of oil shale hydrocarbon generation simulation experiments with close and open systems.

Simulation mode	T(°C)	Total gases (m^3/t_c)	CO_2 (m^3/t_c)	H_2 (m^3/t_c)	HC gases (m^3/t_c)	HC gases (kg/t_c)	Condensed oil (kg/t_c)	Light oil (kg/t_c)	Expulsion oil (kg/t_c)	Residual oil (kg/t_c)	Total oils (kg/t_c)	Total HC (kg/t_c)
yields of HC	sample	0.0	0.0	0.0	0.0	0.0	0.0	0.0	0.0	25.5	25.5	25.5
for	200	30.1	23.8	0.7	0.5	0.9	2.1	5.0	7.2	25.6	32.8	33.7
individual	250	46.6	23.2	2.0	2.3	4.8	5.4	12.7	18.1	36.3	54.4	59.2
simulation	275	65.3	41.9	7.1	8.2	11.9	19.6	42.8	62.4	53.0	115.5	127.4
temperature	300	98.0	57.0	12.8	19.2	33.5	49.1	78.8	127.9	196.8	324.7	358.1
with close	325	119.6	54.5	22.2	33.2	56.3	68.3	137.7	206.1	189.0	395.1	451.4
system	350	126.8	31.9	39.1	48.4	67.9	118.9	108.1	227.0	271.4	498.4	566.3
	400	312.8	58.0	42.0	206.4	308.9	64.8	62.5	127.3	3.7	131.1	440.0
	450	418.5	54.7	19.7	343.3	355.1	66.4	53.5	119.9	1.2	121.1	476.2
	500	617.2	67.4	71.5	474.3	428.0	18.3	26.3	44.6	1.0	45.5	473.6
yields of HC	200	24.2	21.6	0.14	0.14	0.12	0.8	1.6	2.4	40.1	42.5	42.6
at each	250	29.1	21.3	1.4	4.1	5.6	12.7	23.1	35.8	81.2	117.1	122.7
simulation	300	43.8	10.3	11.5	18.7	23.3	43.3	54.3	97.6	272	369.7	393.0
temperature	350	98.7	4.9	30.4	61.3	79.0	57.2	106.3	163.5	100.4	263.8	342.8
stage with	400	89.9	2.5	18.9	68.1	59.3	0.4	11.3	11.7	0.9	12.6	71.9
open system	450	42.6	2.0	17.7	21.6	15.6	0.6	1.5	2.1	0.1	2.2	17.8
	500	73.6	7.7	35.1	26.6	19.1	0.2	2.4	2.6	0.1	2.7	21.8
accumulative	200	24.2	21.61	0.14	0.14	0.12	0.81	1.63	2.44	40.07	42.51	42.6
yields of HC	250	53.3	42.91	1.58	4.21	5.76	13.52	24.76	38.28	81.22	119.49	125.3
at each	300	97.1	53.23	13.12	22.89	29.09	56.81	79.1	135.92	272.02	407.93	473.0
simulation	350	195.8	58.13	43.5	84.2	108.09	114.01	185.38	299.39	100.37	399.76	507.9
temperature	400	285.6	60.64	62.39	152.29	167.37	114.4	196.68	311.08	0.94	312.02	479.4
with open	450	328.2	62.6	80.11	173.86	182.99	115.02	198.17	313.19	0.07	313.26	496.3
system	500	401.8	70.31	115.2	200.43	202.08	115.25	200.55	315.8	0.06	315.87	518.0

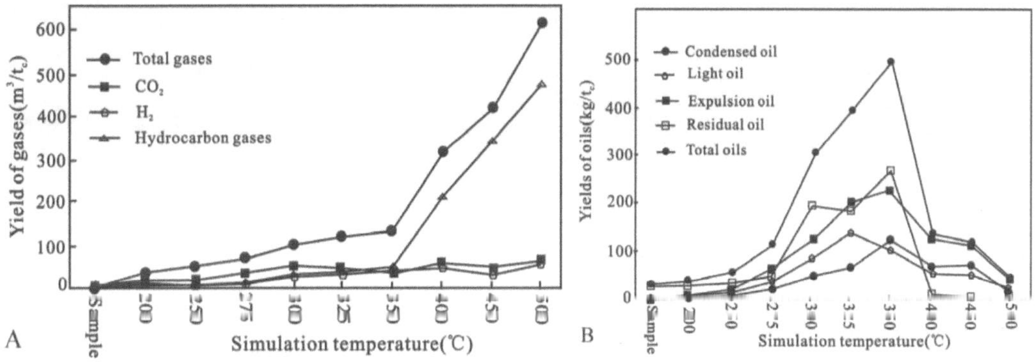

Figure 2. Yields of gases (A) and oils (B) of hydrocarbon generation simulation experiment of the oil shale in close system.

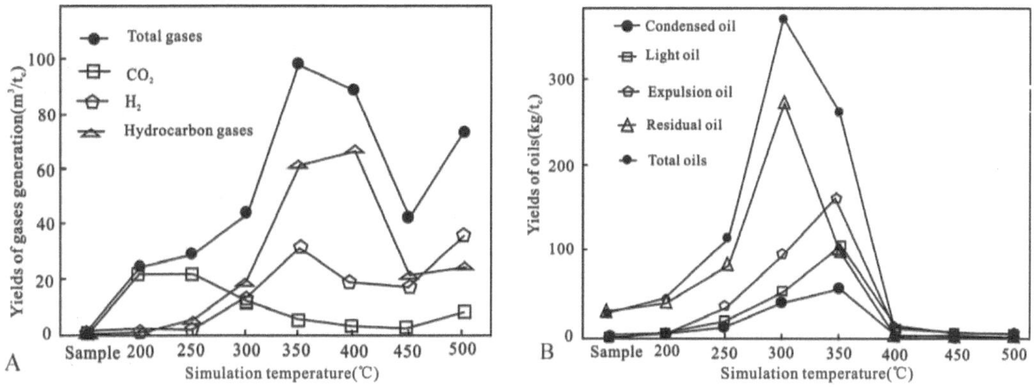

Figure 3. Yields of gases (A) and oils (B) of hydrocarbon generation simulation experiment for oil shale in open system.

4.2 *Results of open system*

For the sample with high abundance of organic matter, high oil generation yield and high content of non-hydrocarbon bitumen in oil family, a part of the generated liquid oil always polymerizes into insoluble kerogen in the process of close system. The shortcoming in close system for such a sample could be compensated for by open system (Wang et al., 2004) because the same original sample was used all along and heated continuously at different simulation temperature stages and the pyrolysis gases and condensate were removed off at every simulation temperature stage in the process of open system simulation experiment.

The oil and gas yields of oil shale hydrocarbon generation simulation at each temperature stage in open system are shown in Table 1 and in Fig. 3 and the accumulative oil and gas yields of oil shale hydrocarbon generation simulation at each simulation temperature stage in open system are shown in Table 1 and Fig. 4. As shown in Table 1 and Fig. 3(A), the total gas yield, gaseous hydrocarbon yield and H_2 yield of open simulation experiment all increased gradually with the simulation temperature increasing at 200~300°C temperature stages, and the yields reached the maximum at 300~400°C temperature stages, then the yields decreased obviously at 450~500°C. But CO_2 yield was relatively higher at 200°C and 250°C, and decreased obviously at 300~500°C. This may be related to decarbonxylation causing higher CO_2 yield at early stage when the maturity of simulation sample is not high.

As shown in Table 1 and Fig. 3(B), the maximum oil generation yield and maximum residual oil yield in open system are 369.7 kg/t_c and 272.0 kg/t_c respectively at 300°C. The maximum oil generation apex of condensate and light oils is at 350°C, the yields of condensate and light oils are 57.2 kg/t_c and 106.3 kg/t_c respectively. This shows the oil shale unit of Lower Jurassic in Qaidam Basin of China can generate and expel oil largely. The oil generation yields decrease sharply when the simulation temperature is higher than 350°C, the total oil generation yield and residual oil yield decrease to 12.6 kg/t_c and 0.9 kg/t_c respectively at 400°C.

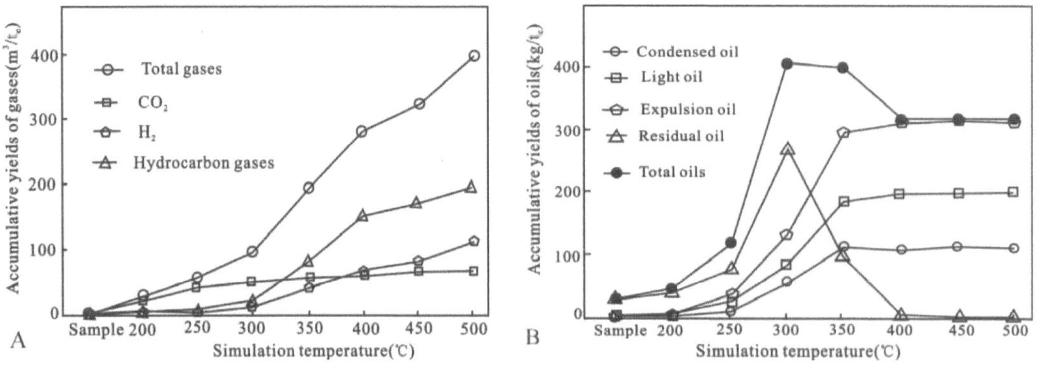

Figure 4. Accumulative yields of gases (A) and oils (B) of hydrocarbon generation simulation.

The accumulative total-gas generation yield, H_2 yield, CO_2 yield and hydrocarbon gas yield all increase with the simulation temperature increasing (see Table 1 and Fig. 4 (A)). The highest accumulative total-gas generation yield is 401.8 m^3/t_c, the highest accumulative hydrocarbon gas yield is 200.4 m^3/t_c at 500°C. They are all lower than the results of close system because the expulsion oil in open system is separated from reaction system, which causes the quantities of hydrocarbon gas and total gases generated from oil cracking are lower. The accumulative CO_2 yield is 21.6 m^3/t_c at 200°C and 70.3 m^3/t_c at 500°C. The accumulative H_2 yield is 115.2 m^3/t_c at 500°C, which are higher than those of close system. This may be due to the participation of hydrogen in second cracking of liquid hydrocarbons in close system.

The accumulative total oil generation yield and total hydrocarbon generation yield in open system increase rapidly with increasing simulation temperature before 300°C(see Table 1 and Fig. 4 (B)). The total oil generation yield and residual oil yield reach the maximum at 300°C, they are 407.9 kg/t_c and 272.0 kg/t_c respectively. But the residual oil yield decreases sharply when the simulation temperature is higher than 300°C. The residual oil yield is only 0.94 kg/t_c at 400°C. This suggests that oil generation reaction of decomposition has finished. The total oil generation yield decreases slightly when the temperature higher than 300°C, whereas the total hydrocarbon yield increases gradually and the accumulative maximum yield is 518 kg/t_c at 500°C.

4.3 *Comparison of oil and gas yields in close and open systems*

The comparison results of hydrocarbon generation simulation in close and open system for oil shale from Lower Jurassic in Qaidam Basin, China are shown in Fig. 5. To study Fig. 5 and the results in Table 1 comprehensively, we can see that the yields of total gases, hydrocarbon gases, CO_2 and H_2 are very close in close and open system when the simulation temperature is lower than 350°C ($R_O = 1.2\%$), but the yields of total gases and hydrocarbon gases in open system are obviously lower than those in close system and the yields of $CO_2$and H_2 in open system are slightly higher than those in close system when the simulation temperature is higher than or equal to 350°C. The yields of expulsion oil and total generation oil in open system are both higher than those in close system, whereas the yields of residual oil in open and close system are very close, but the highest yield occurs at 300°C and 350°C respectively. The yield of total hydrocarbon generation in open system is slightly higher than that in close system on the whole. Therefore, it may be more reasonable to evaluate hydrocarbon generation potentials of a basin by using the average value of hydrocarbon generation simulation results of source rocks in both close and open system

5 CONCLUSION

Hydropyrolysis hydrocarbon generation simulations of an oil shale sampled from Lower Jurassic of the Dameigou profile in the north margin of Qaidam Basin, China were carried out in close and open system. According to the hydrocarbon generation simulating results, following conclusions are obtained.

The yields of hydrocarbon gases, CO_2 and H_2 in both close and open system are very close when the simulation temperature is lower than 350°C, but yields of the total gases and hydrocarbon gases in open system are lower than those in close system when simulation temperature is equal to or higher than 350°C, which indicates liquid hydrocarbons begin to crack into gases when the simulation temperature is higher than 350°C, whereas the yield H_2 in open system is higher than that in close system, which may be due to the participation of hydrogen in second cracking of liquid hydrocarbons in close system.

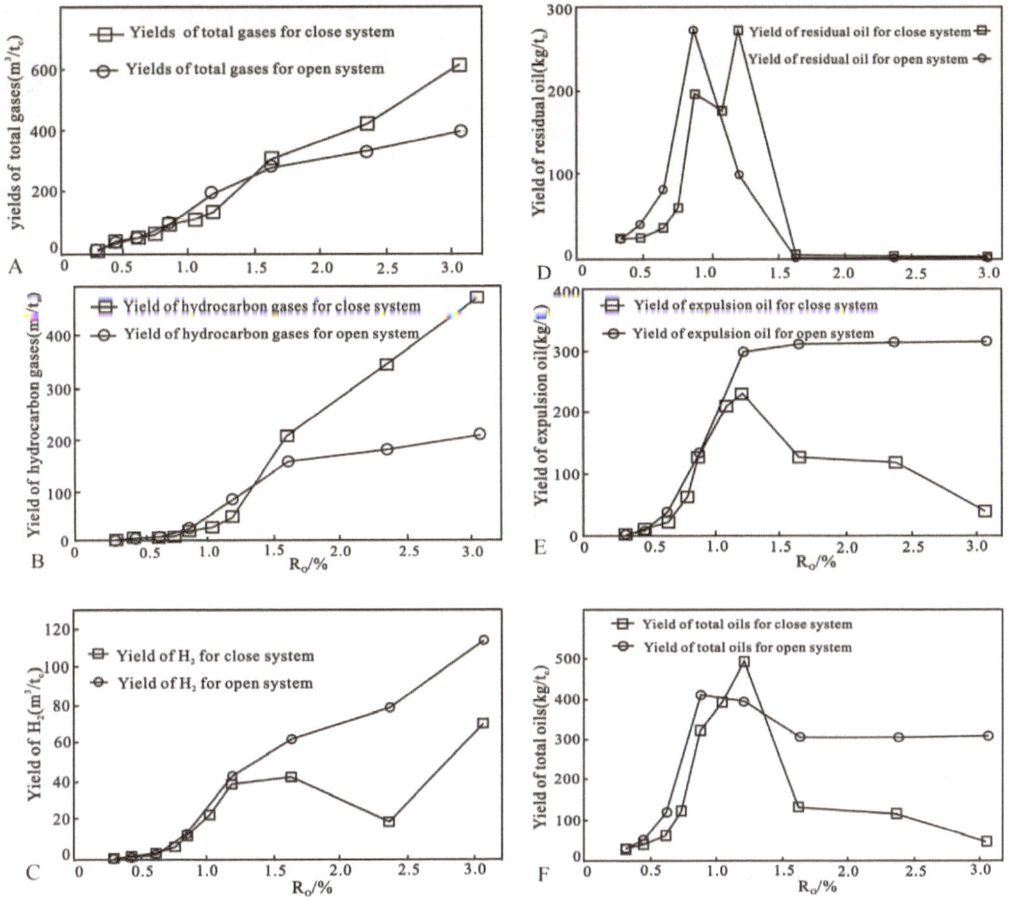

Figure 5. Comparison of oil and gas yields of oil shale hydrocarbon simulation in close and open system A- comparison of total gases; B-comparison of hydrocarbon gases; C-comparison of H_2; D-comparison of residual oil; E-comparison of expulsion oil; F-comparison of total oil.

The yields of expulsion oil and total generation oil in open system are both higher than those of close system, whereas the yields of the residual oil in open and close systems are very close, but the simulation temperature of the highest yield is at 300°C and 350°C, respectively.

The yield of total hydrocarbon generation in open system is slightly higher than that in close system on the whole.

It may be more reasonable to evaluate hydrocarbon generation potential of a basin by using the average value of hydrocarbon generation simulation results of source rocks in both close and open system.

ACKNOWLEDGEMENTS

We are grateful to the staff of AAAPG-2004 secretariate for their careful and constructive reviews of this paper and suggestions on how to improve the clarity of this paper.

REFERENCES

Barth, A., Borgund, A. E., Hopland, A. L. 1989. Generation of compounds by hydropyrolysis of Kimmeridge oil shale-Bulk results and activation energy calculation. Organic Geochemistry, 14(1): 69–76.

Guo, S. Z., 1991. Study on hydrocarbon generation capability of coal maceral of Taiyuan Formation in Shanxi Province. Paleoburial Hill, 3:20–26.

Jia, R. F., Zhou Z. Y., 1981. Thermal simulation experiment of modern algae under about 500 and 1000 barometric pressure. Geochemica, 1:87–94.

Jia, R. F., 1987. The primary study on hydrocarbon generation of resinite in Fushun (I). Science in China (Series B), 1:88–94.

Liu Baoquan, Cai Bin., 1990. Simulation experiment on oil and gas formation of kerogen of shale in Lower Malin

Formation of the Upper Proterozoic. Petroleum Geology and Experiment, 12(2):147–160.
Sofer, Z., 1987. Hydropyrolysis of monterey, asphaltenes. Organic Geochemistry, 13(4–6):939–945.
Stalker, L., Larter, S. R., Farrimond, P., 1998. Biomarker brinding into kerogens: evidence from hydropyrolysis using heavy water (D_2O). Organic Geochemistry, 28(3–4):239–253.
Wang, Y. H., Hou, Q. J., Sun, D. J., et al., 2004. Oil and gas formation and resource evaluation for Mesozoic-Cenozoic strata in the north margin of Qaidam Basin. Beijing: Science Press, pp.35–63.
Weres, O., Newton, A. S., Tsao, L., 1988. Hydropyrolysis of alkanes, alkenes, alcohols and ethers. Organic Geochemistry, 12(5): 433–444.
Zeng Guoshou, Xu Menghong.,1990. Petroleum geochemistry. Beijing: Petroleum Industry Press, pp. 10–150.
Zhao, Z. Z., Qin, J. Z., 2000. Oil and gas formation of marine source rock measures in Qinghai and Xizang plateau. Beijing: Science Press, pp. 56–87.

Petroleum Geochemistry and Exploration in the Afro-Asian Region –
Digang Liang, Darui Wang & Zhenxi Li (eds)
© 2008 Taylor & Francis Group, London, ISBN 978-0-415-44083-7

Carbon isotopic signature of CO_2 in Arthit gas field, Northern Malay basin, the Gulf of Thailand

S. Pisutha-Arnond & A. Sirimongkolkitti
PTT Exploration and Production (Public) Co. Ltd., Bangkok, Thailand

V. Pisutha-Arnond
Geology Department, Science Faculty, Chulalongkorn University, Bangkok, Thailand

ABSTRACT: Carbon dioxide (CO_2) is a critical component found in reservoir gas in the Arthit gas field. After drilling 19 wells, an amount of CO_2 was found to vary from wells to wells. It is necessary to understand its origin and migration in order to accurately predict the CO_2 distribution in the area. The objective of this paper is therefore to discuss the origin and distribution of CO_2 in the Arthit gas field. This study involves an investigation and interpretation of CO_2 contents and carbon isotopic composition ($\delta^{13}C_{CO2}$).

1 INTRODUCTION

The Arthit gas field covers approximately 3900 km^2 and is located on the northwestern margin of the Malay Basin (Figure 1). The basin is an intracratonic basin formed by rifting in early Tertiary time, along North-South (N-S) and Northwest–Southeast (NW-SE) oriented normal faults forming a series of asymmetric half grabens. Sedimentation patterns and environment of deposition were strongly influenced by local sub-basin topography. Major strike-slip tectonic movements in Early Tertiary time are believed to be a main cause of rifting during early Oligocene time. In late Oligocene time, sag subsidence followed the early rifting phase and filled the sub-grabens with alluvial, fluvial and lacustrine sediments. A subsequent N-S transtensional phase associated with continuing NW-SE right lateral strike-slip tectonics dominated during the Neogene. The resulting horsts and grabens developed syn-depositionally from Miocene to the present, and now constitute major hydrocarbon traps in the northern Malay Basin.

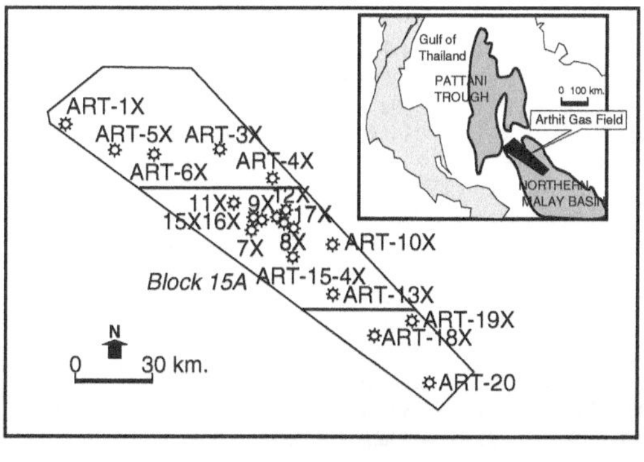

Figure 1. Location map of Arthit gas field, northern Malay basin, the Gulf of Thailand.

The sedimentary history of Arthit gas field commenced in Eocene time and includes four lithostratigraphic units (mega-sequences) deposited during the major stages of basin development. The first unit is syn-rift deposit comprises of fluvial-lacustrine system. The post-rift sequences of Late Oligocene to Middle Miocene age comprise thick fluvial sandstones of regression period and followed by a series of transgressive and regressive cycles influenced by a local marine setting. The upper most formation in the Arthit gas field ranges in age from Late Miocene to Quaternary. This sequence is characterized generally as a shale-rich section.

Carbon dioxide (CO_2) is a critical component found in gas in Arthit gas field. The amount of CO_2 varies from wells to wells ranging from less than 10% to 90%. Though it is believed that the amount of CO_2 increases with depth but in some wells it is reversed. To be able to accurately predict the CO_2 distribution in Arthit gas field it is necessary to understand its origin and migration. The purpose of this study is to discuss the distribution of carbon isotopic data of CO_2 and CO_2 contents. The carbon isotopic compositions and CO_2 contents are from 60 gas samples from the RFT and TST samples from 19 wells drilled during 1999–2002. The carbon isotopic values were reported as per mil (‰) with respect to the PDB standard. It is expected that this work would be able to shed some light on the sources and migrations of CO_2 in the Arthit gas field.

2 SAMPLES AND ANALYTICAL METHODS

During 1999–2002 a total of 20 exploration and appraisal wells were drilled in the concessions. Only one well was abandoned as a dry well. The 19 discovery wells penetrated several zones of gas sandstone reservoirs. The RFT and TST were conducted. The RFT tool (Repeated Formation Tester tool – a trade name) is a pad tool that is pushed against the gas reservoir intervals in the wells. The RFT tool was conducted to investigate the reservoir pressures and collect gas samples. The RFT tool collected gas samples at 300 cc. each. The TST (Tubing Stem Test) is a production test. After the casings were installed and cemented, production tubings were run in. After the perforation gun perforated the tubing and casing into the formation, gases were allowed to flow to surface. Gas samples were collected at the surface in a cylinder of 20 liters capacity. RFT and TST gas samples were sent to PT. Corelab Indonesia for stable isotope analyses.

Carbon isotope ratios of CH_4 and CO_2 in gas samples were measured in a continuous flow GC/stable isotope ratio mass spectrometer. Gas samples were injected into an Agilant 6890 gas chromatograph using helium carrier gas to separate CH_4 and $CO_2$from other components. Each component eluting from the GC was flushed with the helium carrier gas through an oxidizing furnace to convert hydrocarbons to CO_2, then into a Micromass Isoprime stable isotope ratio mass spectrometer. Carbon isotope ratios were measured relative to a working standard, and reported relative to Pee Dee Belemnite (PDB) standard. The analytical uncertainty was better than +/− 0.2 ‰.

Carbon isotope ratios of CH_4 and CO_2 in gas samples were measured in a continuous flow GC/stable isotope ratio mass spectrometer. Gas samples were injected into an Agilant 6890 gas chromatograph using helium carrier gas to separate CH_4 and CO_2 from other components. Each component eluting from the GC was flushed with the helium carrier gas through an oxidizing furnace to convert hydrocarbons to CO_2, then into a Micromass Isoprime stable isotope ratio mass spectrometer. Carbon isotope ratios were measured relative to a working standard, and reported relative to Pee Dee Belemnite (PDB) standard. The analytical uncertainty was better than +/− 0.2 ‰. The results of stable carbon isotopes of methane and carbon dioxide and carbon dioxide contents are displayed in Table 1.

Table 1. Results of stable carbon isotopes of CH_4 and CO_2 and CO_2 contents.

Well	Test Type	Depth (m)	CO_2 (%)	$\delta^{13}C_{CO2}$	$\delta^{13}C_{CH4}$
ART-1X	TST	1680.8	4.88	−13.36	−60.39
ART-1X	TST	3154.0	20.20	−4.53	−40.49
ART-3X	TST	1229.5	6.72	−10.49	−37.11
ART-3X	TST	1229.5	6.72	−10.49	−37.11
ART-3X	RFT	1495.5	63.87	−3.22	−33.8
ART-3X	RFT	1598.0	63.79	−7.18	−33.29
ART-3X	RFT	1683.0	67.70	−5.39	−33.24
ART-3X	TST	2344.5	8.08	−10.35	−36.2
ART-3X	TST	2006.0	49.69	−7.64	−32.88
ART-4X	RFT	1942.0	12.75	−9.60	−44.5
ART-4X	TST	2390.4	22.30	−10.48	−34.06
ART-5X	RFT	2312.0	10.50	−9.95	−48.95
ART-5X	TST	2324.5	11.50	−11.07	−49.6
ART-6X	TST	2801.5	22.80	−10.27	−38.6
ART-7X	RFT	2196.5	13.03	−8.89	−47.95
ART-7X	RFT	2515.3	20.91	−7.54	−43.15
ART-7X	RFT	2661.3	26.25	−8.05	−41.27
ART-7X	RFT	2738.0	37.84	−8.41	−37.27
ART-7X	RFT	2811.8	63.97	−6.81	−30.8
ART-8X	RFT	1958.5	11.43	−8.95	−46.12
ART-8X	RFT	2046.5	13.67	−8.18	−43.09
ART-8X	RFT	2182.5	18.37	−7.83	−42.64
ART-8X	RFT	2454.0	37.52	−7.27	−41.26
ART-8X	RFT	2666.0	72.88	−2.62	−37.32
ART-8X	RFT	2917.5	74.48	−3.39	−32.46
ART-9X	RFT	1899.0	10.49	−8.43	−46.19
ART-9X	RFT	2445.0	25.29	−6.52	−42.53
ART-9X	RFT	2661.5	48.14	−5.66	−38.97
ART-9X	RFT	2711.0	62.77	−5.69	−40.83
ART-9X	RFT	2788.0	75.54	−5.40	−37.74
ART-10X	TST	3595.8	17.50	−4.79	−41.94
ART-10X	TST	4119.5	26.00	−6.81	−40.67
ART-11X	TST	2136.0	21.60	−9.10	−37.84
ART-11X	TST	2084.0	23.30	−8.77	−39.35
ART-11X	TST	2071.0	23.40	−8.72	−38.94
ART-11X	TST	3047.5	46.50	−7.13	−32.01
ART-12X	TST	2378.0	19.40	−7.09	−44.33
ART-13X	TST	2338.0	13.70	−8.64	−41.13
ART-14X	TST	1966.0	17.90	−9.18	−45.22
ART-14X	TST	1623.5	5.40	−11.98	−51.42
ART-15X	TST	2019.5	14.02	−9.16	−45.68
ART-15X	RFT	2226.5	18.72	−7.48	−41.53
ART-15X	RFT	2428.5	28.16	−8.07	−39.96
ART-15X	TST	2461.0	29.54	−8.35	−38.18
ART-15X	RFT	2567.0	36.04	−8.72	−35.78
ART-16X	RFT	2250.0	11.43	−9.63	−47.87
ART-16X	RFT	2571.0	22.98	−7.95	−40.37
ART-16X	RFT	2720.0	31.60	−8.12	−37.46
ART-16X	RFT	3102.5	53.15	−7.91	−27.33
ART-16X	RFT	2805.0	35.17	−9.13	−34.92
ART-17X	RFT	1942.0	8.53	−10.07	−51.63
ART-17X	TST	2232.1	13.50	−10.85	−46.55
ART-17X	RFT	2763.2	26.00	−0.32	−41.98
ART-17X	RFT	3106.5	93.40	−3.43	−30.6
ART-18X	TST	2060.5	10.58	−11.13	−46.24
ART-18X	TST	2664.5	23.83	−5.18	−39.51
ART-18X	TST	2207.8	10.98	−9.62	−40.56
ART-18X	TST	2872.5	45.94	−4.21	−39.02
ART-19X	TST	2157.0	12.60	−9.13	−39.67
ART-20X	TST	1898.0	18.00	−10.40	−43.18
ART-20X	TST	2838.5	36.00	−2.12	−39.18

TST: gas samples collected at surface by TST test.
RFT: gas samples collected at the formation by RFT tool.

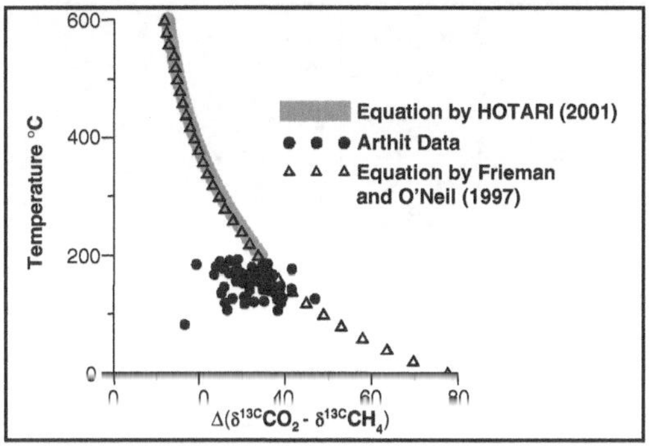

Figure 2. Carbon isotope fractionation factors of CO_2 and CH_4 and the plot of Arthit gas field data (see text for discussion).

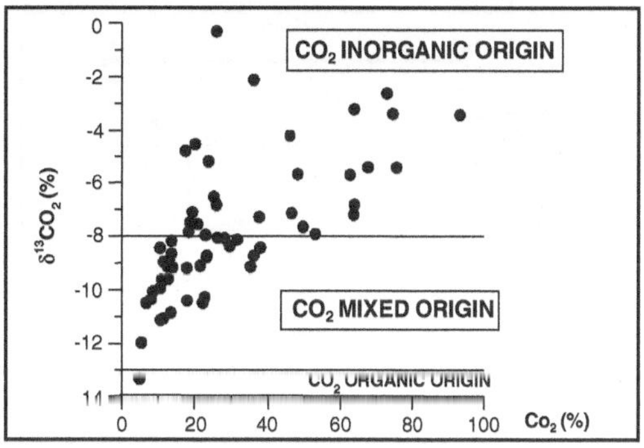

Figure 3. Crossplot between CO_2 contents and carbon isotopic values of CO_2, Arthit gas field.

3 DISCUSSIONS

3.1 *Carbon isotopes of CO_2 an CH_4*

It is normal to speculate that a gas mixture such as CO_2 and CH_4 originated from various sources and accumulated in a reservoir should undergo isotopic re-equilibration particularly at high temperature. Their isotopic values could have been controlled by the isotope fractionation factors and their present values might have been shifted significantly from their original values due to isotopic re-equilibration. Hence their isotopic values might no longer be able to use as a tracer to their sources or origins.

Figure 2 shows the carbon isotopic fractionation factors between CO_2 and CH_4. The triangle symbols are the equilibrium carbon isotopic fractionation curve given by Friedman and O'Neil (1977). The grey circles (packed into grey line) are the equilibrium carbon isotopic fractionation line proposed by Horita (2001; cited in Chacko et al., 2001). The fractionation factor given by Hotari (2001) is in good agreement with that of Friedman and O'Neil (1977).

The $\Delta(\delta^{13}C_{CO2}\text{-}\delta^{13}C_{CH4})$ are plotted against the formation temperatures from Arthit gas field in Figure 2 in order to test whether the CO_2 and CH_4 in the Arthit gas field were in or out of isotopic equilibrium with each other. It is obvious from Figure 2 that most of the points from Arthit gas field were plotted below the equilibrium isotopic fractionation lines. It is therefore likely that the carbon isotopic compositions of CO_2 and CH_4 in Arthit gas field are out of isotopic equilibrium.

It is commonly known that gaseous reactions of CO_2-CH_4 are sluggish; particularly from sedimentary and geothermal systems and that they usually appear to be out of isotopic equilibrium (Ohmoto, 1986; Hotari, 2001; cited in Chacko et al., 2001). Because of the non-isotopic equilibrium and the sluggishness of the CO_2-CH_4 gaseous reaction, the carbon isotopic re-equilibration between CO_2 and CH_4 in Arthit gas field should not have been undergone to a significant degree. It is therefore likely that the carbon isotopes of both CO_2 and CH_4 do maintain their original isotopic signatures. Hence it is possible to use the $\delta^{13}C_{CO2}$ values to interpret the source of CO_2 as well as the $\delta^{13}C$ values of CH_4 for the origin of CH_4 from its own isotopic variation separately. This assumption can be confirmed by the carbon isotopic values of all methane samples (the $\delta^{13}C_{CH4}$ values of 60 samples varying from −26 to −52 ‰, see Figure 4) which fall in a typical range of thermogenic methane even in some gas samples containing small content of CH_4 but large amount of CO_2.

3.2 *Crossplot of carbon isotopes and CO2 contents*

Figure 3 shows the plot of carbon isotopic values of CO_2 against CO_2 contents in the reservoir gas mixture. The light carbon isotopic composition of CO_2 is commonly interpreted to source from organic origin (kerogens) and the heavier carbon isotopic composition of CO_2 is sourced from inorganic origin as shown in Figure 4 (Emery and Robinson, 1993). The carbon isotopic compositions heavier than −8‰ are usually interpreted to originate from inorganic source whereas those lighter than −13‰ are commonly placed for the organic origin. The isotopic composition between −8‰ and −13‰ are therefore interpreted as a mixture between the organic and inorganic origins (Rui Lin, 2001). The distributions of the carbon isotope values of CO_2 lower than −8‰ are in the area where the CO_2 content in the reservoir gas mixture is less than 40%. There are however a number of carbon isotope values of CO_2 above −8‰ that also have CO_2 content in the reservoir gas mixture less than 40%.

The kerogen in Arthit gas field is known to be dominated by sappropelic source. Such kerogen could generate CH_4 with or without minor amount of CO_2

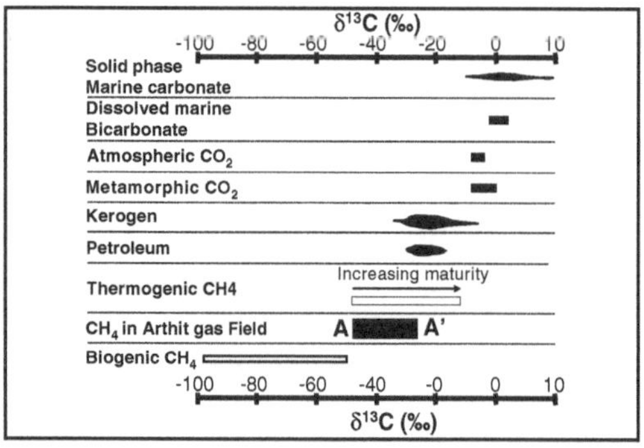

Figure 4. Carbon isotope compositions of major carbon reservoirs in sedimentary basins (from Emery D. and Robinson A., 1993). A-A' represents range of carbon isotope of methane in Arthit gas field.

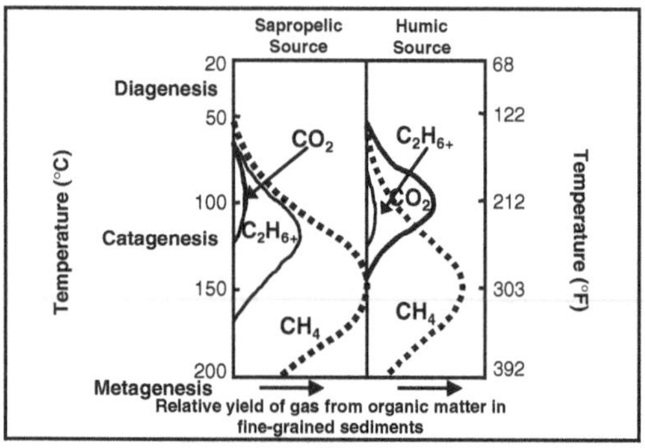

Figure 5. The generation of gases from organic matter with temperature (from Hunt, 1996).

depending on the temperature as shown in Figure 5 (Hunt 1996). Such hydrocarbon gas would have the CO_2 content in the gas mixture closer toward 0% and the $\delta^{13}C$ values of CO_2 below approximately -13‰. On the contrary the pure inorganic CO_2 derived from deep seated sources before mixing with hydrocarbon gases should have the CO_2 content in the gas mixture closer toward 100% and $\delta^{13}C$ values of CO_2 above approximately -8‰ (Figure 6). Mixing of different content of inorganic CO_2 migrated or penetrated into the reservoirs from deep seated origin with hydrocarbon gases dominated by CH_4 without organic CO_2 could produce the reservoir gas mixture having carbon isotopic composition of CO_2 remaining at the same values in any point along the horizontal arrow lines (also due to the sluggishness of gaseous reactions of CO_2-CH_4 as mentioned earlier). However mixing of different content of inorganic CO_2 with hydrocarbon gases with certain amount of organic CO_2 could produce a gas mixture having carbon isotopic composition

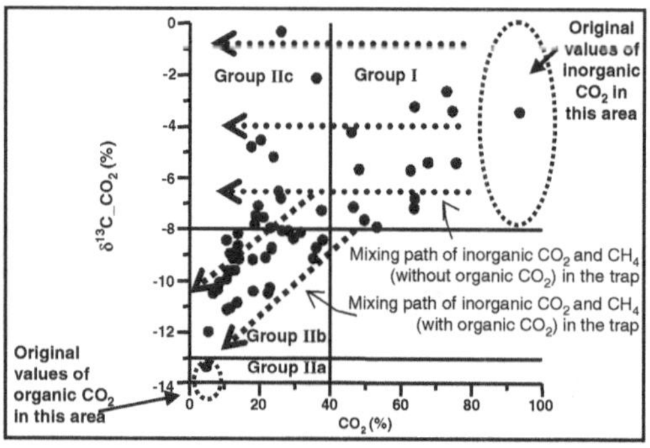

Figure 6. Crossplot between CO_2 contents and Carbon isotopic values of CO_2. Horizontal circle-dotted arrow lines indicate mixing trend of inorganic CO_2 with CH_4 without organic CO_2. Inclined square-dotted arrow lines indicate mixing trend of inorganic CO_2 with CH_4 with certain contents of organic CO_2 (see detailed discussion in text).

of CO_2 being more negative values than the original inorganic CO_2 values depending on mixing ratios along the incline arrow lines (Figure 6).

Therefore inorganic source CO_2 significantly controls the CO_2 contents in gas reservoirs (by its migrated volume) and also controls the present carbon isotopic values in the reservoirs (after mixing). The distribution of the carbon isotopic values against CO_2 contents shown in Figure 3 should therefore be better grouped based primarily on CO_2 contents and the carbon isotopic values as follows (Figure 6):

3.2.1 *Group I (high CO2 contents and enriched carbon isotopic composition)*

This group is characterized by the inorganic dominated source. They show rather high CO_2 content, 40–90% and enriched isotopic values heavier than (-8‰). Because of its inorganic origin, it may generate from deep-seated inorganic sources. The CO_2 from this group is migrated through faults or fractures and accumulated in shallower reservoirs with minor hydrocarbon gas mixing.

3.2.2 *Group II (low to medium CO2 contents and variable $\delta^{13}C$ values)*

This group represent reservoir gases dominated by CH_4 and CO_2 of inorganic, organic (kerogen source) and a mixing origins. The CO_2 contents are generally ranging from 5–40% and variable $\delta^{13}C$ values of CO_2 from 0 ‰ to -14‰. This group can be divided into 3 sub-groups:

3.2.2.1 Sub-group IIa: CO_2 from kerogen source:

This sub-group represents pure kerogen gases dominated by CH_4 and minor amount of organic CO_2 (without the contribution of inorganic CO_2). The carbon isotopic compositions of CO_2 are controlled essentially by kerogen in which their values are below −13‰.

3.2.2.2 Sub-group IIb: CO_2 from mixing sources

This sub-group represents reservoir gases dominated by CH_4 in which the CO_2 is a mixture of both organic and inorganic sources. The carbon isotopic values of CO_2 range from −8‰ to −13‰ and the CO_2 contents range from 5–25%.

3.2.2.3 Sub-group IIc: CO_2 from inorganic dominated source

This sub-group represents reservoir gases dominated by CH_4 without organic CO_2. Varying amount of inorganic CO_2 could migrate into such the reservoirs and their carbon isotopic compositions of CO_2 in the reservoirs are controlled essentially by the isotope values of inorganic CO_2. This sub-group is therefore has the isotopic value similar to inorganic CO_2 which is above −8‰.

4 DISCUSSIONS

4.1 The carbon isotopes of CO_2 and CH_4 in Arthit gas field are out of equilibrium and the carbon isotope compositions of both CO_2 and CH_4 do maintain their original signatures.

4.2 Two sources of CO_2 in Arthit gas field area are recognized as inorganic origin and organic origin by the carbon isotopic values. Based on the content and carbon isotopic values of CO_2, gas reservoirs in Arthit gas field can be grouped as follow:

Group I: High CO_2 content, more than 40%, heavy carbon isotopic values of CO_2 in the range of 0‰ to −8‰. CO_2 is sourced from inorganic origin.
Group II: Low to medium CO_2 content, less than 40%, variable carbon isotopic values of CO_2 from 0‰ to −14‰. CO_2 is from both inorganic origin and organic origin.
Group IIa: Organic dominated source, very light carbon isotopic value, no contribution of inorganic source of CO_2 so the CO_2 (organic source) content is very low.
Group IIb: Mixing sources with minor to moderate CH_4 dilution. Low to medium CO_2 contents (5–40%), carbon isotopic values of CO_2 are ranging from −8‰ to −13‰.
Group IIc: Inorganic dominated source with strong CH_4 dilution. Medium CO_2 contents (15–40%), heavy carbon isotopic values of CO_2 are heavier or more than −8‰.

ACKNOWLEDGEMENTS

The authors thank PTT Exploration and Production Public Co. Ltd., Chevron Thailand Exploration and Production Limited and Mitsui Oil Exploration Co, Ltd. for their permissions to publish this paper. Special thanks are to the Arthit Asset team of PTTEP for their support. We would like to thank John Harrington, PT Corelab Indonesia, for kindly providing the isotope analytical methods.

REFERENCES

Chacko, T.; Cole, D.R. & Horita, J. 2001. Equilibrium oxygen, hydrogen and carbon isotope fractionation factors applicable to geologic systems, In Valley, J.W. & Cole, D.R. (eds) Stable Isotope Geochemistry, Rev Mineral, 43, 1–81.

Emery, D. & Robinson, A. 1993. Inorganic geochemistry applications to petroleum geology, In Blackwell Scientific Publications, Oxford, London.

Friedman, I. & O'Neil, J.R. 1977. Compilation of stable isotope fractionation factors of geological interest, US Geol Surv Prof Paper 440-KK.

Horita, J. 2001. Carbon isotope exchange in the system CO_2-CH_4 at elevated temperatures, Geochim Cosmochim Acta, 65, 1907–1919.

Hunt, J.M. 1996. Petroleum geochemistry and geology, In 2nd edition, Freeman, New York.

Ohmoto, H. 1986. Stable isotope geochemistry of ore deposits. In Valley, J.W.; Taylor, H.P. Jr. & O'Neil, J.R. (eds) Stable Isotopes in High Temperature Geological Processes. Rev Mineral, 16, 492–559.

Rui, Lin 2001. Geochemistry and petroleum systems in the Arthit Area: 1999/2000 Exploration Wells, Unocal Thailand, Ltd., in-house report.

4. Basin modeling, novel geochemical approaches and petroleum strategies

Petroleum Geochemistry and Exploration in the Afro-Asian Region –
Digang Liang, Darui Wang & Zhenxi Li (eds)
© 2008 Taylor & Francis Group, London, ISBN 978-0-415-44083-7

The sedimentary model of $Ng_{(1+2)}$ in Gudao Oilfield and its controls on the remaining oil distribution

Xibin Wang
Earth Resource and Information College, China University of Petroleum, Dongying, China and School of Petroleum Engineering, China University of Petroleum, Dongying, China

Jianhua Zhong, Lin Jiang & Ping Jia
Earth Resource and Information College, China University of Petroleum, Dongying, China

ABSTRACT: The fluvial sedimentary model of the $Ng_{(1+2)}$ sand set is constructed based on the stratigraphic classification and correlation, grain size analysis, morphological feature of fluvial sand bodies and fluvial curvature. According to 16 kinds of fluvial types by Miall, the $Ng_{(1+2)}$ sand set is classified as the fine-grained meander sediment. The study indicates that the main oil layers of $Ng_{(1+2)}$ are the remaining oil-enriched zone and the non-main oil layers are the high remaining oil saturation zones. By waterflooding and gravitational differentiation, the remaining oil is enriched vertically in the middle-upper parts and laterally in the low-permeability units as well as in the poor-injection units within the reservoir. The sedimentary model of $Ng_{(1+2)}$ in Gudao Oilfield controls the inner structure and the macroscopic heterogeneity of the reservoir sands, and the movement rule of the injected water, and thereby controls the distribution of the remaining oil. In a word, the sedimentary model of meandering fluvial facies is the ultimate factor controlling the distribution of the remaining oil in the $Ng_{(1+2)}$ reservoir.

1 INTRODUCTION

The heterogeneity of hydrocarbon reservoirs and the lithofacies variation affects the petroleum-bearing structures. Different fluvial system possesses different characteristics of reservoirs and different distribution of the remaining oil (Liu et al., 2000; Liu and Xu, 2003; Dong et al., 1999; Holtz et al., 1996). In the late stage of oilfield development, it is difficult to keep production steady. So the geological research focuses on defining the sedimentary characteristic of reservoirs, fine-description of reservoir spacial distribution and the distribution characteristic of the remaining oil (Marcu and Clipea, 2001; Virnovsky et al., 1998; Abbots and Van Kuijk, 1997).

Gudao Oilfield, located in the east of the Zhanhua depression of the Jiyang sag in Bohai Gulf Basin, China and bordering Bohai Sea in the east, is a large oil-gas field with a relatively complete north-east anticline structure and drape over the Gudao arch. The main oil-bearing layer is the upper part of Miocene Guantao Formation of Tertiary. The 1 + 2 sand sets of Guantao Formation ($Ng_{(1+2)}$) have fine lithology, they are a set of gray-green or fuchsia mudstone interbedded with muddy siltstone and fine sandstone, the thickness varies from 70 to 100 m. $Ng_{(1+2)}$ belongs to fluvial facies, but the previous researchers haven't reached an agreement on the fluvial classification, some researchers thought it was meandering fluvial deposits (Wang and Qian, 1992) while others held that it was braided fluvial deposits (Li and Zhou, 1982). In the paper, the fluvial sedimentary model of the $Ng_{(1+2)}$ sand set is constructed based on the stratigraphic classification and correlation, grain size analysis, morphological feature of fluvial sand bodies and fluvial curvature. The sedimentary model controlling the distribution of the remaining oil is also discussed from the aspect of distribution rule of the remaining oil.

2 SEDIMENTARY MODELS OF THE $NG_{(1+2)}$ RESERVOIR IN GUDAO OILFIELD

2.1 *Lithologic characteristics of sandstone*

From the core data analysis of Well Bo108, the $Ng_{(1+2)}$ reservoir bed consists of a set of light-brown, dark-brown siltstone and fine sandstone. Thin section observation indicates that the sandstone contains 50~60% quartz, 30~40% feldspar and 5~20% rock fragments; most of feldspars are not weathered, showing fresh

Figure 1. Isopach map for the 8th layer of the $Ng_{(1+2)}$ sand set, Zhong'er District, Gudao Oilfield.

surfaces and distinct cleavages; the content of the rock fragments increases from top to bottom, showing that the deposition occurs after a short transportation. The mudstone, mainly red or brown-yellow and interbedded with thin layers of gray-green mudstone, shows soft to moderate-hardness without lime on the whole, reflecting an oxidizing depositional environment.

2.2 *Grain size characteristics of sandstone*

Sandstone in the $Ng_{(1+2)}$ reservoir presents fine-grained structure, with a median grain diameter of 0.16~0.06 mm and a maximum of 0.64 mm. The probability cumulative curve of the grain size indicates that the sandstone is almost composed of saltated component and suspended component without drag ensemble, which reflects the grain-size characteristics of different structure units in meandering fluvial deposits instead of channel bar of the braided fluvial deposits or anastomosed stream. Slope of the saltated component is steep, about 55~75 degree, the content of suspended matter is up to 30~70%. The coarse and fine tangential points of each curve are 1.5–2.0 ϕ and 2.7–3.5 ϕ, respectively.

2.3 *Morphological features of fluvial sandstone*

The distribution characteristic of the sandstone reflects the channel morphology during the deposition. The isopach map for the sandstone of each layer in Gudao Oilfield represents the characteristic of the meandering river deposition (Fig. 1).

The thickness of a single sand bed in the $Ng_{(1+2)}$ sand set varies greatly, from 0.1 to 14 m, generally from 2 to 5 m, averaging at 2.7 m. The ratio of the sand thickness to the formation thickness averages at 0.21, indicating the mudstone barrier is thick, which, therefore, reflects the characteristic of the meandering river deposition.

2.4 *Channel curvature*

Channel curvature is an important parameter in determining the type of river, which can be calculated by the Schumm formula:

$$P = 3.5F^{-0.27} \quad (1)$$

$$F = 255M^{-1.08} \quad (2)$$

Where, P represents channel curvature, F represents the ratio of width to depth of the river, and M represents aleuropelitic content which is related to the thickness of silty mudstone plus mudstone to the thickness of the formation. Table 1 gives the calculated channel curvature value on the sand-shale content analyzed by the logging result of the selected well in Gudao Oilfield.

According to the classification standard by Miall (Miall, 1996), $P < 1.2$ represents low curvature, $1.2 < p < 1.5$ represents intermediate curvature and $P > 1.5$ represents high curvature. In Table 1, all the channel curvature is more than 2.6, which indicates all the $Ng_{(1+2)}$ sand set was deposited in the high-curvature river.

2.5 *Fluvial sedimentary model*

2.5.1 *Features and vertical cycle of fluvial deposits*

Through the stratigraphic classification and correlation for the $Ng_{(1+2)}$ sand set in the whole Gudao Oilfield, electrofacies analysis and isopach mapping for the sand, the whole sand set is defined as being in one macrocycle, with three epicycles in it (Fig. 2). The dualistic structure of fluvial facies exits apparently in each epicycle. The lower part is coarse grained channel deposit with developed sandstone, represented by the bell- or box-shaped logging curve; the upper part is bank deposit mainly filled with natural levee sediment of lateral or vertical accretion, floodplain sediment of vertical accretion, and interlaid with crevasse splay sediment of fine grains, which are dominated by silty mudstone and mudstone, appearing as flat or dentoid or slightly dentoid crest on the logging curve.

The thickness of the $Ng_{(1+2)}$ sand set in Gudao Oilfield is 70~100 m, and the thickness of each epicycle is about 25~30 m, which is almost consistent with the conclusion of Li et al. (Li et al., 2001), who divided

Table 1. Channel curvature (P) of the Ng (1 + 2) sand set, Gudao Oilfield.

Well no.	X4-12	X4-18	Z8-7	Z8-17	Z25-509	Z28-10	D3-27	D5-17	N8-010	Average
$T_{sand}/T_{formation}$	0.15	0.11	0.35	0.16	0.40	0.04	0.17	0.24	0.21	0.21
$F = 255M^{-1.08}$	2.10	2.00	2.81	2.13	3.06	1.84	2.16	2.37	2.28	2.28
$P = 3.5F^{-0.27}$	2.86	2.90	2.65	2.85	2.59	2.97	2.84	2.77	2.80	2.80

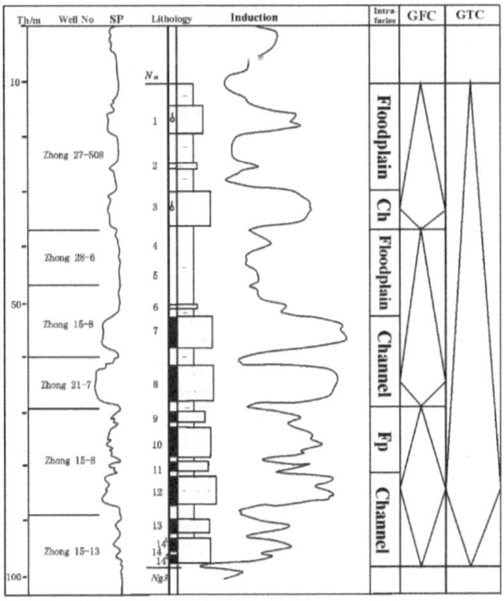

Figure 2. The vertical sedimentary cycle division of the $Ng_{(1+2)}$ sand set in Gudao Oilfield.

the upper part of Guantao Formation in Gudao Oilfield into 10 epicycles and the thickness of each epicycle is about 25 m.

From the reserves distribution in the whole sand sets, the reserves of 12th, 9th and 8th oil layers which developed in the channel subfacies account for 50% of the whole reserves, while the 3rd layer is the main gas-bearing layer, and the other 11 oil layers contain less than 50% of the total reserves, which indicates that the channel subfacies reservoir is the main hydrocarbon accumulation sand body.

2.5.2 *Fluvial sedimentary model*

By the above analysis, the sedimentary characteristic of the $Ng_{(1+2)}$ sand set in Gudao Oilfield is concluded as follows:

1. From the distribution rule of sandbodies, the sand set deposit possesses the characteristic of the meandering river deposit;
2. All the channel curvature values are more than 2.6, indicating a high curvature meandering stream;
3. The grain size is fine-dominated siltstone and well-developed mudstone, medium-coarse sandstone is rare. The top and bottom surfaces of the sand set are conformable contact;
4. According to the grain size analysis, the sediments are transported in the manner of saltation and suspension;
5. The stack pattern of the sequence is very complex in the different depositional period, the flushing and down-cutting by the latter channel sand result in the superposed and incised sands;
6. The sand set can be divided into three epicycles, and each one develops a dualistic structure.

Thereby, compared with the Miall's classification standards (Miall, 1996), the $Ng_{(1+2)}$ sand set of the Gudao Oilfield is classified as the fine-grained meandering river deposit. The fine-grained meandering stream is a kind of low-energy estuary-approached environment (Li et al., 2001) and the thick fine-grained floodplain is well developed, which indicates that the gradient ratio is smaller, the clastics supply is not enough and the depositional rate is slow during the $Ng_{(1+2)}$ deposition. The main sand bodies in $Ng_{(1+2)}$ are typically in slaty, ribbon or sill-like distribution laterally, characterized by fining- and thinning upward pattern. The lag deposit and meander bar are both the products of the lateral accretion when the river channel is transported laterally and transits upwards as the natural levee or mainly as silt with both lateral and vertical accretion. Above the natural levee is a flood plain formed mainly by vertical accretion, whose sediments are mainly silty mudstone and mudstone interbedded with crevasse splay and crevasse channel sands, which constitute the upper unit of the fine-grained meandering stream sequences.

3 SEDIMENTARY MODEL CONTROL ON THE REMAINING OIL DISTRIBUTION

The $Ng_{(1+2)}$ reservoir belongs to a high-curvature channel deposit, with the channel continually migrating and changing, the wide compound meandering river sands of the high-curvature are formed, appearing in a sinuous strip-, ribbon- or branch-like ditribution laterally and lentiform of top-flat and bottom-bulgy or slaty in cross section, surround by the fluvial shaly sediments. It is such a distribution characteristic of fluvial facies reservoir that determines the distribution pattern of the remaining oil in $Ng_{(1+2)}$. According to the statistic data, the average water drive recovery of the fluvial reservoir is 30.2% throughout the whole China (Liu et al., 2000; Liu and Xu, 2003). At present, the recovery of the reserves *in situ* for the $Ng_{(1+2)}$ reservoir in Gudao Oilfield is only 11%, therefore, the potential of the remaining oil in this area is large.

3.1 *Vertical distribution rule of the remaining oil*

3.1.1 *The remaining oil distribution between layers*

The main oil layer is the remaining oil accumulation zone. The dualistic structure of the fluvial facies controls the distribution of the remaining oil, the meander bar and channel lag are the best microfacies developed in the reservoir, and also the main oil layers in $Ng_{(1+2)}$; their big thickness, high initial oil saturations and coarser grain sizes make their porosity, permeability and seepage better than other layers; the oil-bearing thickness and area are greater and the abundance of the ultimate recoverable reserves is higher as well, thus, the absolute amount of the recoverable reserve of the remaining oil is huge. The geological reserve of the 3 main oil layers (12th, 9th and 8th layers) formed in the main developing stage of the $Ng_{(1+2)}$ channel subfacies is up to 50% of the total reserve, even though their recovery efficiency is higher than others, with the amount of 10.52%, 14.44% and 19.45%, respectively, their remaining geological reserves occupy 46.7% of the total remaining amount, so they are still the remaining oil accumulated zones in $Ng_{(1+2)}$.

Non-main oil layers are of high remaining-oil saturation. Plenty of overbank thin sand bodies developed in the host interior, the boundary or the outer of the channel sand (Fig. 3A), are of less thickness, lower porosity, lower permeability and less water drive recovery compared with the channel subfacies (Fig. 3C), thus, their remaining oil saturation is higher than that of the main oil layer (Fig. 3B). The non-main oil layers in $Ng_{(1+2)}$ mainly develop as natural levees, crevasse splay, abandoned channel and flood-plain reservoirs, while the channel subfacies reservoir is poorly developed. At present, the recovery efficiency averages at 6.21% only in these oil layers. In the past, such oil layers were usually perforated with the main layers or untouched due to the limit of techniques, even though some of these layers were exploited together with the main layers, the by-passed oil was still formed in these layers due to their lower permeability. Although the remaining oil saturation in these layers is high, the absolute amount of the recoverable reserves is less than that of the main layers because of their lower initial abundance of the geologic reserves.

It can be obviously seen in Fig. 4 that the saturation of the recoverable remaining oil of the main channel oil layers in the lower part of the $Ng_{(1+2)}$ reservoir is lower than that of the upper non-main layers, which indicates the recovery efficiency of the main layers is higher than that of the non-main oil layers.

3.1.2 *The remaining oil distribution in the layers*

The inner vertical sequence of the sand bodies definitely controls the remaining oil distribution (Fu et al., 2003). The sedimentary sequence in the $Ng_{(1+2)}$ meandering stream-derived reservoir appears mainly in positive rhythm. In the same reservoir, the sand bodies with lower permeability mostly develop in the upper part and are less affected by water injection; the oil layer appears with slight washing or without washing. In addition, the gravitation differentiation occurring during the displacement makes the flushing efficiency decrease in the middle-upper part, thus its remaining oil saturation is higher than that in the lower one, consequently forming the remaining oil enriched zone in the middle-upper part of the reservoir (Fig. 4).

The single meander bar is formed by a series of the crescent lateral-stacking bodies distributed inclinedly and inter-stacked. The upper part between the lateral-stacking bodies is separated by muddy interlayers and the lower part is intercut and interstacked, and consequently connected to form "a semi-connected body"; the muddy interlayers are, in the shape of beveling, transecting, recutting and vertical cutting, distributed within meander bars, which results in lots of barrier stagnant zones that are rich in the remaining oil.

Good micro-positive structures are favorable zones for the remaining oil to accumulate at the later time of the oilfield development (Li and Zhou, 1982). It is found that such microstructures of the meandering river reservoir with the enriched remaining-oil and good production are in different shapes such as top- and bottom-protruding, top-protruding and bottom-flat, top-flat and bottom-protruding and top- and bottom-nose protruding ones. In general, such microstructures can give higher production, while location of wells on the micro-negative structures can yield good flushing efficiency.

3.2 *The lateral distribution rule of the remaining oil*

Sand bodies with different genesis have different production and remaining oil distribution. It can be seen in Fig. 3 that the structural unit with low permeability and incomplete injection-production wells enrich the remaining oil. The $Ng_{(1+2)}$ sand bodies in Gudao Oilfield are sporadically distributed. There are 777 oil-bearing sandbodies in this area and those where oil-bearing area is less than 0.1 km^2 account for 64.5% of the total; it is very hard to establish the regular injection-production well pattern and thereby many reservoir sands are left untouched. These

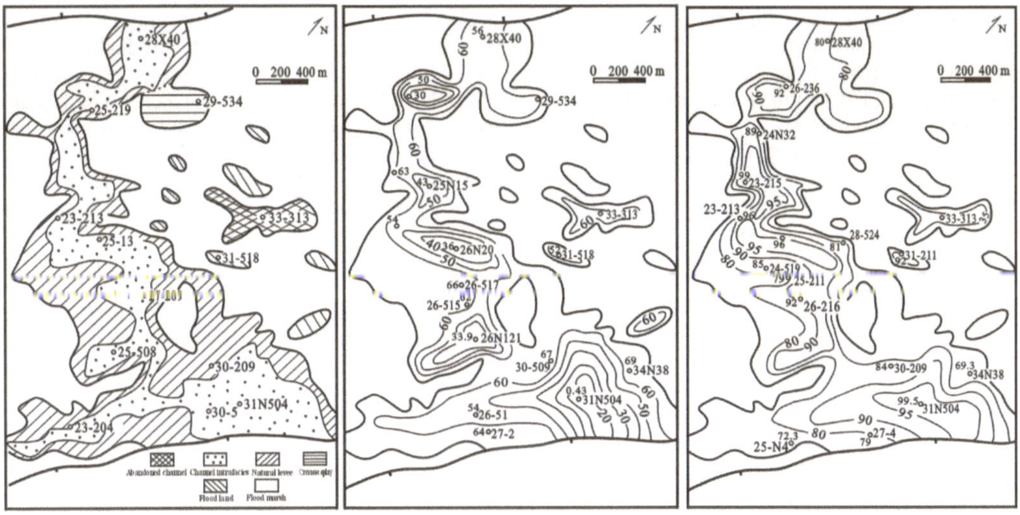

A. sedimentary microfacies ichnography of 9th layer in $Ng_{(1+2)}$ sand set at Zhonger district of Gudao Oilfield

B. remaining oil saturation isoline map of 9th layer in $Ng_{(1+2)}$ sand set at Zhonger district of Gudao Oilfield

C. water content isoline map of 9th layer in $Ng_{(1+2)}$ sand set at Zhonger district of Gudao Oilfield.

Figure 3. Plane sketches showing sedimentary microfacies, remaining oil saturation and water content of production wells in the 9th layer of $Ng_{(1+2)}$ sand set, Zhong'er District, Gudao Oilfield.

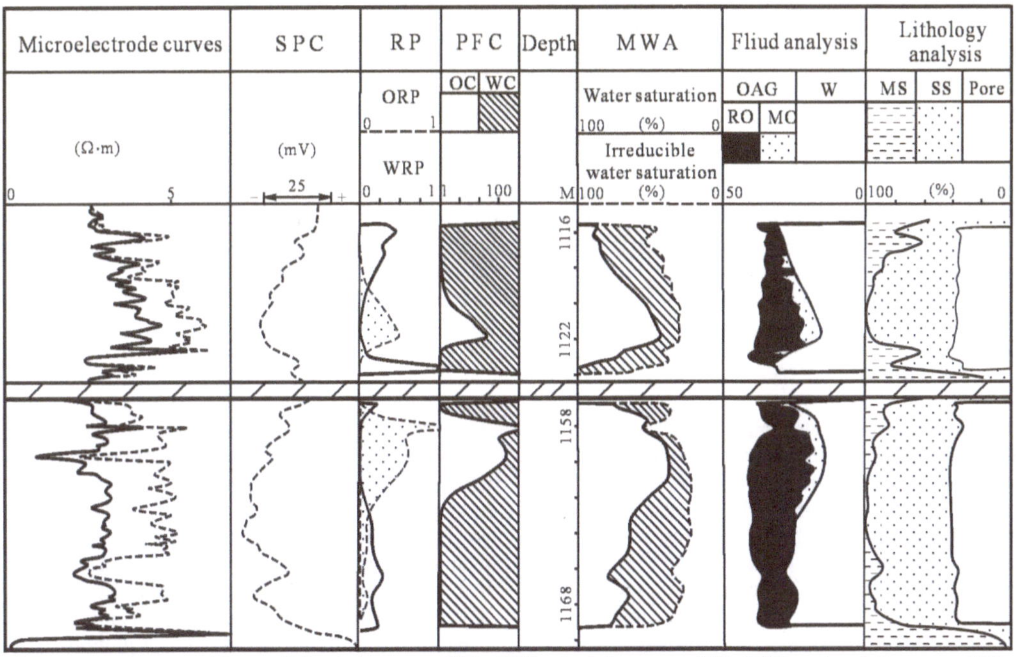

S P C: spontaneous potential curve R P: relative permeability ORP: oil relative permeability WRP: water relative permeability
P F C: produced fluid character OC: oil and gas content WC: water cut M W A: mobile water analysis OAG: oil and gas
RO: residual oil MO: mobile oil W: water MS: mud stone SS: sandstone

Figure 4. Vertical distribution plot of the remaining oil in the $Ng_{(1+2)}$ sand set in Gudao oilfield (Well No. Zhong 30CN515).

reservoir sands with small area and thickness are mainly distributed in the sedimentary structural units of lower-porosity and permeability, such as natural levee, crevasse splay, abandoned channel, floodplain facies and etc., and they form the high-saturation zones of the remaining oil. In addition, the stagnant zones with low pressure difference, the upper parts of partial microstructures, such as the lens and sand body edges independent of the well pattern, the area with incomplete injection-production well pattern, fault barrier and etc., form the high-saturation zones of the remaining oil saturation as well.

3.3 *Non-geological factors resulting in the remaining oil layer*

As development engineering and reservoir management could not be perfect, the pollution of oil layers and interlayer disturbance could result in the formation of the remaining oil layers.

3.4 *Discussions on the relationship between the sedimentary model and remaining oil distribution*

The formation factor is the key factor that controls the remaining oil distribution in the fluvial reservoir (Dong et al., 1999). In the meandering river reservoir, the spacial distribution, geometric shape, scale, continuity of sand bodies, and heterogeneities in interlamination and lateral or inner layers caused by the variation of physical parameters, such as barrier interlayers, porosity and permeability, are mainly controlled by the sedimentary microfacies. The reservoirs in different sedimentary microfacies units possess different porosity, permeability, pore structure and seeping characters as well as different flushing efficiency by water and remaining oil distribution characters. Controlled laterally by the meandering river sedimentary model, sands in the $Ng_{(1+2)}$ reservoir usually appear in sinuous strip, ribbon and branch-like distribution. The physical properties of the reservoir formed in the channel with the favorable porosity-permeability and high initial oil saturation are obviously better than those of the reservoir formed in the flank of natural levee, crevasse splay, abandoned channel and flood-plain. Thus, the injected water preferentially migrates into the reservoir of channel subfacies, and goes along the channel prolongation direction, and then spreads into the two sides of the channel during the water injection of the $Ng_{(1+2)}$ reservoir (Fig. 3C). While the reservoirs of non-channel or channel edge seem to have poor flushing efficiency, low production and thus high saturation of the remaining oil. Vertically, the remaining oil distribution in the single sand is controlled by the sedimentary rhythm and the impermeable interlayers in reservoirs, which are actually caused by the vertical variation of the sedimentary microfacies of the meandering river (i.e. controlled by the dualistic structure of the meandering river). From the formation of the $Ng_{(1+2)}$ meandering river, channel lag deposit, channel sand deposit, marginal bank or meander bar deposit are formed successively from bottom to top, and overbank and flood-plain deposits appear upward, resulting correspondingly in the physical properties of the reservoir getting worse and the thickness and the frequency of interlayers increasing from bottom to top. Thus, the injected water will break preferentially through along the lower zone of high-permeability, while the poor flushing efficiency, low production and high saturation of the remaining oil appear at the upper zone of low-permeability.

Therefore, the sedimentary model of the $Ng_{(1+2)}$ controls the distribution of the remaining oil in the following ways.

The sedimentary model of the meandering river of high-curvature and fine-grained controls the architecture and macroscopic heterogeneity of the $Ng_{(1+2)}$ reservoir sands. It is the vertical difference between the layers and sharp facies change on the lateral that makes the reservoir units have different porosity-permeability and seeping characters, different displacement efficiency and distribution characters of the remaining oil, and consequently form the different types of the special distribution of the remaining oil. The main oil layers developed in the channel subfacies possess the characteristics of the high-production and low-saturation of the remaining oil, while the non-main oil layers developed at the edge facies of the channel appear to have the low-production and high-saturation of the remaining oil. In the same oil layer, the remaining-oil saturation of the main channel reservoir is lower than that of the thin beds developed at the edge and outer of the channel facies.

The sedimentary model of the fine-grained meandering river controls the spacial motion rule of the injected water. The distribution rule of the water content of the producing wells in the $Ng_{(1+2)}$ reservoir indicates that the injected water always migrates preferentially into the channel and breaks thorough along the channel in the direction of the high-pressure gradient until the pressure gradient decreases or the drainage is blocked, then it moves upward or into the two sides of the channel. The lateral change of the sedimentary microfacies causes the lateral heterogeneity of the seepage capacity as well as the heterogeneity of water displacing oil, so as to result in the lateral heterogeneity of the remaining oil distribution.

The fine-grained meandering river sedimentary model determines the sedimentary rhythm and interlayers distribution characters, and consequently controls the distribution of the remaining oil *in situ*. The thinning-upward rhythm of the single sand vertically controls the volume of the injected water, which results in water breakthrough at the bottom of the channel sand and the formation of the remaining-oil enriched zone in the upper part. The interlayers *in situ* influence and control incoordinately the seepage of water and oil in the oil layers, and the influence extent is related with the thickness, the prolongation and the position of the interlayers. The interlayers in the upper oil layer affect lightly the seepage of the water and oil but the middle or middle-upper parts affect it highly.

Different from common fluvial deposits, the high curvature fine-grained meandering stream deposit makes the distribution of the remaining oil in the $Ng_{(1+2)}$ reservoir more complex. First, there are lots of oil sandbodies formed in $Ng_{(1+2)}$ and controlled by the sedimentary model, and most of them are thin with small oil-bearing areas, it is hard to establish the reasonable well pattern. Thus, many remaining oil-enriched zones are formed laterally. Secondly, the commingled production for most of the $Ng_{(1+2)}$ reservoirs makes reservoirs at the channel edge or in the upper unit hard to produce oils, even become into by-passed oil zones. Thereby, it is important but also difficult how to improve the recovery in the $Ng_{(1+2)}$ reservoirs.

4 CONCLUSION

In view of the lithological character of sandstone, grain size character and spacial distribution of sandbodies, the $Ng_{(1+2)}$ sand set in Gudao Oilfield belongs to the meandering river deposit, and the value of curvature P suggests that it is a high-curvature of the meandering river deposit. The $Ng_{(1+2)}$ sand set can be divided into 3 epicycles, which are also in accordance with the fine-grained meandering stream model by Miall.

The main oil layers are still remaining oil-enriched zones, while the non-main oil layers are of high remaining oil saturation; the remaining oil enriched zones are formed vertically in the reservoirs, and on the lateral, the remaining oil is enriched in the structural units of low permeability and less developed zones.

The sedimentary model of the meandering river of the $Ng_{(1+2)}$ sand set in Gudao Oilfield controls the inner structure and macroscopic heterogeneity of the reservoir sandbodies as well as the spacial movement rule of the injected water, and consequently the meandering river sedimentary model is the ultimate factor that controls the distribution of the remaining oil in the $Ng_{(1+2)}$ reservoir.

ACKNOWLEDGEMENTS

We thank Shujun Song, Xinming Duan, Zhaojie Xie, Dongmei Wang for providing samples and geological data

REFERENCES

Abbots, F.V. & Van Kuijk, A.D. 1997. Using 3D geological modeling and connectivity analysis to locate remaining oil targets in the Brent reservoir of the mature Brent Field. *SPE Offshore Europe Conference Proceedings* PP. 104–114 (SPE-38473), Aberdeen

Dong, D., Chen, J. & Qiu, M.W. 1999. Residual oil types and distribution rule in fluvial facies reservoir. *Oil & Gas Recovery Technology* 6(3): 39–46 (in Chinese with English abstract)

Fu, G.M., Li, Y.J. & Shi, J.P. 2003. The heterogeneity model of the fluvial facies reservoir sandbody and the distribution of the remaining-oil. *Journal of Chang'an University (Earth Science Edition)* 25(1): 15–19 (in Chinese with English abstract)

Holtz, M., Knox, P., Mcrae, L., Hentz, T. & Chang, C. 1996. Revitalizing a mature oil play: strategies for finding and producing uncovered oil in Frio fluvial-deltaic sandstone reservoirs of South Texas: Annual report. *US DOE Fossil Energy Report*, No. DOE/BC/14959-17

Li, S.Y., Li Z., Wang, Z.C., Liu, J.M., Wu, C.D. & Li, X. 2001. The study of sedimentary model in the upper member of Guantao Formation in Gudao Oilfield of Shengli Petroleum Province. *Acta Sedimentologica Sinica* 19(3): 387–393 (in Chinese with English abstract)

Li, X.G. & Zhou, X.C. 1982. The two types of fluvial facies reservoirs and the analysis for their production effects in Gudao Oilfield. *Petroleum Exploration and Development* 9(2): 44–51 (in Chinese with English abstract)

Liu, J.M., Li, Y. & Yan, J.X. 2000. The discussion on remaining-oil distribution and it's controlling factors in fluvial reservoirs. *Oil & Gas Recovery Technology* 7(1): 50–53 (in Chinese with English abstract)

Liu, J.M. & Xu, S.Y. 2003. Reservoir sedimentary model of fluvial facies and it's control on the remaining oil distribution. *Acta Petrolei Sinica* 24(1): 58–62 (in Chinese with English abstract)

Marcu, D. & Clipea, M. 2001. The revealing and exploitation of un-drained areas from old oil fields. *Petroleum Society of CIM/CAN International Petroleum Conference Proceedings* Paper No. 2001-111, Calgary

Miall, A.D. 1996. *The Geology of Fluvial Deposits*. Berlin: Springer Verlag

Virnovsky, G.A., Helset, H.M., Skjaeveland, S.M. & Skauge, A. 1998. A model to quantify residual saturation distribution in heterogeneous reservoirs. *SPE et al India Oil & Gas Conference Proceedings* PP. 307–314 (SPE-39538), New Delhi

Wang, B.H. & Qian, K. 1992. *Geological study and exploration practice in Shengli Oilfield*. Dongying: Petroleum University Press (in Chinese)

Petroleum Geochemistry and Exploration in the Afro-Asian Region – Digang Liang, Darui Wang & Zhenxi Li (eds)
© 2008 Taylor & Francis Group, London, ISBN 978-0-415-44083-7

Organic and trace elements geochemistry of western Niger delta oils

A.A. Mustafa & J.I. Nwachukwu
Department of Geology, Obafemi Awolowo University, Ile-Ife, Nigeria

ABSTRACT: Thirteen crude oil samples from on-shore western swamp fields in the Niger Delta were analyzed using standard organic and inorganic geochemical techniques (column chromatography, gas chromatography and inductively coupled plasma spectrophotometry) to permit their characterization. The results show that API gravity ranges from 23.5–43°; CPI ranges from 0.99–1.48; Pr/Ph ratio ranges from 1.25–2.65; Pr/nC_{17} ratio ranges from 0.41–3.90; and Ph/nC_{18} ratio ranges from 0.24–4.15. These indicate mixed organic sourced oils. The concentration of transition metals (V, Cr, Mn, Fe, Co, Ni, Cu) in these oils ranges from 0.04–16.69 ppm. Statistical analysis of the trace metals contents and organic geochemical data enabled the correlation and classification of the oils into two groups (A & B) similar to those of earlier workers. Star diagrams and cross plots (Co/Ni versus V/Ni) confirm that the two groups of oils are genetically related with group A oils being the biodegraded equivalents of group B oils. The threshold temperature for biodegraded oils was found to be 79° C, which is in agreement with earlier published data. In general, our results support the presence of a single petroleum system in the western swamp of the Tertiary Niger Delta.

1 INTRODUCTION

The Niger delta petroleum province has been extensively studied such that its source rocks, reservoir rocks and hydrocarbon occurrence are fairly well known (see Evamy et al., 1978; Ekweozor & Okoye, 1980; Nwachukwu & Chukwura, 1986; Bustin, 1988; Doust & Omatosola, 1990; Ekweozor & Daukoru, 1994; Tuttle et al., 1999; Haack et al., 2000; Matava et al., 2003; Eneogwe & Ekundayo, 2004 among others).

Most of these authors are of the opinion that we have a single Tertiary petroleum system in the delta. However, the possibility of a Cretaceous petroleum system has been reported (Haack et al., 2000). Since this has not been unambiguously established partly because of dearth of reliable data and partly because of conflicting geochemical evidence in the literature (Tuttle et al., 1999; Haack et al., 2000), there is the need to further re-evaluate Niger delta crude oils and source rocks.

In this study we have analyzed crude oil samples from the onshore western swamp of the delta, using inductively coupled plasma spectrophotometry (for trace metals) and standard organic geochemical techniques. Previous trace elements studies used Instrumental neutron activation analysis (Nwachukwu et al., 1995) and Atomic absorption spectrophotometry (Udo et al., 1992).

2 GEOLOGIC SETTING

The Niger delta is located in the Gulf of Guinea and covers an area of over 70,000 km^2. Basin formation began in the Early Cretaceous with the formation of the Benue – Abakaliki trough as a failed arm of the rift triple junction associated with the separation of the African and South American continents and subsequent opening of the south Atlantic (Tuttle et al., 1995). The Santonian folding, faulting and uplift of the sediment filled Benue-Abakaliki trough caused the subsidence of the Anambra basin to the west, and Afikpo syncline to the east. Marine deposition continued in these basins up to the Paleocene (Bustin, 1988). In the Eocene, the Benin and Calabar flanks were uplifted leading to the merging of the Niger-Benue and Cross- river delta systems in Middle Miocene, and progressive outbuilding of the delta southwards.

The stratigraphy of the Tertiary Niger delta is well known and consists of three diachronous units which show an overall upward transition from dark-gray to black, marine, prodelta shales (Akata Formation) through the paralic Agabda Formation, an alternating sequence of sands and shale to continental sands and gravels of the Benin Formation (Weber & Daukoru, 1975; Evamy et al., 1978; Fig. 1). About five submarine canyons are present in the delta (Tuttle et al., 1999).

Five offlapping siliciclastic sedimentation cycles (depobelts) are also present in the delta with each

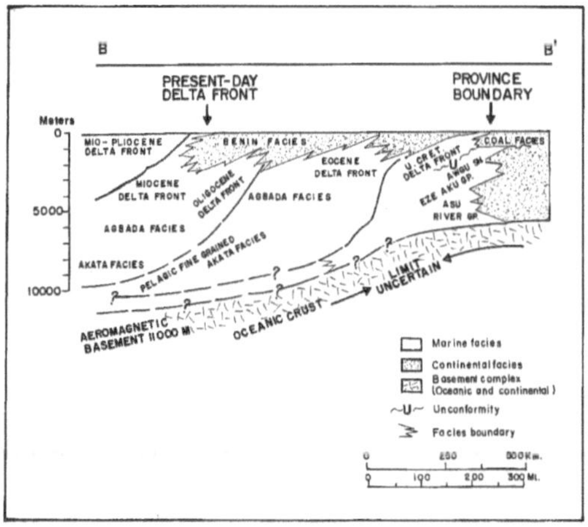

Figure 1. Stratigraphic section of the Niger Delta (After Tuttle et al., 1999).

having its own separate sedimentation, deformation and petroleum history (Evamy et al., 1978; Ejedawe, 1981; Doust & Omatsola, 1990).

3 MATERIALS AND METHODS

Thirteen oil samples from three fields (Abiteye, Makaraba and Utonana) in the western swamp of the Niger delta (Fig.2) were analyzed. The oils were fractionated by column chromatography on a Varian 500 mg pre-packed bond eluate cartridge (i.e. pre-packed with silica gel) and eluted with n-hexane. The nitrogen, sulphur and oxygen containing compounds (NSO's) were retained in the column while the alkanes and aromatics were fractionated on a Waters HPLC fitted with a 600 multi-solvent delivery system. The detector system includes a 410 Differential refractometer and a 486 Tunable absorbance detector. The column used was a Whatman Partsil 10 while the solvent was hexane. At a flow rate of 2.05 ml per minute, the saturate fraction eluted at about 15 to 18 minutes while the aromatic fraction eluted at about 28 to 30 minutes. Both fractions were monitored on a Waters 746 Data module coupled to the HPLC.

Gas chromatography was achieved on a Hewlett Packard 5890 Series 11 Plus G.C., equipped with a 50 m × 0.2 mm × 0.5 um film thickness PONA (paraffins, olefins, naphthenes and aromatics) cross-linked methyl siloxane capillary column, for the whole oil, saturate and aromatic fractions. A flame ionization detector was used and the carrier gas was hydrogen flowing at 300 ml/sec. The oven was programmed from 30 to 305°C at 1°C/min. The final temperature was held isothermally for 20 minutes. The peak areas were electronically integrated with the H P Chemstation

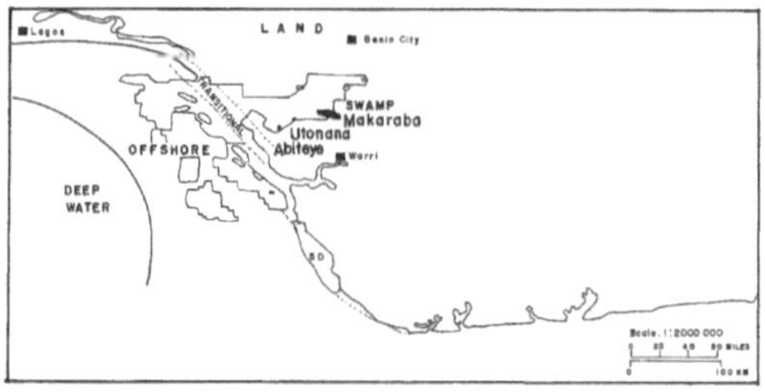

Figure 2. Map of Niger Delta showing oil field locations used for this study.

Software and identification of the n-alkanes was based on retention times and comparison with authentic standards.

The inductively coupled plasma spectrophotometer (ICPS) was used for the detection of trace metals concentration in the oils. Prior to the ICPS analysis, the oil samples were digested with tetraoxosulphate V1 acid and hydrogen peroxide under intense heating for the removal of the hydrocarbon compounds and concentration of the trace metals.

The product was a colourless liquid often called "water sample" but concentrated with trace metals. Approximately 1 g of the oil sample was digested and diluted to 50 cm^3 by the reagents (tetraoxosulphate V1 acid and hydrogen peroxide). The "water sample" was then loaded into the ICPS for analysis. The results of the analysis were corrected to parts per million by multiplying the ICPS value by the dilution factor which in this case is the volume to which sample was diluted (i.e. 50 cm^3) divided by the weight of the sample.

4 RESULTS AND DISCUSSION

It is well known that the physical and chemical properties of Niger delta oils are highly variable even down to reservoir level (Tuttle et. al. 1999). Such is also the case with the oils we analyzed. API gravity of the oils ranged from 23.5 to 43°.

Inspection of the whole oil chromatograms showed that the oils were biodegraded to various degrees. Six were normal, 4 had minor biodegradation while 3 were moderately biodegraded (Table.1; Fig.3). Detailed gravimetric data are shown in Table 2.

Carbon Preference Index (CPI) ranged from 0.99 to 1.48 averaging 1.05. This is consistent with abundance of terrestrial organic matter in the source rocks. Pristane/Phytane (Pr/Ph) ratio ranged from 1.25 to 2.65 averaging 2.03. Pr/nC_{17} ranged from 0.41 to 3.90 averaging 1.20 while Ph/nC_{18} ranged from 0.24 to 4.15 averaging 0.86. A plot of Pr/nC_{17} against Ph/nC_{18} (Fig.4) shows the oils plotting within the oxidizing peat-coal environment and mixed organic source facies. One sample plotted within the marine

Table 1. Bulk geochemical data for oils.

1	Reservoir sand	Reservoir age	Reservoir depth m(ft) top	Reservoir depth m(ft) bottom	Density g/cm^3	API gravity	V/Ni ratio	Co/Ni ratio	Pr/Phy ratio	Pr/nC_{17} ratio	Phy/nC_{18} ratio	CPI C_{21-29}	Bacterial degradation
UT-1	E-05/UT-01	Early	2185(7283)	2199(7330)	0.85	35	1.42	0.68	1.67	0.63	0.40	1.03	0 (Normal)
UT-1D	D-15/UT-01	Miocene	2074(6912)	2159(7198)	0.86	32.5	1.35	0.63	1.83	1.01	0.59	1.03	1 (Minor)
UT-4D	D-03/UT-04		1969(6562)	2018(6726)	0.91	23.5	0.89	0,49	1.25	1.99	1.62	1.21	4 (Moderate)
UT-8	E-07/UT-01		2470(8232)	2482(8272)	0.83	39.6	1.37	0.70	2.13	0.55	0.28	1.03	0 (Normal)
UT-8D	D-02/ut-05		2018(6727)	2033(6778)	0.91	23.9	1.02	0.61	1.80	2.35	1.75	1.03	3(Moderate)
UT-9	E-07/UT-01		2647(8822)	2662(8874)	0.83	39.9	1.03	0.64	2.29	0.52	0.25	1.03	0 (Normal)
UT-9D	D-03/UT-04		2091(6970)	2155(7182)	0.90	26.4	0.59	0.43	1.78	3.90	4.[illegible]5	1.20	4 (Moderate)
MK-1D	C-01/MK-01		1961(6536)	1979(6598)	0.90	27.5	1.59	0.70	1.89	0.41	0.24	0.99	0 (Normal)
MK-3	G-01/MK-01	Oligocene	3150(10500)	3158(10528)	0.81	43	0.93	0.60	2.65	0.60	0.26	1.02	0 (Normal)
MK-11	C-03/MK-01	Early	1999(6664)	2053(6842)	0.88	30.3	1.32	0.71	2.14	1.29	0.47	1.02	1 (Minor)
AB-7D	C-10/AB-03	miocene	1925(6418)	1934(6448)	0.86	33.6	1.15	0.63	2.17	0.81	0.41	1.02	1 (Minor)
AB-19	C-12/AB-03		1971(6571)	1988(6628)	0.86	33.5	1.40	0.73	2.28	0.85	0.42	1.04	1 (Mnor)
AB-26	C-09/AB-13		2097(6990)	2106(7020)	0.83	40	1.45	0.71	2.45	0.63	0.29	1.02	0 (Normal)

0 = No obvious degradation; 1 = Lower homologs of n-paraffins depleted; 3 = Only traces of n-paraffins remain; 4 = Non-paraffins, acyclic isoprenoids intact; 5 = Acyclic isoprenoids absent; 6–8 = Selected steranes and hopanes partly degraded; 9 = Steranes and hopanes heavily degraded.

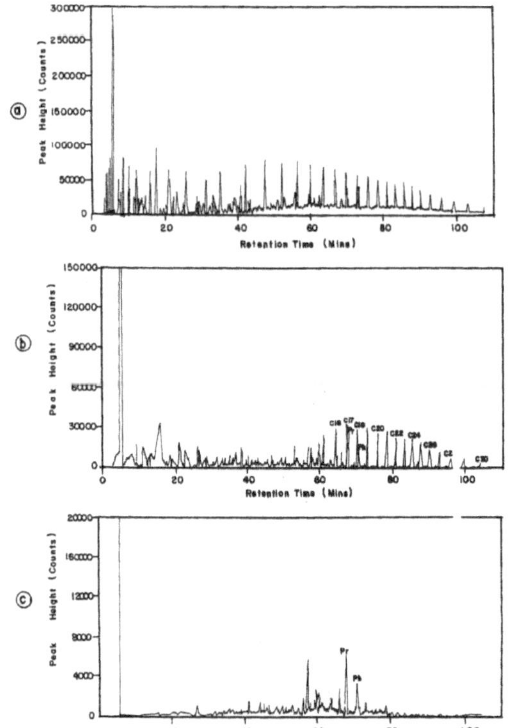

Figure 3. Gas chromatograms of oils: (a) UT-1 (b) AB-7D and (c) UT-9D.

organic matter field and is a moderately, biodegraded oil (UT-9D). Two groupings are observed, reflecting the biodegraded and normal crudes.

The normal crudes appear to be more mature. The two groups are identical with those of Nwachukwu et al. (1995). Group B oils are undegraded and correspond to the Bonny light stream (Evamy et al., 1978; Doust & Omatsola, 1990) which is paraffinic and waxy. Group A oils are largely biodegraded and correspond to Bonny medium crude oils (Evamy et al., 1978; Doust & Omatsola, 1990) which are naphthenic and non waxy. All the oils are early Miocene oils, with the exception of one (MK-3), which is Oligocene in age (Table 1).

Comparison of the data with other Miocene Niger delta oils shows that the non-biodegraded oils have Pr/Ph ratio of 1.67–2.65 averaging 2.18, which is

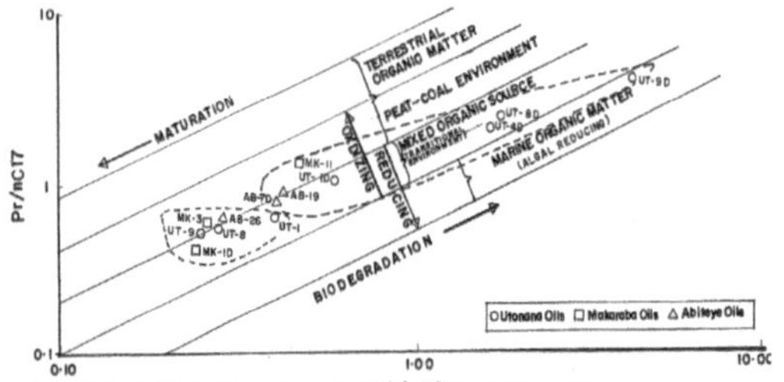

Figure 4. Pr/nC_{17} versus Ph/nC_{18} plot for oils (After Orr: In: Shanmugam, 1985).

Table 2. Gravimetric data of oils.

	UT-1	UT-1D	UT-4D	UT-8	UT-8D	UT-9	UT-9D	MK-1D	MK-3	MK-11	AB-7D	AB-19	AB-26
C_{15}	9.2	9.6	5.5	10.0	9.2	9.9	5.7	8.9	10.5	1.3	6.6	6.9	11.7
C_{16}	8.7	8.5	5.1	9.3	6.4	9.3	5.5	8.8	9.4	3.0	8.3	8.6	10.8
C_{17}	8.2	8.5	4.5	8.8	8.2	8.8	8.6	8.5	8.7	5.4	8.9	9.0	9.6
PR	5.1	8.6	9.0	4.8	19.1	4.5	33.4	3.5	5.2	6.9	7.2	7.6	6.0
C_{18}	7.7	7.9	4.4	7.9	6.1	8.0	4.5	7.7	7.6	7.0	8.0	8.0	8.4
PH	3.1	4.7	7.1	2.3	10.6	2.0	18.8	1.9	2.0	3.2	3.3	3.3	2.5
C_{19}	7.3	7.8	5.7	7.4	7.4	7.7	8.4	7.8	7.2	8.3	7.9	7.9	7.7
C_{20}	6.6	6.9	5.3	7.0	5.4	7.1	4.0	7.4	6.7	8.3	7.3	7.1	6.9
C_{21}	6.6	6.6	5.4	6.7	4.2	6.7	1.7	7.4	6.7	8.7	7.3	6.9	7.0
C_{22}	6.1	6.3	5.3	6.1	3.9	6.4	1.2	7.1	6.0	7.9	6.3	6.1	5.8
C_{23}	6.3	6.0	4.8	6.1	3.8	6.1	1.1	6.9	6.0	8.1	6.1	6.0	5.2
C_{24}	5.7	5.0	5.8	5.7	3.8	5.5	1.4	6.5	5.7	7.7	5.9	5.5	4.9
C_{25}	5.0	4.4	11.5	5.0	3.2	5.0	1.7	5.4	5.0	6.8	5.0	4.8	4.0
C_{26}	4.1	3.3	4.8	4.1	2.4	3.9	0.7	4.4	4.1	5.5	4.0	3.8	3.1
C_{27}	3.5	2.6	3.9	3.3	2.1	3.3	0.6	3.3	3.4	4.4	3.1	3.2	2.5
C_{28}	2.7	1.9	4.5	2.4	1.6	2.5	0.6	2.2	2.6	3.3	2.2	2.3	1.8
C_{29}	2.3	1.5	3.9	1.9	1.5	2.0	0.9	1.4	2.0	2.6	1.6	1.8	1.4
C_{30}	1.7	0.1	3.4	1.3	1.3	1.4	1.2	0.8	1.3	1.8	1.0	1.2	0.9
Pr/Ph	1.67	1.83	1.25	2.13	1.80	2.29	1.78	1.89	2.65	2.14	2.17	2.28	2.45
Pr/nC_{17}	0.63	1.01	1.99	0.55	2.35	0.52	3.90	0.41	0.60	1.29	0.81	0.85	0.63
Ph/nC_{18}	0.40	0.59	1.62	0.28	1.75	0.25	4.15	0.24	0.26	0.47	0.41	0.42	0.29
CPI	1.03	1.03	1.21	1.03	1.03	1.03	1.20	0.99	1.02	1.02	1.02	1.04	1.02

Carbon Preference Index (CPI) (after Bray & Evans, 1961).

comparable to 2.62 of Geomark Research Inc. (1988) and 2.25 of Eneogwe & Ekundayo (2004).

This indicates mixed marine and terrigenous organic source facies. The biodegraded oils had Pr/Ph ratio ranging from 1.25 to 2.28 averaging 1.85. This is radically different from 3.74 of Geomark Research Inc. (1988), which probably reflects higher terrestrial organic matter input and a more oxic depositional environment of source facies. Matava et al. (2003) reported Pr/Ph ratios of 2–5 for Niger delta oils in OML 67–70. It thus appears that the oils, irrespective of biodegradation and other alteration effects have the same source facies. Minor variations reflect migration and biodegradation effects (Matava et al., 2003).

4.1 *Star diagram*

A star diagram (also referred to as a polygon plot by Chambers & Kleiner, 1982) is a polar plot of selected G.C.-peak ratios. It has been found very useful in geochemical fingerprinting of oils and oil-oil correlation (Kaufman et al., 1990; Ganz et al., 2004). Usually, the oil peak ratios are plotted and the points connected, creating a star-shaped or polygon-shaped pattern characteristic of each oil. The ratios of peak numbers (peak ratios) selected for this study are listed in Table 3. Since no G.C.-Mass spectrometry was done, identity of these compounds could not be ascertained. The star diagram (Fig.5) shows two groups of oils as earlier indicated by the Pr/nC_{17}versus Ph/nC_{18} plot (Fig.4).

4.2 *Trace metals*

Cluster analysis of trace metals data was carried out using a number of variables ranging from API gravity to total transition metals concentration (TTM). It was observed that the dendrogram using API + TTM as variable (Fig.6) had the best agreement (85%) with the geochemical grouping (Fig.4) as it only misclassified two oils (UT-1 and MK-1D). Two major groups are also apparent (Groups A & B). The group A oils could be seen to be subdivided into two sub-groups, $A_1$and A_2depending on their degree of biodegradation. A_1's show minor degradation while A_2's show moderate degradation. MK-1D was misclassified due to its low

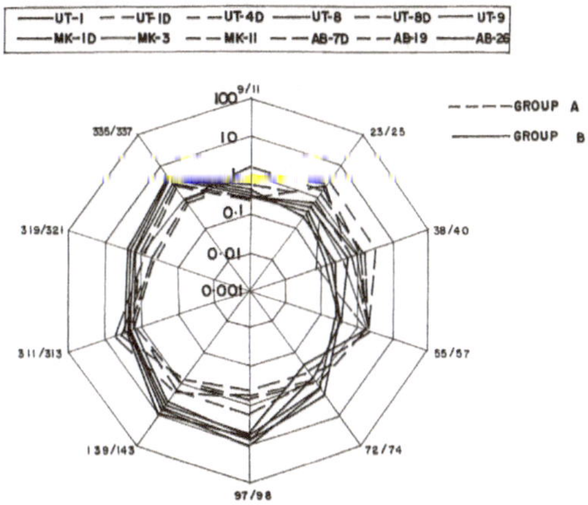

Figure 5. Star diagram of oils. The identity of the peaks could not be ascertained since no G.C.-Mass spectrometry was carried out.

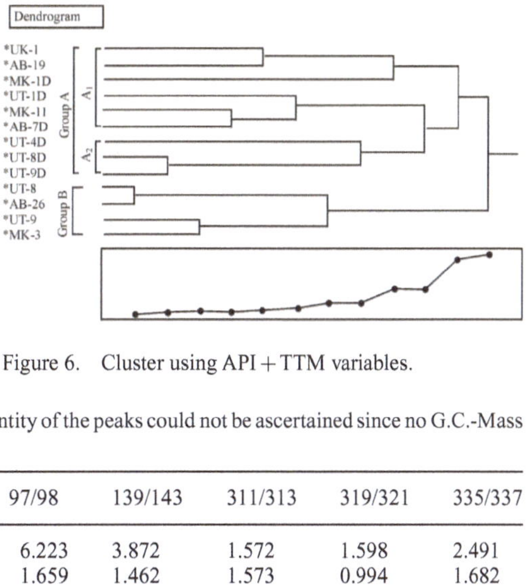

Figure 6. Cluster using API + TTM variables.

Table 3. Star diagram peak ratios (using peak numbers). The identity of the peaks could not be ascertained since no G.C.-Mass spectrometry was carried out.

Sample	9/11	23/25	38/40	55/57	72/74	97/98	139/143	311/313	319/321	335/337
UT-1	0.367	0.337	0.187	0.318	1.334	6.223	3.872	1.572	1.598	2.491
UT-1D	0.338	0.593	0.690	2.157	0.777	1.659	1.462	1.573	0.994	1.682
UT-4D	1.610	1.603	1.144	1.590	0.835	0.635	0.884	1.523	0.503	0.618
UT-8	0.300	0.425	0.357	2.042	0.840	7.036	5.084	2.380	1.828	3.525
UT-8D	0.938	2.545	3.322	2.159	0.399	0.372	0.709	3.396	0.426	0.573
UT-9	0.289	0.472	0.509	0.375	0.730	7.501	6.309	2.582	1.941	4.028
MK-D	0.369	0.237	0.215	0.221	1.402	10.956	6.675	2.067	2.413	4.174
MK-3	0.361	0.263	0.129	0.312	2.383	7.996	4.679	2.551	1.679	3.891
MK-11	0.207	1.945	0.651	2.024	0.896	0.495	1.852	4.962	0.773	2.147
AB-7D	0.574	0.970	1.336	2.056	0.603	0.720	0.715	3.081	1.236	2.412
AB-19	0.840	1.124	0.063	1.970	0.661	0.743	0.856	3.102	1.182	2.387
AB-26	0.228	0.578	0.927	1.805	0.244	4.815	9.064	2.437	1.593	3.428

API gravity, which wrongly suggests biodegradation instead of low thermal maturity.

The dendrogram using V/V + Ni fraction as variable (Fig.7) shows 62% agreement with the geochemical grouping (Fig.4). This dendrogram correctly classified the two oils earlier misclassified by Fig.6. Also 75% of the minor biodegraded oils (i.e. sub-group A_1) were classified as Group B while two normal oils were misclassified as degraded oils. The minor degraded oils may suggest a link (or relationship) between the normal Group B oils and the moderately biodegraded oils of sub-group A_2. Thus indicating that Group A oils are the biodegraded equivalents of Group B oils as earlier observed by Nwachukwu et al. (1995).

4.3 *Transition metals concentrations*

Table 4 shows the concentration of the transition metals (V, Fe, Ni, Cu, Cr, Co, Mn) and metal ratios for the oils. Individual element values range from 0.04 (Mn) to 16.69 ppm (V) while overall average values range from 0.33 (Mn)–5.79 ppm (Fe).

Of particular importance is the concentration and metal ratios of V, Ni and Co, which have been used in characterizing and correlating oils (Barwise, 1990; Nwachukwu et al., 1995). Vanadium concentration ranges from 0.38–16.69 ppm, averaging 4.13 ppm. The values are in agreement with 3.70–40.0 ppm of Udo et al. (1992) but are higher than values obtained by Ndiokwere (1983) and Nwachukwu et al. (1995). Nickel was found to range from 0.37–10.52 ppm, averaging 3.08 ppm, which is in agreement with values reported in the literature (Ndiokwere, 1983; Udo et al., 1992; Nwachukwu et al., 1995; Geomark Research Inc., 1998). The concentration of Co in the oils ranged from 0.23–7.35 ppm, averaging 2.02 ppm. The values are in agreement with data of Ndiokwere (1983) and Udo et al. (1992) but higher than 0.003–1.174 ppm obtained by Nwachukwu et al. (1995).

The V/V + Ni fraction ranges from 0.37–0.61 averaging 0.54. This is higher than 0.01–0.41 (Average: 0.12) obtained by Nwachukwu et al. (1995) and is in agreement with 0.69–0.74 of Geomark Research Inc. (1998) for Niger Delta Miocene oils and 0.46 of Oluwole et al. (1985) for Cretaceous tar sand oils. Ekweozor & Nwachukwu (1989) and Ekweozor (1990) have demonstrated that the tar sands bitumen reservoired in Cretaceous sediments are not insitu generated. Therefore the similarity in V/V + Ni ratios of the oils and the bitumen probably supports a common origin. In addition, Katz (2002) in his study of ODP leg 159 sediments confirmed a similarity between Oligo-Miocene intervals and deepwater oils of the delta. According to him, "No such similarity exists between the Cretaceous organic rich intervals and these oils".

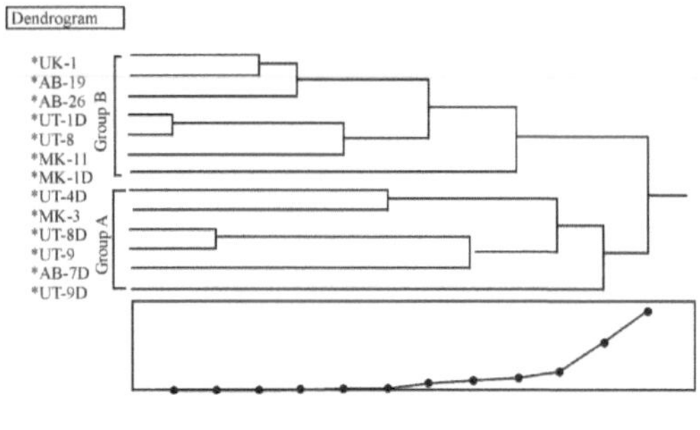

Figure 7. Cluster using V/V+Ni variable.

4.4 *Cross-plots*

Figure 8 shows the plot of API gravity versus Depth for all oils. The correlation coefficient is high (r = 0.699) which shows that the plot is useful as oil type discriminant. Based on the pattern observed a modified cross-plot (Quadrant cross-plot) was developed to give a proper view of the relationship between the oils.

Table 4. Transition metals concentration and ratios.

Sample	V	Fe	Ni	Cu	Cr	Co	Mn	TTM$^+$	V/ Ni	Ca/ Ni	V/ V+Ni	Fe/ V
UT-1	9.79	12.5	6.88	7.97	3.56	4.68	0.78	46.16	1.42	0.68	0.59	1.28
UT-1D	4.27	5.14	3.15	3.44	1.59	2.00	0.33	19.91	1.35	0.63	0.58	1.20
UT-4D	0.40	7.14	4.94	4.25	1.68	2.40	0.37	25.18	0.89	0.49	0.47	1.62
UT-8	2.92	4.81	2.13	2.77	1.13	1.49	0.24	15.49	1.37	0.70	0.58	1.65
UT-8D	1.34	4.62	1.32	1.18	0.54	0.80	0.12	9.92	1.02	0.61	0.50	3.45
UT-9	0.38	1.02	0.37	0.39	0.18	0.23	0.04	2.60	1.03	0.64	0.51	2.71
UT-9D	0.57	3.05	0.96	0.54	0.26	0.41	0.06	5.86	0.59	0.43	0.37	5.32
MK-1D	16.69	12.9	10.52	12.99	5.78	7.35	1.19	67.42	1.59	0.70	0.61	0.77
MK-3	0.50	2.42	0.54	0.76	0.24	0.32	0.06	4.83	0.93	0.60	0.48	4.86
MK-11	1.46	1.86	1.11	1.27	0.58	0.78	0.12	7.18	1.32	0.71	0.57	1.27
AB-7D	1.02	2.76	0.89	1.11	0.42	0.56	0.10	6.87	1.15	0.63	0.53	2.71
AB-19	6.53	11.5	4.66	5.97	2.41	3.39	0.57	35.02	1.40	0.73	0.58	1.76
AB-26	3.80	5.53	2.63	3.31	1.42	1.87	0.30	18.86	1.45	0.71	0.59	1.45
Average	4.13	5.53	2.63	3.31	1.42	1.87	0.30	18.86	1.45	0.71	0.59	1.45

The mean of both API and Depth values of all the oils was used as the origin of the plot on X and Y axes (respectively) instead of using zero origin.

Quadrant 1 shows normal mature oils probably derived from deep source rocks. Quadrant 2 shows minor biodegraded to normal mature oils that have probably migrated upward from deeply buried source rocks. Quadrant 3 shows minor to moderately biodegraded oils and /or immature oils expelled from source rocks at probably low thermal maturity. For example MK-1D was found to be normal but misclassified as degraded because of its API gravity which might be as a result of low thermal maturity. A CPI of 0.99 for MK-1D confirms its low thermal maturity. Quadrant 4 is empty and probably signifies immature oils. However thermal maturity parameters for Niger delta oils are not related to depth and do not vary greatly (Matava et al., 2003). The sickle shaped pattern formed by the oils (Fig. 8) may reflect a common origin for these oils.

Consideration of oil reservoir temperatures (Table 5) will show that Quadrant 1 oils have a temperature range of 81–107°C at depths of 2482–3158 m (8272–0528ft). Oils in quadrant 2 have a temperature range of 74–79°C at depths of 1934–2199 m (6448–7330 ft) while oils in quadrant 3 have a temperature range of 66–75°C at depths of 2018–2159 m (6726–7198 ft). The threshold of oil biodegradation was found to be 79°C at a depth of 2199 m (7330 ft). This is consistent with earlier observed data. For example Dickey et al. (1987) observed heavy oils occurring in shallow reservoirs at temperatures of 66–82° C. Their study also showed no clear relationship between the degree of oil degradation and the presence of meteoric water

Table 5. Reservoir temperature data for oils.

Well ID	Reservoir/ sand	Reservoir/ depth(ft)	Formation temperature (°)	API gravity
UT-1	E-05/UT-01	7283–7330	79	35
UT-1D	D-15/UT-01	6912–7198	75	32.5
UT-4D	D-03/UT-04	6562–6726	66	23.5
UT-8	E-07/UT-01	8232–8272	81	39.6
UT-8D	D-02/UT-05	6727–6778	66	23.9
UT-9	E-07/UT-01	8822–8874	82	39.9
UT-9D	D-03/UT-04	6970–7182	68	26.4
MK-1D	C-01/MK-01	6536–6598	71	27.5
MK-3	G-01/MK-01	10,500–10,528	107	43
MK-11	C-03/MK-06	6664–6842	72	30.3
AB-7D	C-10/AB-03	6418–6448	74	33.6
AB-19	C-12/AB-03	6571–6628	76	33.5
AB-26	C-09/AB-13	6990–7020	77	40

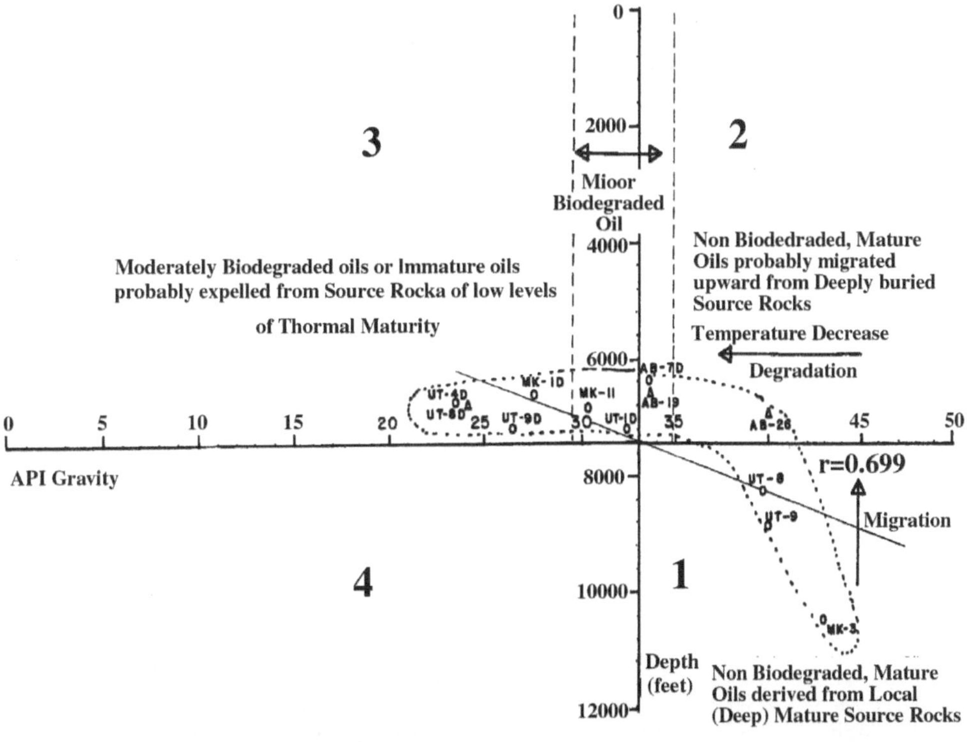

Figure 8. API versus Depth cross-plot.

as degraded oils were sometimes found associated with connate and meteoric waters, and undegraded oils with meteoric water. This they opined suggests extensive secondary migration of oil from one structure to another through either meteoric or connate water, or invasion of meteoric water into reservoirs containing undegraded oil accumulation. Also Ejedawe & Wagbatsoma (2002) observed that in the Greater Ughelli depobelt, the depth to the 80°C temperature level is about 9000 ft and corresponds to the change from heavy to light crude and that there is no clear relationship between API gravity and water salinity.

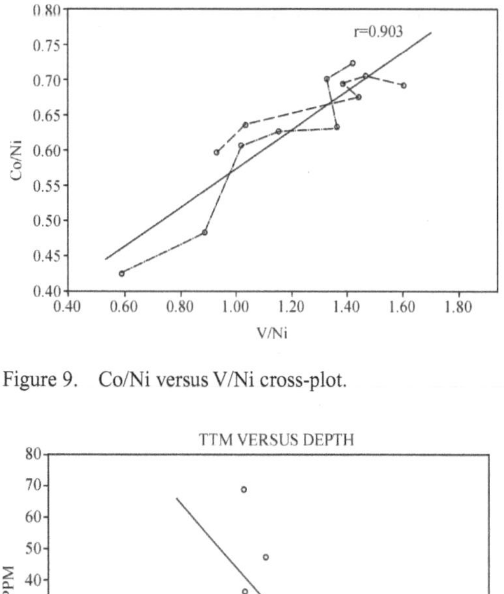

Figure 9. Co/Ni versus V/Ni cross-plot.

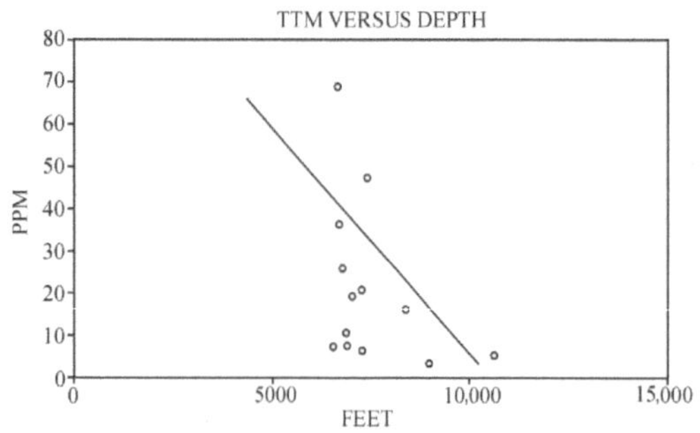

Figure 10. TTM versus Depth plot for oils.

4.5 *Co/Ni versus V/Ni cross-plot*

Figure 9 shows the cross-plot of Co/Ni versus V/Ni for the oils. This plot shows that Group B oils are distinct from group A oils. The two plots have similar trend, which indicates a genetic relationship as earlier observed by Nwachukwu et al. (1995).

4.6 *TTM versus depth cross-plot*

This plot (Fig.10) was also largely able to show two groupings although there is some overlap. In general, the concentrations (TTM) decrease with increasing burial depth. This decrease could be a biodegradation effect with the normal oils at greater depths having lower concentrations while the biodegraded shallower oils have higher TTM.

4.7 *Inter metal correlations*

Table 6 is a trace metals regression Table for the 13 oils. A unique feature of this is the strong correlation of all the metals with one another which may strongly indicate a common origin or close genetic relationship.

Co correlates strongly with Cu and Cr (r = 0.999 for both). The least strong positive correlation is Fe with V (r = 0.901) while the strongest positive correlation is Cr with V (r = 1.000). V also correlates strongly with Ni (r = 0.983).

As would be expected API gravity shows negative correlation with all transition metals because their effects on oils are opposite (Nwachukwu et al., 1995).

5 CONCLUSIONS

It is here concluded that:

- Organic geochemistry coupled with statistical analysis of trace metals data have been used to characterize 13 western Niger delta crude oils.
- API gravity ranges from 23.5–43°; CPI ranges from 0.99–1.48; Pr/Ph ratio ranges from 1.25–2.65; Pr/nC_{17} ranges from 0.41–3.90; and Ph/C_{18} ranges

Table 6. Regression matrix for oils.

Sample	V	Fe	Ni	Cu	Cr	Co	Mn	TTM^{+}	API
V	1								
Fe	0.901	1							
Ni	0.983	0.923	1						
Cu	0.997	0.923	0.989	1					
Cr	1.000	0.911	0.986	0.998	1				
Cu	0.997	0.927	0.990	0.999	0.999	1			
Mn	0.996	0.933	0.987	0.999	0.998	0.999	1		
TTM	0.991	0.948	0.991	0.997	0.994	0.998	0.998	1	
API	−0.214	−0.231	−0.302	−0.209	−0.213	−0.228	−0.208	−0.235	1

from 0.24–4.15. These indicate mixed organic sourced oils.
- The V and Ni concentrations are less than 100 ppm, which indicate abundance of terrestrial organic matter in the source rocks.
- Pr/nC_{17}versus Ph/nC_{18} and Co/Ni versus V/Ni cross-plots indicate the presence of two groups of oils, with Group A oils being the biodegraded equivalents of Group B oils as earlier observed by Nwachukwu et al. (1995).
- The similarity of the V/V+Ni fractions of the oils and tar sands bitumen may indicate a close relationship. Since the tar sands bitumen is non-indigenous, a single petroleum system is therefore supported by our data.
- API gravity is negatively correlated with all transition metals as earlier observed by previous workers.
- The threshold temperature for oil biodegradation was found to be 79°C, which is in conformity with earlier published data.

ACKNOWLEDGEMENT

We are grateful to Chevron Nigeria Limited and her joint venture partners for samples and analytical support and permission to publish the outcome of this research, which reflects the views of the authors.

REFERENCES

Barwise, A.J.G. 1990. Role of nickel and vanadium in petroleum classification. *Energy and Fuels* 4: 647–652.

Bustin, R.M. 1988. Sedimentology and characteristics of dispersed organic matter in Tertiary Niger Delta: Origin of source rocks in a deltaic environment. *AAPG Bull.* 72: 277–298.

Chambers, J.M. & Kleiner, B. 1982. Graphical techniques for multivariate data and for clustering. In: Handbook for statistics, P.R. Krisnaiah and L. N. Kanal (eds) *Classification pattern recognition and reduction of dimentionality.* New York, North – Holland, 2: 162.

Dickey, A.P., George, G.O. & Barker, C. 1987. Relationships among oils and water compositions in Niger delta. *AAPG Bull.* 71 (10): 1319–1328.

Doust, H. & Omatsola, E. 1990. Niger delta. In: J.D. Edwards & P.A. Santogrossi (eds.), *Divergent/Passive margin basins.* AAPG Memoir 48: 201–238.

Ejedawe, J.E. 1981. Patterns of incidente of oil reserves in Niger Delta Basin. *AAPG Bull.* 65: 1574–1585.

Ejedawe, J.E. & M. Wagbatsoma. 2002. Salinity variation and Heavy Oil Occurrence in Greater Ughelli depobelt. *NAPE Bull.* 16 (2): 106–120.

Ekweozor, C.M. 1990: Geochemistry of oil sands of Southwestern Nigeria. In: B.D. Ako & E.I. Neu (eds.). *Occurrence, Utilization and economics of Nigerian Tar sands.* NMGS Ibadan Chapter Publication: 50–62.

Ekweozor, C.M. & Daukoru, E.M. 1994. Northern delta depobelt portion of the Akata-Agbada(!) petroleum system, Niger Delta, Nigeria. In: Magoon, L.B. & Dow, W.G. (eds.) *The Petroleum System—From Source to Trap.* AAPG Memoir 60: 599–614.

Ekweozor, C.M & Nwachukwu, J.I. 1989. The origin of Tar sands of Southwestern Nigeria. *NAPE Bull.* 4 (2): 82–94.

Ekweozor, C.M. & Okoye, N.V. 1980. Petroleum source bed evaluation of Tertiary Niger Delta: *AAPG Bull.* 64: 1251–1259.

Eneogwe, C. & Ekundayo, O. 2004. Application of Mango's C_7 hydrocarbon parameters to the correlation of oil in the western Niger Delta, Nigeria. *Journal of Mining and Geology* 40 (1): 17–28.

Evamy, B.D., Haremboure, J., Kamerling, P., Knaap, W.A., Molly, F.A. & Rowlands, P.H. 1978. Hydrocarbon habitat of Tertiary Niger Delta. *AAPG Bull.* 62: 1–39.

Ganz, H., Dolapo, M.A., Okpoto, V. & Davidson, C. 2004. Geochemical Fingerprinting Technology in SPDC: Improving Reservoir Models and Field Development. SPE Paper 88884, *Technical Journal, Nigerian Annual International Conference & Exhibition, Abuja, Nigeria* 2nd–4th August 2004.

GeoMark Research, Inc. 1998. OILS: Oil Information Library System, version 1.0: Houston, Texas, GeoMark Research, Inc., [database available from GeoMark Research, Inc., 9748 Whithorn Drive, Houston, TX 77095].

Haack, R.C., Sundararaman, P., Diedjomahor, J.O., Xiao, H., Gant, N.J., May, E.D. & Kelsch, K. 2000. Niger Delta Petroleum Systems, Nigeria. In: M.R. Mello & B.J. Katz (eds) *Petroleum Systems of South Atlantic Margins.* AAPG Memoir 73: 213–231.

Katz, B. 2002. Significance of ODP Results on Deepwater Exploration – Greater Niger Delta Region. *NAPE Bull.* 16 (2): 93–105.

Kaufman, R.L., Ahmed, A.S. & Elsinger, R.J. 1990. Gas Chromatography as a Development and Production Tool for fingerprinting Oils from individual Reservoirs: Applications in the Gulf of Mexico. *GCSSEPM Foundation Ninth Annual Research Conference Proceedings:* 263–282.

Matava, T., Rooney, M.A., Chung, H.M., Nwankwo, B.C. & Unomah, G.I. 2003. Migration effects on the composition of hydrocarbon accumulations in the OML 67–70 areas of the Niger Delta. *AAPG Bull.* 87 (7): 1193–1206.

Ndiokwere, CH.L. 1983. Analysis of Nigerian petroleum for trace elements by Neutron activation. *Radiochem. Radioanal. Lett.* 59 (4): 201–212.

Nwachukwu, J.I. & Chukwura, P.I. 1986. Organic matter of Agbada Formation, Niger Delta, Nigeria. *AAPG Bull.* 70 (1): 48–55.

Oluwole, A.F., Adegoke, O.S., Kehinde, L.O, Borisade, A.B., Ako, B.D., Ajayi, T.R., Enu, E.I., Nwachukwu, J.I., Coker, S.J.L., Wallace, D., Asubiojo, O.I. & Ogunsola, O. 1985. Chemical composition of Bitumen Extracts from Nigerian Tar sands. In: *Proceedings of 3rd Int. Conf. On Heavy Crude and Tar sands. Long Beach, California*: 373–379.

Orife, J.M & Avbovbo, A.A. 1982. Stratigraphic and Unconformity Traps in the Niger delta. In: M.T. Halbouty (ed.). *The Deliberate Search for the Subtle Trap.* AAPG Memoir 32: 251–265.

Shanmugam, G. 1985. Significance of coniferous rain forests and related organic matter in generating commercial quantities of oil, Gippsland Basin, Australia. *AAPG Bull.* 69: 1241–1254.

Tuttle, M.L.W., Charpentier, R.R. & Brownfield, M.E. 1999. The Niger delta Petroleum System: Niger delta Province, Niger, Cameroon, and Equatorial Guinea, Africa. *USGS Open-File Report* 99-50-H: 64 pp.
Udo, O.T., Ekwere, S. & Abrakasa, S. 1992. Some trace metals in selected Niger Delta crude oils: Application in oil-oil correlation studies. *J. Mining and Geology* 28(2): 289–292.
Weber, K.J. & Daukoru, E.M. 1975. Petroleum Geological aspects of the Niger Delta. *Proc. 9th World Petr. Congress.* 2: 209–222.

Petroleum Geochemistry and Exploration in the Afro-Asian Region – Digang Liang, Darui Wang & Zhenxi Li (eds)
 ISBN 978-0-415-44083-7

Predictive models of the formation volume factor and density of the saturated crude oil

Haitao Xue, Shuangfang Lu, Bo Wang & Xiaoyan Liu
Daqing Petroleum Institute, Daqing, P R China

Xiaochun Fu
Chongqing Medicine Industry Academe, Chongqing, P R China

ABSTRACT: In this paper, the dissolving process of natural gas in crude oil is investigated theoretically. It is found that the volume change of crude oil is mainly related to two factors: ① the volume expansion caused by temperature increase, which can be described by temperature coefficient $f\ (t)$; ② the volume expansion caused by dissolution of gas in crude oil, which can be described by liquid apparent molar volume (V_{gmi}) of each constituent of gas and its molar solubility in oil (C_i) or gas-oil ratio (R_{Si}). Based on the volume addition principle of ideal solution, we deduced the theoretical equation in which the volume factor (B) and density (d_p) of the saturated crude oil can be figured out. The results of the study show that the new models in which the volume factor and density of the saturated crude oil are expressed as the function of temperature and solubility of each constituent of gas or gas-oil ratio are more reasonable than other models where a relative density of gas and a total gas-oil ratio are used only. And new models are appropriate for normal gas and unconventional natural gas rich in non-hydrocarbon gas constituents.

Keywords: saturated crude oil; formation volume factor; density; solubility; gas-oil ratio

1 INTRODUCTION

The formation volume factor (FVF), gas-oil ratio and density of the saturated crude oil are important and common parameters in petroleum engineering calculations (Elsharkawy, 1997) and assessment of traps (Chen and Xu, 1993). Empirical PVT correlations have been the subject of researches since the beginning of reservoir and production engineering. Several empirical PVT correlations have been published for crude oil from various geographical locations around the world (Standing, 1947; Lasater, 1958; Galsø, 1980; Vasquez and Beggs, 1980; Obomanu and Okpobiri, 1987; Al-Marhoun, 1988, 2004; Dokla and Osman, 1992; Kartoatmodjo and Schmidt, 1994; Elsharkawy, 1997). Some of the most widely used PVT correlations can be found in Standing (1947), Vasquez and Beggs (1980). The FVF and gas-oil ratio correlations of Vasquez and Beggs have been appraised to be the most accuratio models (Bradley, 1996; Al-Marhoun, 2004). Tentative application reveals that Vasquez-Beggs' model shows discontinuousness when API equals 30° due to adopting two sets of constant coefficients. Elsharkawy model of FVF (1997) has the same problem. Both Standing's and Vasquez and Beggs' models can neither fit for the natural gas with high N_2 and CO_2 contents nor for methane. Among the input variables (such as gas-oil ratio, crude oil API, gas gravity, reservoir pressure and temperature) in any PVT correlation, gas gravity is the most questionable property because it is rarely measured due to high expense and its dependence on surface separation conditions (Elsharkawy, 1997). In addition, it is unreasonable to express the property of gas with a relative density. Based on the experimental data, we investigated theoretically the dissolving process of gas in crude oil and established the theoretical predicting models for FVF and density of the saturated crude oil.

2 THE DEFECTS OF STANDING'S AND VASQUEZ AND BEGGS' MODELS

Although Standing's and Vasquez and Beggs' models are used widely, the authors noticed these correlations have the following defects.

(1) Vasquez and Beggs' correlations adopt two sets of constant coefficients. That is, when the API of crude oil is ≤30° or >30°, a different set of constant coefficients is used. We gained the high-pressure property data of 8 oils from test wells in the Daqing oilfield (see Table 1). Using the input

Table 1. High-pressure property data of 8 oils from test wells in Daqing Oilfield.

Temperature/ °C	Pressure/ ×0.1 MPa	Ground oil gravity/ g·cm^{-3}	Ground oil API/°	Gas specific gravity (air = 1.0)
108.3	67.5	0.8619	32.7	0.8528
62.8	71.5	0.8514	34.7	0.8529
57.2	55.6	0.8715	30.9	0.8603
85	97	0.8715	30.9	0.8603
85.6	40	0.8754	30.1	0.8592
81	70	0.8503	34.9	0.8164
85	75	0.8427	36.4	0.8476
71.6	130	0.8385	37.3	0.6167

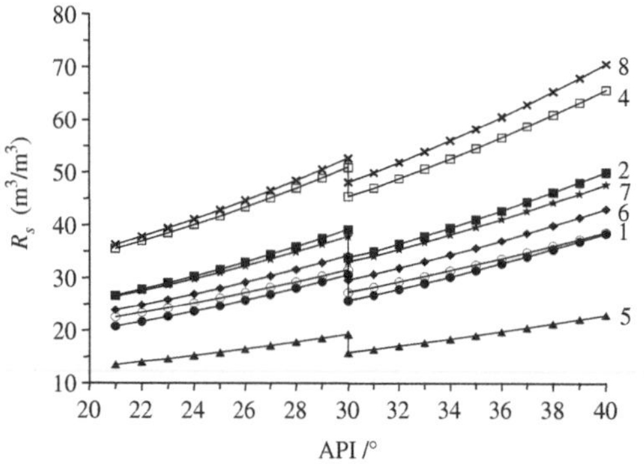

Figure 1. Relationship between gas-oil ratio (R_s) calculated from Vasquez-Beggs' model and crude oil API. When the API = 30° (equivalent to 0.8762 g/cm^3) the curves of gas-oil ratio arise discontinuously. So in a certain range of API, it causes the wrong result that gas-oil ratios of the high-gravity oil are lower than those of the black oil.

parameters provided by Table 1 such as temperature, pressure and gas gravity, and supposing the given gas dissolved in different API crude oils, we calculated FVF and gas-oil ratio of the crude oil by Vasquez and Beggs' model. When the API = 30° (equivalent to 0.8762 g/cm^3) the curves of FVF and gas-oil ratio arise discontinuously (Fig. 1 and Fig. 2). This is the biggest defect of Vasquez and Beggs' model. Elsharkawy (1997) model has the same defect. From Fig. 1 and Fig. 2 it is obviously seen that R_s ~API and B_{ob} ~API curves arise with breakpoints. So in a certain range of API, it causes the wrong result that FVF and gas-oil ratio of the high-gravity oil are lower than those of the black oil. The API for most crude oils in China is just around 30°, so the application of these models is restricted.

(2) They can't fit for the natural gas with high N_2 and CO_2 contents. In different regions, constituents of natural gases are obviously different. The above-mentioned models have a general character: the bigger the gas specific gravity, the greater the gas-oil ratio. The gas specific gravity of N_2 is close to 1 while that of CO_2 is greater than 1, and that of methane is equal to 0.55. The solubility of methane in the crude oil is higher than that of N_2 and lower than that of CO_2. Therefore, when natural gas contains plenty of N_2, the model calculation value will be higher than the actual value. There will be divagation for natural gases with high CO_2 content, too.

(3) They can't fit for pure methane. Gas specific gravity of methane is 0.55. But the allowed range of gas specific gravity of Standing's and Vasquez-Beggs' models is 0.59~0.95, 0.56~1.118, respectively.

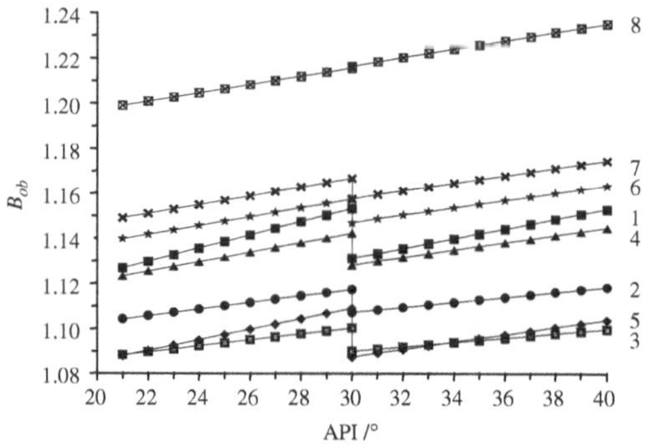

Figure 2. Relationship between FVF (B_{ob}) calculated from Vasquez-Beggs' model and crude oil API. When the API = 30° (equivalent to 0.8762 g/cm^3) the curves of FVF arise discontinuously. So in a certain range of API, it causes the wrong result that FVF of the high-gravity oil is lower than those of the black oil.

3 PREDICTING MODEL OF FORMATION VOLUME FACTOR

3.1 *Formation volume factor of the crude oil with single dissolved component*

Assuming the V_0 volume of surface oil becomes V_2 under the formation temperature and pressure conditions when dissolving gas, the formation volume factor of the reservoir oil is defined as follows:

$$B = \frac{V_2}{V_0} \tag{1}$$

The ground conditions are generally defined as 20°C of temperature and 101.3 kPa.of pressure. As a result, there is a difference in temperature between the stratum condition and ground condition, which makes the volume of crude oil change. If the volume of some

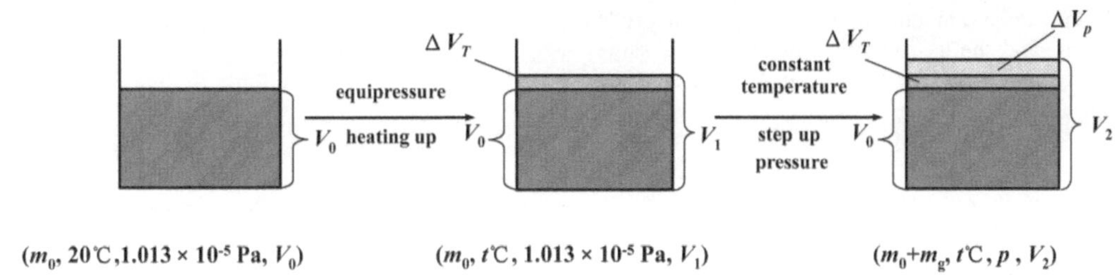

Figure 3. Volume changes of the crude oil from the ground state to the stratum state.

Table 2. Apparent liquid density and apparent liquid molar volume of components of the natural gas under the normal pressure.

Components	Molecular weight	Specific gravity of liquid[William *et al.*, 1984] (15.6°C/15.6°C)	d^{20}[Wang and Qin, 1990] ($g \cdot cm^{-3}$)	V_{gm} ($cm^3 \cdot mol^{-1}$)
CH_4	16.043	0.30	–	53.42*
C_2H_6	30.070	0.3564	0.3399	88.47
C_3H_8	44.097	0.5077	0.5005	88.11
n-C_4H_{10}	58.124	0.5844	0.5788	100.42
i-C_4H_{10}	58.124	0.5631	0.5572	104.42
i-C_5H_{12}	72.151	0.6247	0.6197	116.43
n-C_5H_{12}	72.151	0.6310	0.6262	115.22
CO_2	44.010	0.827	–	53.16*
N_2	28.013	0.808	–	34.69*
H_2S	34.076	0.79	–	43.18*

Note: Data marked with "*" are attained through conversion of specific gravity of liquid.

quantities of crude oil that dissolves no gas under certain temperature t (°C) is V_1, the ratio of V_1 and its volume at ground temperature, V_0, is $f(t)$ which is called temperature coefficient

$$f(t) = \frac{V_1}{V_0} \tag{2}$$

Compared with the volume of the crude oil under ground condition, the increase under stratum condition is due to two factors: temperature increase and the solution of gas. The whole process of the condition changes can be described in two steps (Fig. 3).

If there is a certain quantity of crude oil, its mass is m_0, its volume under ground condition is V_0, and then B can be defined according to Figure (3) and Formula (1):

$$B = \frac{V_2}{V_0} = \frac{V_0 + \Delta V_t + \Delta V_p}{V_0} = \frac{V_1}{V_0} + \frac{\Delta V_p}{V_0} \tag{3}$$

Formula (2) is put into the above one:

$$B = f(t) + \frac{\Delta V_p}{V_0} \tag{4}$$

For the calculation of ΔV_p, the conception of apparent liquid density of gas is introduced, which is a supposed state, that is to assume the gas dissolved in crude oil has the characters of liquid, then the density of this kind of liquid gas is called the apparent liquid density of gas. If the temperature is certain, the mass of gas dissolved in a given amount of crude oil is m_g (g), and the apparent liquid volume of the gas is V_{gl} (cm^3), then the definition formula of apparent liquid density of gas can be defined as:.

$$d_{gl} = \frac{m_g}{V_{gl}} \tag{5}$$

If the molar mass of gas is M ($g \cdot mol^{-1}$), its apparent liquid molar volume (under ground condition) is

$$V_{gm} = \frac{M}{d_{gl}} \ (cm^3 \ . \ mol^{-1}) \tag{6}$$

V_{gm} is a constant related to the types of gas. On the basis of data reported (William et al., 1984; Wang and Qin, 1990), the apparent liquid density and liquid molar volume of the main components of natural gas are listed in Table 2.

If the molar amount of gas dissolved in crude oil is n_g, and the gas dissolved in oil has the same temperature coefficient as oil phase, then at t(°C) temperature:

$$\Delta V_p = f(t) \cdot V_{gm} \cdot n_g \tag{7}$$

Formula (7) is put into (4), we get

$$B = f(t) + \frac{f(t)V_{gm}n_g}{V_0} = f(t) + \frac{V_2}{V_0}\frac{f(t)V_{gm}n_g}{V_2}$$

$$= f(t) + B \cdot f(t) \cdot V_{gm} \cdot C \tag{8}$$

from Formula (8), we get

$$B = \frac{f(t)}{1 - f(t)V_{gm}C} \tag{9}$$

Formula (9) just is the theoretical model of formation volume factor when single component gas is dissolved in crude oil. In this formula, C is molar solubility of gas, $C = n_g/V_2$, and its unit is mol · cm^{-3}. From Formula (9), we can know that B is the function of temperature t and C, and C is related to temperature and pressure. So long as the expressions of C and f (t) are determined, the FVF of the saturated crude oil can be figured out, and the calculation of C is studied in other papers (Xue et al., 2005).

3.2 *FVF of the multi-component saturated crude oil*

We suppose there are n components in the mixed gas and there is no interaction between each other. The dissolution of each component of gas makes the volume of oil phase increase, and the increasing quantities are $\Delta V_{p1}, \Delta V_{p2}, \ldots, \Delta V_{pn}$; the molar amounts of each gas component in oil phase are $n_{g1}, n_{g2}, \ldots, n_{gn}$; and the apparent liquid molar volumes of each gas component are $V_{gm1}, V_{gm2}, \ldots, V_{gmn}$. If the volume change of the oil caused by the temperature change is ΔV_t, then

$$B = \frac{V_0 + \Delta V_t + \Delta V_p}{V_0} = f(t) + \frac{\Delta V_p}{V_0}$$

$$= f(t) + \frac{\Delta V_{p1}}{V_0} + \frac{\Delta V_{p2}}{V_0} + \cdots + \frac{\Delta V_{pn}}{V_0} \tag{10}$$

Put Formula (7) into (10), then

$$B = f(t) + \sum_{i=1}^{n} \frac{f(t)V_{gmi}n_{gi}}{V_0}$$

$$= f(t) \cdot (1 + \sum_{i=1}^{n} \frac{V_{gmi}n_{gi}}{V_0} \cdot \frac{V_2}{V_2})$$

$$= f(t) \cdot (1 + \sum_{i=1}^{n} V_{gmi} \cdot \frac{n_{gi}}{V_2} \cdot \frac{V_2}{V_0})$$

$$= f(t) \cdot (1 + \sum_{i=1}^{n} V_{gmi} \cdot C'_i \cdot B) \tag{11}$$

C′i is the solubility of No.i component in the reservoir oil.

From Formula (11), we get:

$$B = \frac{f(t)}{1 - f(t) \cdot \sum_{i=1}^{n} V_{gmi} \cdot C'_i} \tag{12}$$

This is the FVF formula that is expressed by the solubility of multi-component gas.

FVF is slightly small when the formation pressure is greater than the saturated pressure; at present, for the relevant rectified formulas cited widely (see the reference by Bradley, 1996).

On the basis of experimental data, we have studied the relationship between the temperature coefficient f and temperature t (°C) from 20°C to 150°C, and obtained a regression equation:

$$f(t) = 0.97974 + 7.31378 \times 10^{-4} t + 7.8749 \times 10^{-7} t^2 \tag{13}$$

4 THE CALCULATION MODEL OF DENSITY OF CRUDE OILS

4.1 *The calculation model of density of crude oils with single dissolved component gas*

We suppose the density of crude oils is d_0 under ground condition. If the temperature and pressure are given, the crude oil gets saturated when gas m_g(g) dissolves in crude oil m_0(g), and the corresponding pressure and density of the saturated oil are p and d_p, respectively. At this time, the volume of crude oil is changed from V_0 (under ground condition) to V_2, and then the density of the saturated crude oil is

$$d_p = \frac{m_0 + m_g}{V_2} = \frac{m_0}{V_2} + \frac{m_g}{V_2} = \frac{m_0}{V_2} + C_m \tag{14}$$

C_m (g · cm^{-3}) is the mass concentration of gas. Rewrite Formula (14) into:

$$d_p = \frac{m_0}{V_0} \cdot \frac{V_0}{V_2} + C_m = \frac{d_0}{B} + C_m \tag{15}$$

For $C_m = M \cdot C$, then Formula (15) can be rewritten into

$$d_p = \frac{d_0}{B} + C \cdot M \tag{16}$$

where C is molar solubility ($\mathrm{mol \cdot cm^{-3}}$), M is molar mass of gas, d_0 is density of the degassed crude oil under ground condition. If FVF Formula (9) is put into (16), this can be deduced:

$$d_p = \frac{d_0}{f(t)} \cdot (1 - f(t) \cdot V_{gm} \cdot C) + MC$$

$$= \frac{d_0}{f(t)} + (M - d_0 \cdot V_{gm})C \tag{17}$$

Formula (17) shows the theoretical calculation result of density of crude oils under the bubble point pressure when single-component gas dissolves in crude oils, which suggests the density d_p of the reservoir oil is related to the temperature, solubility and type of gas. If the solubility and temperature of a gas are known, the density of the saturated crude oil can be figured out theoretically.

4.2 *The calculation model of density of the saturated crude oil with multi- component gas*

The calculation model for the multi-component gas can be deduced by the similar method of single-component gas. It can be supposed that the masses of each gaseous component dissolved in crude oils are $m_{g1}, m_{g2}, \ldots, m_{gn}$, then

$$d_p = \frac{m_0 + m_{g1} + m_{g2} + \cdots + m_{gn}}{V_2}$$

$$= \frac{d_0}{B} + C_{m1} + C_{m2} + \cdots + C_{mn}$$

$$= \frac{d_0}{B} + C_1' M_1 + C_2' M_2 + \cdots + C_n' M_n$$

$$= \frac{d_0}{B} + \sum_{i=1}^{n} C_i' M_i \tag{18}$$

Formula (12) is put into (18), then

$$d_p = \frac{d_0(1 - f(t) \cdot \sum_{i=1}^{n} V_{gmi} \cdot C_i')}{f(t)} + \sum_{i=1}^{n} C_i' M_i$$

$$= \frac{d_0}{f(t)} - \sum_{i=1}^{n} V_{gmi} \cdot C_i' + \sum_{i=1}^{n} C_i' M_i \tag{19}$$

5 THE RELATIONSHIP OF THE GAS-OIL RATIO, MOLAR SOLUBILITY, FVF AND DENSITY

If the volume of a certain mass of crude oils is V_0 under the ground conditions, which is saturated with dissolved gas at certain formation temperature and under the given pressure, then it is degassed completely under the ground conditions. If the volume of degassing is V_g, and the gas-oil ratio is defined as:

$$R_S = \frac{V_g}{V_0} \tag{20}$$

And the volume of 1 mol gas is 22400 cm^3 under the standard condition (0°C, 101.3 kPa), then

$$V_g = 22400 n_g \quad (\mathrm{cm^3}) \tag{21}$$

Formula (21) is put into (20)

$$R_S = \frac{V_g}{V_0} = \frac{22400 n_g}{V_0} = 22400 \cdot \frac{V_2}{V_0} \cdot \frac{n_g}{V_2} = 22400 B \cdot C$$

$$(\mathrm{cm^3/cm^3} \text{ ground crude oil}) \tag{22}$$

Formula (22) just shows the relationship between the gas-oil ratio and the molar solubility when single-component gas is dissolved in crude oils.

It is supposed that there isn't interaction among various components of gases, when the mixed gas is dissolved in crude oils.

If R'_{Si} is the gas-oil ratio of the ith component of the mixture gases dissolved in crude oils under the partial pressure p_i, R_{Si} is the gas-oil ratio of single gas i dissolved in an equal quantity of crude oils under the pressure p_i, according to the above hypothesis, the following can be reached.

$$R'_{Si} = R_{Si} \tag{23}$$

At a certain temperature, when the multi-component mixture dissolves in crude oils, its gas-oil ratio equals the sum of the gas-oil ratios when each component of gases dissolves in crude oils, respectively, under the relevant partial pressure, which is

$$R_S = \sum_{i=1}^{n} R_{Si} = \sum_{i=1}^{n} R'_{Si} = 22400 \cdot \sum_{i=1}^{n} C_i \cdot B_i$$

$$= 22400 \cdot \sum_{i=1}^{n} C_i' \cdot B \tag{24}$$

Formula (24) shows the basic relationship of the gas-oil ratio, component molar solubility and FVF when the multi-component gas dissolves in crude oils. In this formula, C_i and B_i are molar solubility and FVF of the ith component of the mixed gas dissolved in crude oils, respectively. And B, C_i' are the FVF of crude oils and molar solubility of component i when the multi-component gas dissolves in crude oils, respectively.

Based on Formula (24), Formula (12) can be rewritten into a more practical one

$$B = f(t)\cdot(1+\sum_{i=1}^{n}V_{gmi}\cdot\frac{R_{Si}}{22400}) \tag{25}$$

For the single-component gas is dissolved in crude oils, B and R_S can be expressed in a more simple formula

$$B = f(t)\cdot(1+\frac{R_S}{22400}\cdot V_{gm}) \tag{26}$$

The above two formulas are practical for integrating B and R_S. Based on Formula (24), Formula (18) can be rewritten into a more practical one

$$d_p = \frac{d_0}{B}+\frac{1}{B}(\frac{R_{S1}}{22400}M_1+\frac{R_{S2}}{22400}M_2+\cdots+\frac{R_{Sn}}{22400}M_n)$$

$$=\frac{1}{B}(d_0+\sum_{i=1}^{n}\frac{R_{Si}}{22400}M_i) \tag{27}$$

Formula (27) is the density equation of crude oils with the multi-component dissolved gas under the saturated pressure. By the comparison of Formula (27) with the empirical formula (28) reported by Chen (1991), it can be found that the both have similar form. However, the former expresses the density of liquid through the solubility of multi-component gas and the molar mass of each component, which is more reasonable than the one that shows the density of crude oils with dissolved gas by a constant coefficient ($C_1 = 1.0$ and $C_2 = 1.2237 \times 10^{-3}$ in International System, respectively) and a relative density of gas in Equation (28) (Chen, 1991).

$$d_p = (C_1 d_0 + C_2 d_g R_S)\cdot\frac{1}{B} \tag{28}$$

When the formation pressure is higher than the saturation pressure, the density of crude oils is larger under the former condition. At present, one can find out the rectified formula is widely used in references (Bradley, 1996).

6 RESULT AND COMPARISON

The solubility of natural gas, R_{Si} and the temperature coefficient in Formula (13) are put into Formula (25), and FVF B is calculated. The relative density of gas can be figured out by the composition of the dissolved gas, and then it can be put into Standing's model (1947) and Vasquez-Beggs' model (1980) to get FVF B. Under the given temperature and pressure conditions, the values of FVF and density of crude oils, which are obtained by each model, are showed in Table 3.

Table 3. Comparison between the values of calculation and experiments from various models of FVF and density of the saturated crude oil ($d_0 = 0.8609\text{g}\cdot\text{cm}^{-3}$).

			FVF			Density of crude oils ($\text{g}\cdot\text{cm}^{-3}$)	
Temperature (°C)	Pressure (10^5 Pa)	Experimental gas-oil ratio ($\text{m}^3\cdot\text{m}^{-3}$)	Model in this paper	Model I	Model II	Model III	Model in this paper
35	55.7	24.69	1.0743	1.0677	1.0817	0.8170	0.8239
	99.0	44.09	1.1273	1.1202	1.1330	0.7953	0.7988
	143.3	62.63	1.1749	1.1640	1.1822	0.7756	0.7808
	171.6	78.32	1.2239	1.2132	1.2223	0.7658	0.7661
	221.8	100.8	1.2836	1.2781	1.2821	0.7446	0.7448
45	53.2	22.7	1.0782	1.0739	1.0849	0.8131	0.8196
	97.5	41.9	1.1324	1.1247	1.1350	0.7938	0.7971
	141.4	58.62	1.1739	1.1633	1.1809	0.7734	0.7794
	180.3	77.69	1.2315	1.2214	1.2296	0.7603	0.7610
	217.6	95.89	1.2860	1.2744	1.2772	0.7464	0.7453
55	53.9	21.28	1.0815	1.0777	1.0912	0.8064	0.8159
	99.5	38.68	1.1281	1.1237	1.1383	0.7859	0.7938
	140.8	56.67	1.1808	1.1735	1.1838	0.7714	0.7748
	184.4	74.11	1.2283	1.2156	1.2300	0.7554	0.7599
	212.8	87.48	1.2691	1.2504	1.2645	0.7460	0.7497
75	55.8	20.1	1.0958	1.0985	1.1053	0.7954	0.8030
	100	35.87	1.1395	1.1436	1.1489	0.7769	0.7820
	141.2	51.47	1.1847	1.1874	1.1887	0.7636	0.7652
	176.5	64.23	1.2202	1.2211	1.2226	0.7518	0.7532
	213	76.93	1.2538	1.2423	1.2565	0.7406	0.7460

Note: Model I is Standing's Model; Model-II is Vasquez-Beggs' Model; Model-III comes from the reference (Bradley, 1996).

From Table 3, it can be seen that the values of FVF, which are figured out by Standing's and Vasquez-Beggs' models, are quite different. In general, the difference is 0.02 or so. While the values calculated from the model put forward in this paper are nearly between Standing's and Vasquez-Beggs' models. Therefore, it is evident that the calculated results of the model in the paper are reasonable.

As for the calculation model of density of the saturated crude oil, the typical one is put forward in the reference (Bradley, 1996). And the comparison of the calculated results from the two models between this paper and the typical one can be seen in Table 3. From which, one can see that the calculated results from the two models are very similar, and the maximum absolute bias is less than 0.01. For this experimental system, no matter which model it is, the calculated results are all acceptable. But the model in this paper is more reasonable because the composition of gases is considered. When the content of non-hydrocarbon gases is high, Model III may cause big error owing to representing gas composition by its relative density only.

7 CONCLUSIONS

(1) In this paper, the changes of volume are investigated deeply in the dissolving process of gases in crude oils. The concept of the apparent liquid density of gases is introduced. On the basis of this concept, the theoretical formula is deduced, in which the FVF of the saturated crude oil is regarded as the function of molar solubility of gases.
(2) According to the relation between molar solubility of gases and the gas-oil ratio, the theoretical formula of FVF and gas-oil ratio are deduced, it has a simple form and the composition of gases is considered.
(3) Theoretically, the formula is deduced to show the relationship between the density of the saturated crude oil and molar solubility of gases or gas-oil ratio.

All the formulas mentioned above consider the factors of the multi-component gas and temperature factor, and pressure factor is concealed in solubility. In principle, as long as the molar solubility or gas-oil ratio of gaseous components in crude oils is known at a given temperature, the FVF and density of the reservoir oil can be calculated accurately by the formula put forward in this paper.

REFERENCES

AlMarhoun, M.A., 1988. PVT Correlations for Middle East crude oils. *Journal of Petroleum Technology* 40, 650–666.
AlMarhoun, M.A., 2004. Evaluation of empirically derived PVT properties for Middle East crude oils. *Journal of Petroleum Science and Engineering* 42, 209–221.
Bradley, H.B., 1996. *Petroleum engineering handbook (the last part)*. Beijing: Petroleum Industry Publishing House (in Chinese).
Chen Yuanqian, 1991. *Calculating methods for reservoir engineering*. Beijing: Petroleum Industry Publishing House (in Chinese).
Chen Zhangming, Xu Jingzhen, 1993. *Evaluating methods of oil and gas trap*. Harbin: Harbin Industry University Publishing House (in Chinese).
Dokla, M.E., Osman, M.E., 1992. Correlation of PVT properties for UAE crudes. *Society Petroleum Engineer*, Mar, 41–46.
Elsharkawy, A.M., Alikhan, A.A., 1997. Correlations for predicting solution gas/oil ratio, oil formation volume factor, and undersaturated oil compressibility. *Journal of Petroleum Science and Engineering* 17, 291–302.
Galsø, O., 1980. Generalized pressure-volume-temperature correlations. *Journal of Petroleum Technology* 32, May, 785–795.
Kartoatmodjo, T., Schmidt, Z., 1994. Large data bank improves crude oil physical property correlations. *Oil and Gas Journal* 92, July 4, 51–55.
Lasater, J.A., 1958. Bubble Point Pressure Correlation. *Trans. AIME* 213, 379–381.
Obomanu, D.A., Okpobiri, G.A., 1987. Correlating the PVT properties of Nigerian crudes. *Journal of Energy Resource Technology* 109, 214–217.
Standing, M.B., 1947. A pressure-volume-temperature correlation for mixtures of California oils and gases. *API Drill Prod Pract*, May, 275–287.
Vasquez, M.E., Beggs, H.D., 1980. Correlations for fluid physical property prediction. *Journal Petroleum Technology*, June, 968–970.
Wang Wenhu, Qin Yanlong, 1990. Hydrocarbon physical chemistry handbook. Beijing: Hydrocarbon Processing Publishing House (in Chinese).
William, D., McCain, J.R., 1984. Petroleum liquid character. Beijing: Petroleum Industry Publishing House (in Chinese).
Xue Haitao, Lu Shuangfang, Fu Xiaotai., 2005. Forcasting model of solubility of CH_4, CO_2 and N_2 in crude oil. *Oil & Gas Geology* 26, 444–449 (in Chinese).

Petroleum Geochemistry and Exploration in the Afro-Asian Region – Digang Liang, Darui Wang & Zhenxi Li (eds)
© 2008 Taylor & Francis Group, London, ISBN 978-0-415-44083-7

The application of rock pyrolysis technique to geochemistry evaluation of reservoirs in China

Liyan Wu & Zhenling Zhang
Research Institute of Petroleum Exploration and Development, PetroChina, Beijing, China

ABSTRACT: The rock pyrolysis analysis technique has been applied originally to geochemistry evaluation of source rocks, which is evolved into an extensive application to the assessment of reservoirs in China. PetroChina Research Institute of Petroleum Exploration and Development has invented a quantitative and qualitative analysis method to measure the hydrocarbon within reservoir rocks (Chinese Patent No 89107286.1). By pyrolytic analysis of the cuttings of reservoir rocks, petroleum inside is divided into five group compounds. As a consequence of the quantification of each group of compounds, the oil density is distinguished. And the total quantity of oils and gases, and the porosity within reservoir rocks are measured by this method. Thus, parameters of hydrocarbon quantity, oil density and porosity of rocks are combined to calculate the oil saturation of reservoir rocks and to evaluate the characters and oil grade of reservoirs. In which the probability of rock samples polluted by oil mixed mud of drilling is diagnosed, so that analysis results of polluted samples can be removed. This analysis method is generally used in drilling site and laboratory in China to discover new oil fields, particularly light oil reservoirs, and it has become a normal item of geochemistry logging in petroleum exploration in China.

1 A QUANTITATIVE ANALYSIS METHOD OF OILS AND GAS IN RESERVOIR ROCKS

Chinese petroleum technicians started to extensively apply the rock pyrolysis technique to geochemistry evaluation of reservoirs in1990, which had been applied originally to geochemistry evaluation of source rocks. Through more than 15 years of diligent practice, Research Institute of Petroleum Exploration & Development (RIPED), PetroChina has worked hard to manufacture successively the OGE workstation and geochemistry logging instrument, which is now used in well sites or laboratories of petroleum exploration for reservoir rock characterization all over China. Through the analysis of cuttings or cores the direct diagnosis of oil reservoirs and oil & gas shows has become a normal item of geochemistry logging in China petroleum exploration.

The invention patent "A Quantitative Analysis Method of Oil and Gas in Reservoir Rocks" (Chinese patent No 89107286.1) of RIPED, PetroChina has been applied in the OGE workstation and geochemistry logging instrument. World Intellectual Property Organization and Patent Bureau of the Peoples Republic of China awarded this technique the Patent for its social and economical effect in November 1997. By thermal distillation and pyrolysis of petroleum inside reservoirs, hydrocarbons are divided into five group compounds, each compound is quantitatively identified as shown below (Table1 and Figure1).

Table 1. Analysis conditions of "Quantitative Analysis Method of Oil And Gas in Reservoir Rocks".

Parameters of analysis	Temperature (°C) Initial	Temperature (°C) Terminal	Isothermal (min)	Heating rate (°C/min)
Gas S'_0 (mg/g)	90	90	2	—
Gasoline S'_{11} (mg/g)	200	200	1	—
Diesel S'_{21} (mg/g)	200	350	1 (350°C)	50
Wax & heavy fraction S'_{22} (mg/g)	350	450	1 (450°C)	50
Cracked hydrocarbons of resin & asphalt S'_{23}	450	600	1 (600°C)	50

2 DETERMINATION OF OIL DENSITY

The oil density is an important parameter for calculating the oil saturation of reservoir rocks. According to the ratio of oil compounds of reservoirs, the oil density is distinguished in an abas (Figure 2).

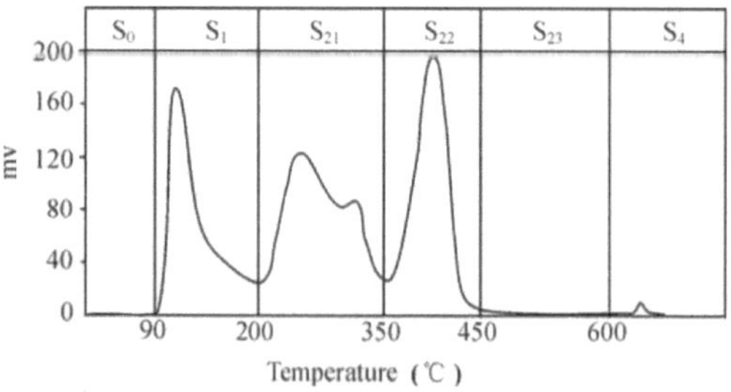

Figure 1. Analysis diagram of a sandstone.

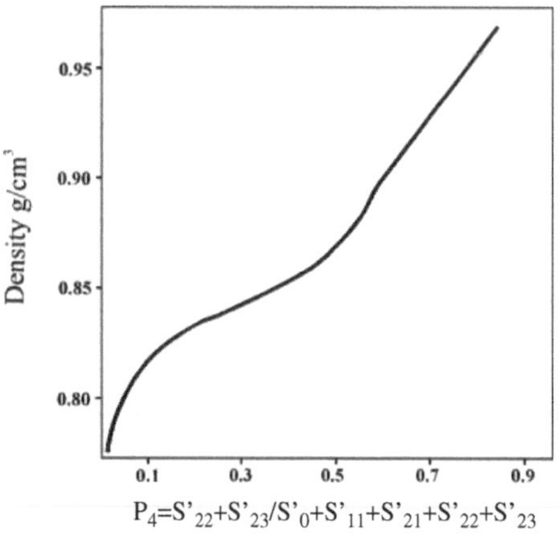

Figure 2. Abas of oil density.

Table 2. Comparison of oil density values (after Dongsheng LANG, 1995).

Well	Depth (m)	S'22 + S'23/ST*	Measurement by densimeter in well testing	Abas density value (g/cm^3)
Da 404	555-558	0.11	0.8212	0.827
Da 420	2378-385	0.13	0.8200	0.830
Yan 142	1878-881	0.41	0.8546	0.859
Yan 241	1685-690	0.49	0.8720	0.878
Po 431	1629-663	0.69	0.9120	0.932

* $S'_0 + S'_{11} + S'_{21} + S'_{22} + S'_{23}$.

The comparison of oil density values between pyrolysis method and densimeter resulted in a better effect in the Daqing oil field, China (Table 2).

3 THE MEASUREMENT OF POROSITY OF RESERVOIR ROCK

The porosity of reservoir rocks is very important for calculating the oil saturation of reservoir rocks and characterizing reservoirs. The process of reservoir rock pyrolysis includes the evaporation and cracking of fluid (gas, oil and water) in rock samples. After pyrolysis, the rock is deducted all fluid, its residua is mineral matrix. The difference in values between mass of rock and mass of mineral matrix is the fluid mass. Because pores of rocks are charged by fluid, the volume of fluid is the volume of pores. So the formula for computing the porosity Φ is as follows:

Table 3. Comparison of porosity values of sandstone.

Well	Depth (m)	Lithology	Measurement by helium porosimeter	Pyrolysis Φ(%)
Liu139	1938.0	trace oil sandstone	8.40	8.05
Liu137	1971.2	trace oil sandstone	10.10	10.95
W5-3	1108.48	impregnated oil sandstone	15.20	15.47

$$\Phi = (1 - \frac{\rho_r}{\rho_m} \times \frac{W_m}{W_r}) \times 100\ \% \tag{1}$$

Φ—Porosity of reservoir rock, %;
Wr—Mass of rock, g;
Wm—Mass of mineral matrix , g;
ρ r—Density of rock, g/cm^3;
ρ m—Density of mineral matrix, g/cm^3;

The coincidence of porosity values of sandstone between the pyrolysis method and helium porosimeter was good in the Jiangsu oil field, China (Table 3).

4 CALCULATION OF OIL SATURATION IN RESERVOIR ROCKS

The parameter of oil saturation in reservoir rocks is important for characterization of oil reservoirs. The formula to calculate the oil saturation in reservoir rocks is as follows:

$$So = \frac{S_T \times \rho_r}{\rho_0 \times \Phi} \times 10 \tag{2}$$

So—Oil saturation, %;
ST—Total hydrocarbon yields in reservoir rocks, mg/g;
ρ r—Density of reservoir rocks, g/cm^3;
ρ_0—Density of oils, g/cm^3;
Φ—Porosity of reservoir rocks, %.

5 THE EVALUATION OF RESERVOIR CHARACTER

The porosity, oil saturation and total hydrocarbon content of reservoir rocks are applied to evaluate the

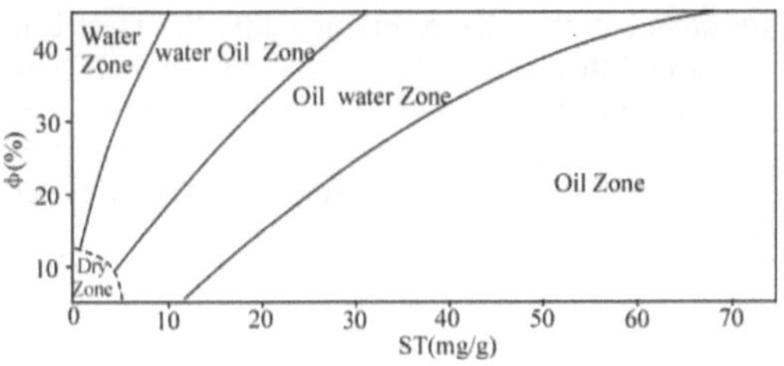

Figure 3. Abas evaluating the characters of reservoirs (after Lianhua Ding, 2000).

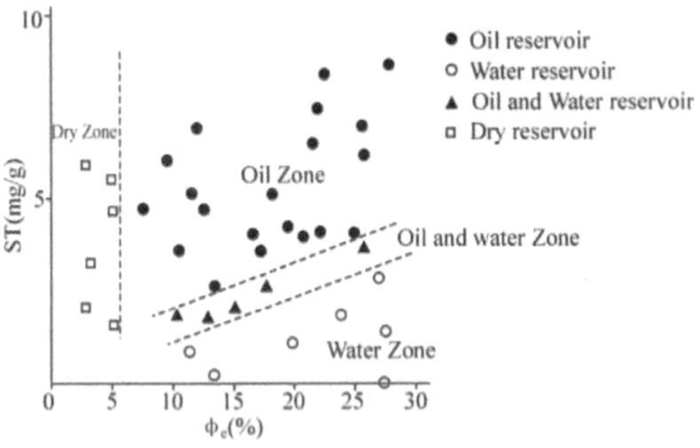

Figure 4. Abas evaluating the characters of light oil reservoirs (after Dongsheng Lang, 2004).

characters and oil grade of reservoirs. Senior engineer Lianhua Ding from the Shengli oil field, China proposed a abas for evaluating the characters of reservoirs (Figure3), which is applied to evaluate normal and heavy oil reservoirs. By using this abas several oil fields have been discovered in China.

Dr. Dongsheng Lang from the Daqing oil field, China proposed a abas to evaluate the light oil reservoir (Figure 4), with which some light oil fields are discovered in Hailaer Basin, Northeast China.

6 OGE WORKSTATION

The pyrolysis instrument OGE workstation is specially designed to evaluate oil reservoir rocks for petroleum exploration. By pyrolytically analysing cuttings of reservoir rocks, five groups of hydrocarbon compounds and the oil content in rocks can be gained. The OGE workstation integrated with software interprets data, diagnoses oil density, rock porosity, total hydrocarbon content and oil saturation, respectively, and characterizes reservoirs as well (see the analysis report on reservoir rocks by the OGE workstation in Table 4). There exist six analysis cycles: 1) Reservoir cycle, applied to oil reservoirs, pollution of drilling mud and bitumen; (2) Source rock cycle, hydrocarbon potential, maturity and type of organic matter are evaluated; 3) Carbonate rock cycle, with high sensitivity for analysing low organic matter and high maturity carbonate rocks in China and dealt by a multi-peaks processing software to illustrate Tmax and calculation percentage of each peak area; 4) Thermo-simulation cycle, measuring the relation between hydrocarbon generation rate and maturity of source rocks; 5) Activation energy cycle, measuring the relation between hydrocarbon generation rate and activation energy; and 6) User-defined cycle, all running parameters (temperature gradient and isothermal time of pyrolysis, etc) can be determined by users.

Table 4. Analysis report on reservoir rocks.

Depth (m)	2105	2635
Age	E	E
Lithology	sandstone	sandstone
Oil & gas mg/g	79.3	12.31
Oil density g/cm^3	0.8546	0.9122
Porosity %	29.9	24.8
Oil saturation %	48.9	11.86
Oil character	normal oil	heavy oil
Oil grade	impregnated	trace
Reservoir character	oil reservoir	water reservoir

Table 5. Relation between the hydrocarbon content of cuttings and oil characters (after Lianhua Ding, 2000).

Reservoir characters	S'_0 (mg/g)	$S'_{11}+S'_{21}$ (mg/g)	$S'_{22}+S'_{23}$ (mg/g)	ST* (mg/g)
Light oil reservoir cuttings	>0.05	>6	>2	>8
Normal oil reservoir cuttings	>0.05	>14	>12	>26
Heavy oil reservoir cuttings	>0.05	>20	>27	>47
Extra heavy oil reservoir cuttings	>0.05	>22	>30	>52

* $S'_0+S'_{11}+S'_{21}+S'_{22}+S'_{23}$

7 RESULTS

7.1 *Characterization of reservoirs*

The hydrocarbon content in reservoir rocks depends directly on the porosity of reservoirs and the density of oils. For example, the hydrocarbon content in reservoir rocks from the Shengli oil field is >8 mg/g for light oil reservoir cuttings only, but >47 mg/g for heavy oil reservoir cuttings (Table 5). High porosity reservoir cuttings are generally with high hydrocarbon content (Table 6). With these relation principles oil reservoirs can be characterized (Table 7).

7.2 *Diagnosis of pollution*

Through the analysis of five groups of oil compounds, the diagnosis on the pollution of oil mixed mud of drilling was made so as to remove the polluted data. For example, Well Yu 502 of the Daqing oil field was drilled by diesel mixed mud, it was found by the rock pyrolysis that the cuttings contaisned high quantity of S'_{21} (mg/g) compounds probably caused by diesel pollution. Therefore, the rock pyrolysis logging interpreted it as a dry reservoir while the electrical log showed that as an oil reservoir. At last well testing assured that it was a dry reservoir (Table 8), and according to two intervals (1521.2 ~ 1522.6 m and 1549.4 ~ 1556.2 m), their ST (mg/g) looks as diesel pollution by 2.4 mg/g.

Table 6. Relation between the hydrocarbon content of cuttings and porosity (after Lianhua Ding, 2000).

Porosity of oil reservoirs	S'_0 (mg/g)	$S'_{11}+S'_{21}$ (mg/g)	$S'_{22}+S'_{23}$ (mg/g)	ST* (mg/g)
>25%	>0.1	>23	>21	>44
10% ~ 25%	>0.1	>17	>14	>31
<10%	>0.05	>17	>12	>29

$^*S'_0+S'_{11}+S'_{21}+S'_{22}+S'_{23}$

8 CONCLUSIONS

After 15 years of experience and practice, now thousands of OGE workstations and derivative instruments are running in well sites of petroleum exploration and in laboratories in China for oil reservoir characterization, which makes a contribution to the discovery of new oil fields. The OGE workstation provides: 1) Determination of the hydrocarbon content in reservoir rocks; 2) Estimated quantity of gas and four oil compounds; 3) Measurement of porosity of reservoir rocks; 4) Calculation performance of oil density; 5) Calculation of oil saturation of reservoirs; 6) Performance by abas software for reservoir characterization.

Table 7. Rock pyrolysis logging of a well in the Shengli Oil Field (after Lianhua Ding, 2000).

Depth Interval (m)	S'_0 (mg/g)	$S'_{11}+S'_{21}$ (mg/g)	$S'_{22}+S'_{23}$ (mg/g)	ST (mg/g)	So (%)	Φ (%)	Oil character	Oil producing capacity (t/d.m)	Reservoir character
2412–2413	0.09	17.5	6.25	23.75	40	15.78	normal oil	3.81	oil reservoir
2414–2415	0.09	25.47	8.49	33.96	58	15.68	normal oil	20.00	oil reservoir
2416–2418.5	0.01	25.17	8.68	33.85	40	22.11	normal oil	3.51	oil reservoir
2419–2420	0.01	4.47	1.79	6.27	8	21.21	normal oil	0.00	dry reservoir
2421–2423	0.05	15.87	5.12	28.99	32	17.99	normal oil	2.21	oil-water reservoir
2426–2432	0.01	22.35	7.87	30.67	40	20.91	normal oil	3.60	oil reservoir
2709–2710	0.05	15.6	5.38	20.98	34	17.31	normal oil	2.22	oil-water reservoir
2715–2716	0.03	18.18	6.02	24.2	48	13.3	normal oil	7.61	oil reservoir
2717–2719	0.01	24.69	8.99	33.68	51	18.1	normal oil	8.38	oil reservoir
2720–2724	0.09	25.39	8.58	33.97	44	20.93	normal oil	4.96	oil reservoir
2965–2967	0.04	1.8	4.63	6.52	16	10.17	heavy oil	0.00	dry reservoir
3035–3036	0.02	4.16	20.2	24.4	41	14.2	heavy oil	0.20	dry reservoir
3037–3039	0.05	12.36	19.35	31.73	61	12.68	heavy oil	3.43	oil reservoir
3040–3041	0.04	9.24	17.89	27.18	55	12.55	heavy oil	1.76	oil reservoir
3042–3044	0.01	9.17	19.82	29.03	56	12.87	heavy oil	1.63	oil reservoir

Table 8. Reservoir interpretation on Well Yu 502 of the Daqing oil field (after Dongsheng Lang, 1995).

Depth interval (m)	S'_{11} (mg/g)	S'_{21} (mg/g)	S'_{22} (mg/g)	S'_{23} (mg/g)	ST (mg/g)	ST-2.40 (mg/g)	Interpretation of electrical log	Interpretation of rock pyrolysis logging
1521.2-1522.6	0.02	1.96	0.07	0.51	2.56	0.16	Dry reservoir	Dry reservoir
1536.6-1544.2	1.59	3.52	0.79	0.67	6.57	4.17	Poor oil reservoir	Dry reservoir
1549.4-1556.2	0.01	1.93	0.03	0.49	2.46	0.06	Poor oil reservoir	Dry reservoir
1562.0-1566.0	2.01	3.47	0.95	0.79	7.22	4.82	Oil reservoir	Dry reservoir

ACKNOWLEDGEMENTS

We are grateful to the Geo-Logging Company of Shengli Oil Field, SINOPEC and Geo-Logging Company of Daqing Oil Field and Xinjing Oil Field, CNPC for their support. We thank Professors Digang LIANG and Shuichang ZHANG of Key Laboratory of Petroleum Geochemistry, CNPC for their assistance at AAAPG-2004 Conference.

REFERENCES

Clementz. D. M. 1979. Effect of oil and bitumen saturation on source rocks pyrolysis. *AAPG Bull*, Vol. 63, No12, PP2227–2232

Dongsheng Lang, 1995. Brief presentation YQIII oil gas show evaluation instrument. *Logging Technology Communication*, Vol,6, No2, PP19–22 (in Chinese).

Dongsheng Lang, Daohua Jiang, 2004. Pyrolysis Quantitative Evaluation on the Light crude Oil-Producing Layer in HAILAER Basin Northeast China. *Petroleum Exploration and Development*. Vol.31, No.1, PP79–81 (in Chinese).

Espitalie, J. 1984. Geochemical logging, Analytical pyrolysis techniques and application. *Butterworth*. London.

Liyan Wu, Lianhua Ding & Bin Li, 2000. *The Qualitative and Quantitative Evaluation of the Oil & Gas Reservoir by Quick Pyrolysis Analysis of Rock*. Petroleum Industry Press, Beijing (in Chinese).

Petroleum Geochemistry and Exploration in the Afro-Asian Region –
Digang Liang, Darui Wang & Zhenxi Li (eds)

Study of fault sealing by geochemical techniques: Fault sealing evaluation in petroleum migration and reservoir continuity

Dujie Hou, Jinchuan Zhang & Zulie Long
School of Energy Resource, China University of Geosciences, Beijing, China

Junzhang Zhu & Youjun Tang
Geochemistry Research Center, Yangtze University, Hubei, Jingzhou, China

Xinde Xu & Baojia Huang
CNOOC Nanhai West Research Institute, Zhanjiang, Guangdong, China

ABSTRACT: The fault sealing and reservoir continuity of different fault blocks are very important issues for the study of oil filling histories and production. The petroleum reservoir geochemistry characteristics near faults are mainly involved in the present study that can give us a hint to have information on fault sealing in the geological history and hydrocarbons expulsion stage. The two areas have been chosen for the study of fault sealing in petroleum migration and reservoir continuity. One is in the sag in the south west of Weizhou, Beibuwan Basin near the South China Sea and another is in the Oulituozi region, Liaohe Depression, Bohai Gulf Basin in the East of China. In the first example, the fault has been found by using geochemical chromatographic fingerprints. The main supporting evidence is poor reservoir continuity although it is a small fault. The geochemical and geological analysis of the Oulituozi region shows that the oil pool near the fault has relatively high maturity and is abundant in pyrrolic nitrogen compounds. These results approve that the fault did play an important role for petroleum migration in the geological history. Two typical research examples show advantages of geochemical techniques for documenting fault sealing. Moreover, these techniques can provide more useful information than other methods.

1 INTRODUCTION

Fault sealing means the ability of preventing fluid from continuing flowing, which is caused by the difference of drainage pressure between the rocks of hanging wall and footwall or that between the rocks of the fractured zone and the hanging wall and footwall. Fault sealing has always been an important subject for research that petroleum geologists have paid close attention to (Mcgregor, 1996), because it can control the amount of the trapped petroleum (whether it is present or not and the percentage of oil filling), the migration route of the petroleum, the vertical and horizontal distribution of hydrocarbon compounds, and hydrocarbon compound movement during oil field development (Downey, 1984; Jian, 1995; Yielding, 1997; Xiang et al., 2004). Therefore it is meaningful and interesting to study fault sealing both from the perspective of petroleum exploration and development (Allan., 1989; Knott, 1993; Gibson, 1994).

Traditionally, the study of fault sealing often concentrates on the contact of the faces at both sides of the fault (Smith, 1966). If sandstone of the hanging wall and footwall joins together, the fault is thought to open, and if the permeable rock and impermeable rock of the hanging wall and footwall join together, the fault is thought to close (Fig. 1). But the actual conditions are not as simple as that. In fact, though the faults may occur as a plane, they mostly develop a fault fracture zone, and the fault sealing is influenced by the lithologic character and mudstone and clay contents. However, geochemical method can provide information on fault sealing from the aspect of the chemical properties of the oil, gas and water of the hanging wall and footwall of the fault, and the aspect of the geochemical characteristics of the oil reservoir nearby the fault, it is even more advantageous than other methods. First, it can reflect the geological history, and especially fault sealing during the period of hydrocarbon migration according to the geochemical characteristics of the oil reservoir adjacent to the fault. Second, according to the chemical composition of the crude oil on both sides of the fault, it can help to judge if reservoirs are interconnected via the same lithology

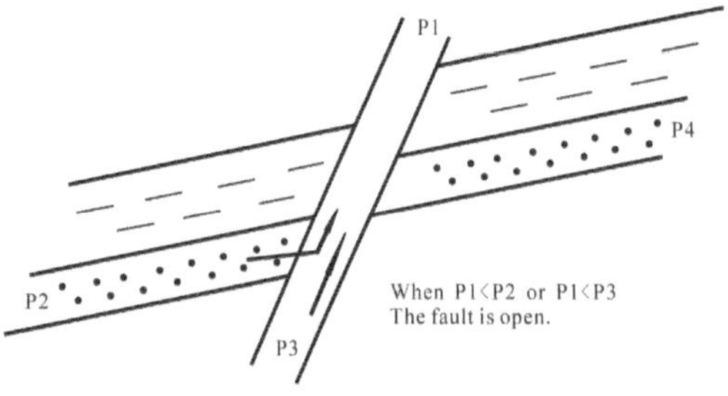

Figure 1. Schematic diagram showing the fault sealing.

or disconnected for fluids, and can offer quantitative appraisal on the connectedness of reservoirs, thus to guide the petroleum exploration (Hu, 2002).

2 INFLUENTIAL FACTORS OF FAULT SEALING AND THE RESEARCH APPROACH

Faulting is an important adjunct to structural deformation because it establishes the stress orientations that develop the geometry of tectonic provinces and facilitate their exploration. One of important issues related to the oil accumulation is fault sealing.

The fault sealing can be divided into the vertical fault sealing and lateral fault sealing. The vertical sealing often determines the migration direction of petroleum (Montel, 1993; Wang 1997; Wang, 2004), the reallocation of oil reservoirs and the amount of the trapped petroleum. The lateral sealing determines the distribution range of petroleum, the exploitation deployment of oil reservoirs and exploration targets.

2.1 *Influential factors of vertical sealing*

The following aspects mainly determine the vertical sealing of a fault:

1) Mechanical properties of the fault
   Generally speaking, the sealing of a tension fault is uncommon, while that of a compression fault is more likely.
2) Hade
   The greater the hade is, the less the pressure of the overlying formation is on the fault zone. On the contrary, the lower the hade is, the better the fault sealing is.
3) The property of the matrix of the ruptured zone
   The higher the mudstone and clay content are, the denser the matrix is, and the better the sealing is.
4) Fracture
   If the cap rock is fractured, sealing is less likely, and if the cap rock is not fractured, sealing is more likely.
5) Contemporaneous fault or growth fault
   The growth fault is commonly associated with flowing mudstone, which makes the fault sealing possible. Meanwhile, factors such as fragmentation, diagenesis and the change of the permeability of the sandstone, the activity and structural type of fault in different basins etc., have influences on the sealing of a fault.

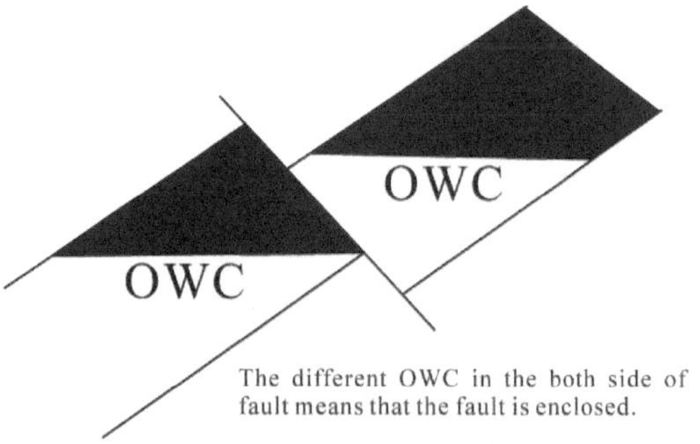

Figure 2. Conceptual diagram showing the fault sealing by different Oil Water Contact.

2.2 *Influential factors of lateral sealing*

Besides the above-mentioned factors, sealing in the lateral direction is closely related with the lithology of the fault of the hanging wall and footwall. Generally speaking, if the permeable rock stratum of the hanging wall and footwall remain in contact, lateral sealing is good, but it depends on other factors as well. Meanwhile, it should be pointed out that the sealing of a fault is conditional. For example, a fault may be impermeable to oil but permeable to natural gas. The height of the oil column is limited corresponding to fault sealing, when the height is exceeded, the fault may discharge hydrocarbons.

2.3 *Common research method of the fault*

The most common research approach to studying the behavior of faults is to use the distribution of the oil-water contact in different blocks. If the oil-water contact of the hanging wall and footwall of the fault is different, it is certain that the fault is closed. But if it is the same, the fault can both be closed and not be closed (Fig. 2). According to the data of well testing, if the oil reservoir of two blocks of the fault accords with a uniform decline curve of pressure, then the fault may be laterally interconnected. In addition, fault sealing can be evaluated according to acoustic logging data or the smearing coefficient of mudstone (Knott, 1993). But it has to be pointed out that all the approaches have certain advantages and limitations. The sealing of one fault is also variable. For example, it may be sealed at certain stratum, but leaky at another, and the fault sealing will also change with time. Therefore, as petroleum geologists, we mainly want to know about the migration period of the petroleum and the lateral sealing of the existing oil reservoir, and for this reason

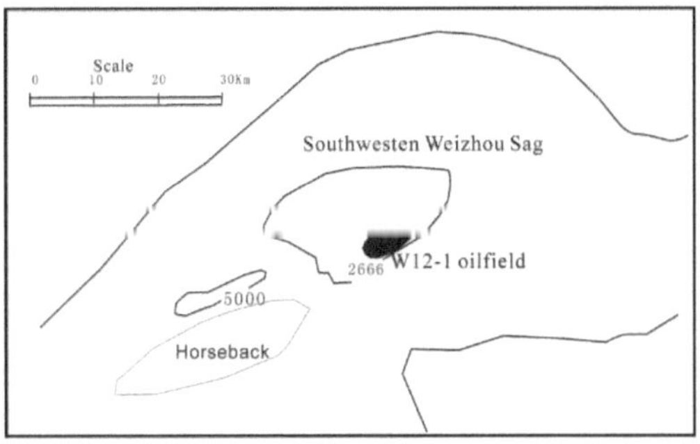

Figure 3. Location of research area.

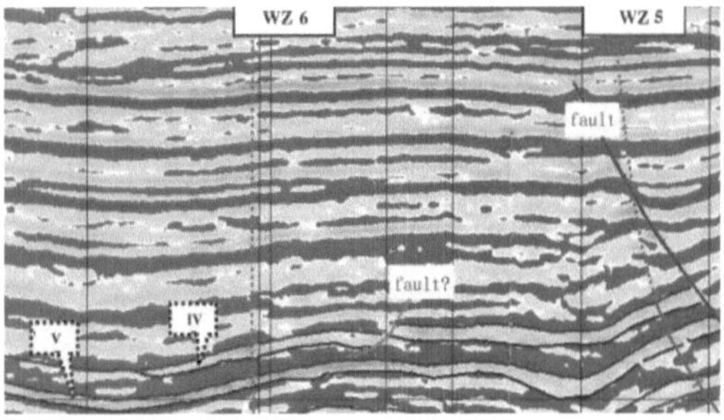

Figure 4. Regional structure map of the Weizhou 12-1 oilfield.

the geochemical analysis has important value both in theory and practical application.

3 LAYOUT OF TEXT

3.1 *Text and indenting*

The WZ 12-1 oil field in the west of the South China Sea is taken as an example (Fig. 3). This study is mainly a research project carried on according to the need of the development department of the oil field who identified a small fault in the formation group IV of WZ 6 oil well and WZ5 oil well according to the geophysical prospecting data (Fig. 4). The fault displacement is relatively small, it seems that it should be relatively obvious that if the existence of the fault is uncertain, then the relative continuity of beds across a fault that may not be present cannot be known. Because this oil field will undergo development soon, the existence of this fault needs to be reevaluated, and it has to know whether there is fault sealing or reservoir continuity of the formation group IV.

During the research, the GC peak ratios of the oil samples from the formation group IV and V were carefully analyzed and studied. By the orthogonal analysis approach, over ten parameters were optimized and selected from lots of peak ratios (Fig. 5). These ten parameters cover the maximum difference of the fluid peak ratios of the oil pool in different wells, as well as the difference of the available parameters of different formations within each well. Therefore, they are representative and can truly reflect the continuity and sealing of the fluid of oil reservoirs.

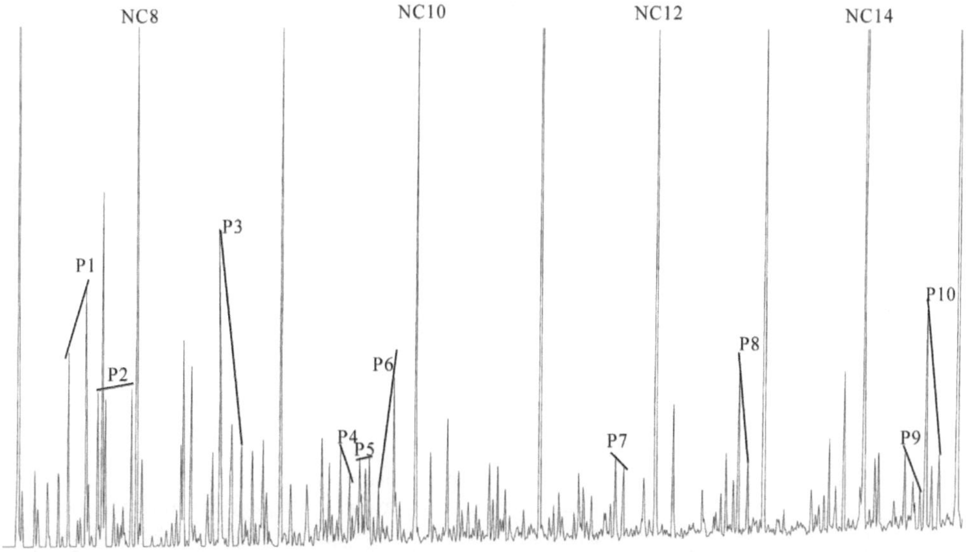

Figure 5. Ten GC peak ratios used to determine reservoir continuity.

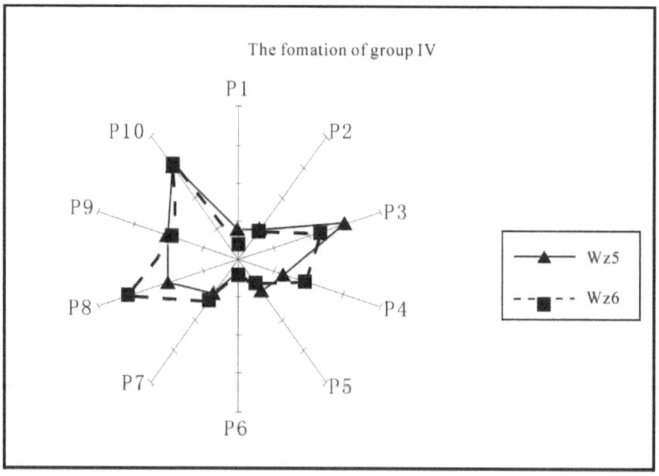

Figure 6. Radar or star plot of selected peak ratios from gas chromatographic analysis of oils in the formation group IV of Wells WZ5 and WZ6.

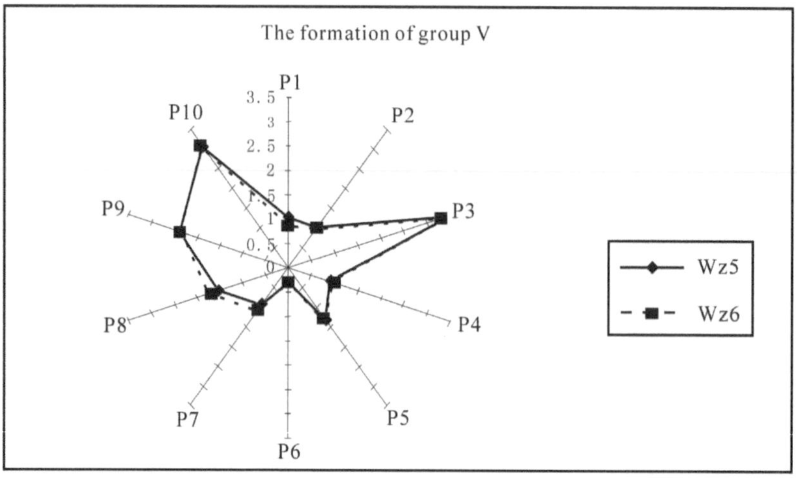

Figure 7. Radar or star plot of selected peak ratios from gas chromatographic analysis of oils in the formation group V of Wells WZ5 and WZ6.

In order to study the fault plane continuity in the formation group IV and V, the 10 selected peak ratios from P1 to P10 are posted in a radar or star plot, as shown in Fig. 6 and Fig 7. From Fig 7, the GC peak ratios of the oil samples from the formation group V of the WZ5 well and WZ6 Well are almost the same. Since there is no fault in group V formation sandbody between the WZ6 and WZ5 wells, it would seem that similar chemistry could indicate that continuity of the fluid of the formation group V is very good.

From Fig. 6, there is a difference between the GC peak ratios of the two wells of the formation group IV, which reflects that the continuity of the fluid of the formation group is limited. It should be noted that all these samples were analyzed at the same time and two samples were re-analyzed and showed the reproducibility for the GC analysis. In order to describe the reservoir continuity quantitatively, according to the standard of continuity evaluation (based on the similarity of chemical compositions) and the ΔPx data table (Table 1), the coincidence rate of the GC peak ratios of the formation group V is 90%, and the continuity of the fluid of the oil reservoir is class I (excellent). While the coincidence rate of the GC peak ratios of the formation group IV is 30%, and the continuity of the fluid of the oil reservoir is class V. It shows that the materials of the formation group IV of the WZ5 and WZ6 wells are different. The seismic data also show that the

Table 1. Data of ΔPx parameters of the formation V and IV of the WZ 12-1 oil field.

ΔPx Horizon	ΔP1 (%)	ΔP2 (%)	ΔP3 (%)	ΔP4 (%)	ΔP5 (%)	ΔP6 (%)	ΔP7 (%)	ΔP8 (%)	ΔP9 (%)	ΔP10 (%)	Coincidence rate (%)	Continuity grade
Formation V	8.29	0.16	2.43	4.65	1.57	2.39	5.11	5.01	0.86	0.64	90	Good
Formation IV	30.87	2.09	13.41	18.37	13.14	11.99	8.47	22.69	5.04	0.51	30	Bad

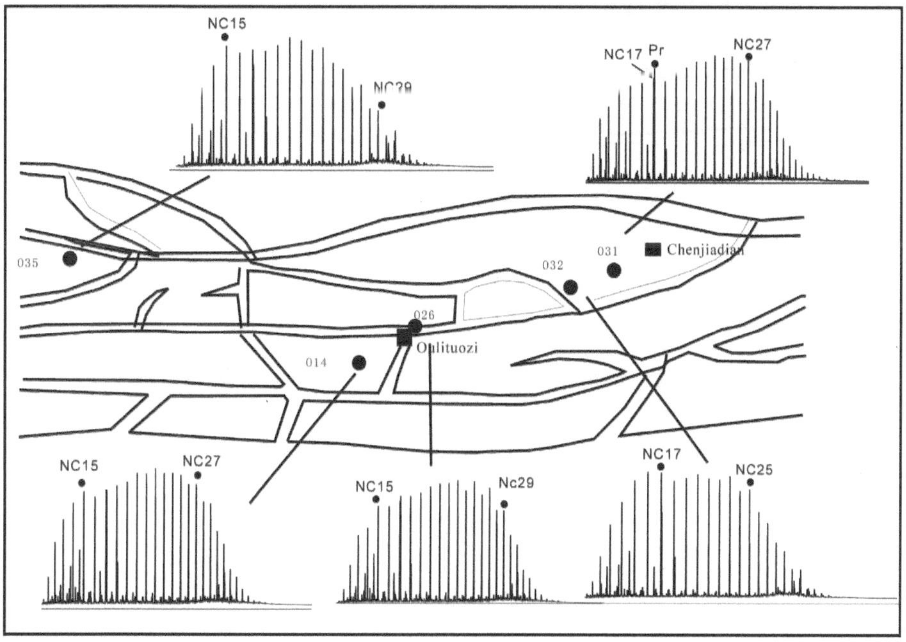

Figure 8. Diagram showing the distribution of n-alkanes in oils of the Oulituozi region.

physical properties of the reservoir of the formation group V are good for petroleum continuity, and there is heterogeneity in formation group IV. The reservoir continuity is identical with the geological section. The above-mentioned results have proved that there really is a small fault in the formation group IV, and it has certain influence on the reservoir continuity. The continuity is not very good, therefore, the impact on fluid of the fault should be taken into consideration in future oil development.

3.2 *Evaluation on the function of a fault during the migration of petroleum*

The midsection in the east of the Liaohe depression has also been of recent research interest. Within this area, there is a significant amount of petroleum in the fractured anticline belt in the area of Oulituozi. Based on the geological data of the area, the function of the fault during the migration of petroleum is analyzed. The position of the sampled well, the detailed geochemistry analysis and the research results of the characteristics of the crude oil source have been published by Hou(2002).

Generally speaking, the crude oil of the area of Oulituozi belongs to normal crude oil, the density of the crude oil differs from 0.83 to 0.89, the viscosity from 2 to 20 mPa.S, and also the content of NSO varies from 8% to 22%. Those features appear in such an area of a few square kilometers of Oulituozi, reflecting the difference of oil pools and possible difference of source input and maturity of crude oils. The nature of the crude oils O 14 and O 15 are relatively similar based on analysis. Although the geographic positions of Wells O 32 and O 31 are near in distance, the nature of the crude oil is very different. The result of the study in the past verified that the crude oil of O 26 is a typical oil of mixing source, while the crude oil of O 32 and O 31 derived from different source rocks.

The distribution of GC fingerprints of oils in the Oulituozi region is shown in Fig. 8. It can be seen that the O32 and O35 oils close to the fault have relatively mature n-alkane distribution. Compared with

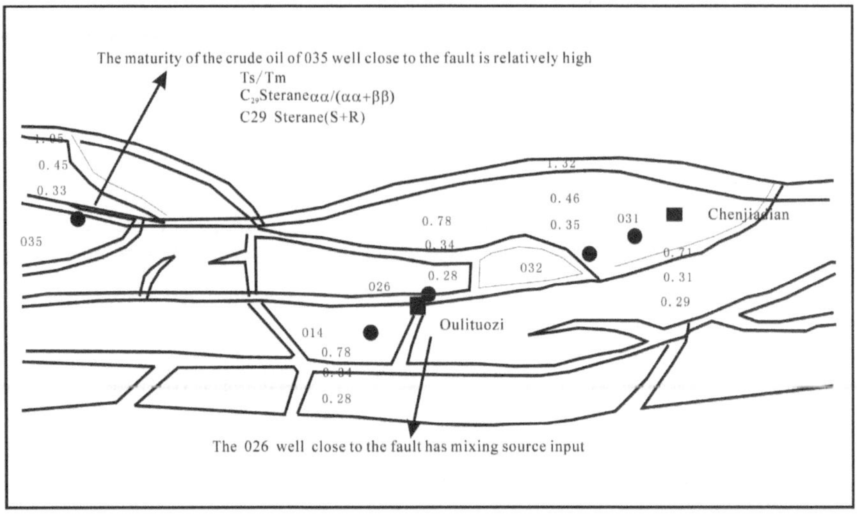

Figure 9. Diagram showing the fault which plays an important role on petroleum migration by using maturity parameters.

Table 2. Parameters of pyrrolic nitrogen compounds of the crude oil in the Oulituozi region.

Well	O 31	O 26	O 14	O 35	O 32
Depth	2405.9–2412.7	2176.3–2262	2178–2182	3000.5–3015.6	2294–2335.3
Formation	Es3	Es3	Es3	Es3	Es3
Benzo[a]carbazole/ Benzo[c]carbazole	0.78	1.06	0.96	1.04	1.47
Carbazole	2.02	3.46	3.36	2.91	1.20
C_1—Carbazole	6.64	11.80	11.77	8.21	7.51
C_2—Carbazole	24.61	29.40	29.68	27.44	29.82
C_3—Carbazole	66.73	55.35	55.18	61.45	61.47
1,8-Dimethyl Carbazole/ 1,7- Dimethyl Carbazole	1.08	1.05	1.07	0.96	1.11
1,8- Dimethyl Carbazole/ 1,6- Dimethyl Carbazole	0.91	1.24	0.97	1.28	1.25
C_3-Carbazole/C_1-Carbazole	10.05	4.69	4.69	7.49	8.19
C_3-Carbazole/C_2-Carbazole	2.71	1.88	1.86	2.24	2.06
Carbazole/Benzocarbazole	0.57	1.16	0.98	0.97	0.96

oils of O32 and O31, the distribution of n-alkane is also different from other oils because Well O32 is near the fault and more mature oils are filling into trap. The biomarker maturity parameters showed the big difference between oils in the Oulituozi region. The oils of O32 and O35 are mature and have parameters of higher maturity (Fig. 9) while the oil of O31 is typical immature oil (Hou, 2002). Oils of O14 and O16 are mixing ones and the maturity parameter is between oils of O32 and O31. It should be known that Wells O32 and O35 are near the main fault of this region. It seems that the fault may be the main conduit for the migration and filling of mature oils.

The analysis results of nitrogen compounds of the crude oils in this area are shown in Table 2. Results showed that the absolute concentration of pyrrolic nitrogen compounds of the crude oils of O 32 and O 35 is relatively high compared with the other oils in this area (Fig. 10), which shows that they are more close to the source rock kitchen than the other oils and meanwhile verifies that the fault functioned as a migration conduit for petroleum during the course of the petroleum filling (Li et al., 1998). The composition of carbazole series compounds such as C_1, C_2, C_3, and carbazole/benzocarbazole also support that oils in Wells O32 and O35 mainly migrated through fault, therefore, there is no obvious differentiation process happening to the pyrrolic nitrogen compounds.

Therefore, we can get some information about fault sealing by the oil of traps near fault or far from fault. The maturity and distribution of pyrrolic nitrogen compounds of oils in the Oulituozi region do indicate which fault can play an important role on oil migration and filling.

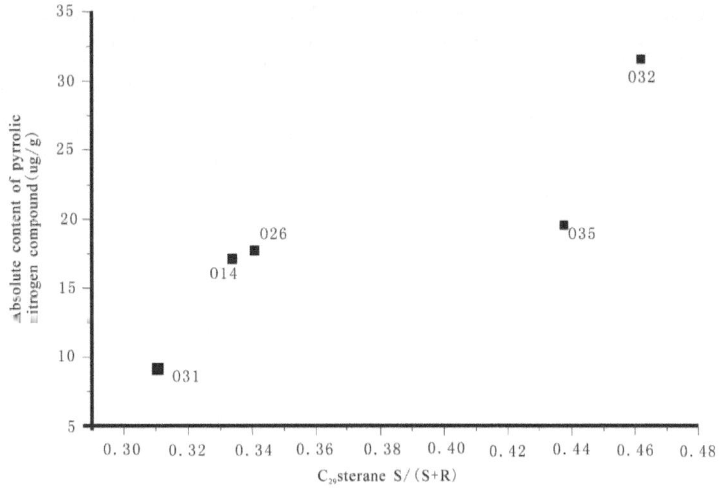

Figure 10. Relationship between concentration of pyrrolic nitrogen compounds in the oils of the Oulituozi region with maturity parameters of C_{29} steranes.

4 CONCLUSIONS

Through the analysis on the chemical composition and biomarker distribution of petroleum in the fault of two different blocks, it is reasonable that the geochemical analysis of petroleum can also offer important information for the study on fault sealing, and help us obtain the knowledge about fault sealing.

1) The analysis on geophysics data of the small fault of the southwest oilfield of Weizhou shows that the small fault has had little influence on the continuity of fluid, but the geochemistry analysis of oils on opposite sides of the fault shows that it still has great influence on the continuity of reservoir fluid.
2) According to the data of pyrrolic nitrogen compounds and combined with the distribution of reservoirs, fault and the maturity of crude oils, it can be determined whether the fault has functioned as a migration conduit for petroleum during the course of the petroleum filling.
3) Geochemical analysis has bright prospects in the study on the continuity of fault.

ACKNOWLEDGEMENTS

This present research work is supported by the natural gas program of the 973 National Key Basic Research Projects. We are grateful to Zhang Minqiang, the manager of the exploring department of the oil company of the East China Sea (the former vice-president of the research institute of the western oil company of the South China Sea), for a lot of constructive advice and suggestion on the research work, to Lin Jincheng the chief engineer, Xie Peiyong the director of the development department of the research institute of the western oil company of the South China Sea. for their strong support and care during the research about the petroleum in the west of South China Sea. We thank Senior Engineers Huang Baojia and Li Li, Su Wenhui, the director, and Mr. Xu Xinde, Xu Yueming, Huang Yiwen and so on for having offered beneficial suggestion and help. The research on the Liaohe depression was directed and helped by Wu Tiesheng, the ex-chief geologist of the research institute of the Liaohe oilfield administration bureau. Xiao Qianhua, the chief of the research institute, Li Jinyou, the director of the regional department, Mr Xu Kun and his colleagues of the well logging company of the Liaohe oilfield administration bureau are sincerely thanked for their great help to us. The authors acknowledge and appreciate Dr. L. R. Snowdon and anonymous reviewers for the help in improving the clarity of the manuscript.

REFERENCES

Allan, A.S., 1989. Model for hydrocarbon migration and entrapment within faulted structures. American Association of Petroleum Geologists Bulletin 73, 803–811.

Downey, M.W., 1984. Evaluating seals for hydrocarbon accumulations. American Association of Petroleum Geologists Bulletin 68, 1752–1763.

Gibson, R.G., 1994. Fault-zone seals in siliciclastic strata of the Columbus Basin, offshore Trinidad, American Association of Petroleum Geologists Bulletin 78, 1372–1385.

Graham Yielding, Brett Freeman, D. Tim Needham, 1997, Quantitative fault seal prediction, American Association of Petroleum Geologists Bulletin 81, 987–997.

Hou D.J., Tang Y.J., Li X.Q., Zhu J.Z., 2002. The identification and quantitative analysis of mixing oil in Oulituozi Region, Symposium on The 4th National Petroleum geology experiment technology and administration, Petroleum Industry Press, Beijing.74–82 (in Chinese).

Hu, C.Y., Kong, Z.P., 2002. The Principle of the Formation of Oil and Gas Reservoir. Petroleum Industry Press, Beijing.111–113 (in Chinese).

Jian C.Z., Wei G.Q., Yao H.J., 1995, Structural Evolution and Regional Structural Geology of Tarim Basin. Petroleum Industry Press, Beijing.1–20 (in Chinese).

Kissin, Y.V., 1987, Catagenesis and composition of petroleum: Origin of n-alkanes and isoalkanes in petroleum crude. Geochimica et Cosmochimica Acta 51, 2445–2457.

Knott S.D., 1993, Fault seal analysis in the North Sea, American Association of Petroleum Geologists Bulletin 77, 778–792.

Li M., Lin R.Z., Wang P.R., 1998, Study on the oil and gas resource of condensate oil-gas field in Kekeya. Hydrocarbon prospecting and reservoir geochemistry. Petroleum Industry Press, Beijing. 17–32 (in Chinese).

Macgregor, D.S., 1996, Factors controlling the destruction or preservation of giant light oilfields, Petroleum Geoscience 2, 197–221.

Smith D.A., 1966, Theoretical consideration of sealing and nonsealing faults, American Association of Petroleum Geologists Bulletin 50, 363–374.

Wang X.B., 2004, The formation and customization of the tectogenetic movement since Neogene Period of China gas reservoir. Oil and gas geology 25, 126–132 (in Chinese)

Wang Z.X., Xin Q. L., 1997, The discussion on underground fault sealing, Geological Journal of University 3/3, 293–300 (in Chinese).

Xiang C.F., Xia B., Jie X., 2004, Main Pathway of oil and gas migration and transport of the west incline of Songliao basin. Oil and gas geology 25, 204–208 (in Chinese).

Petroleum Geochemistry and Exploration in the Afro-Asian Region –
Digang Liang, Darui Wang & Zhenxi Li (eds)
© 2008 Taylor & Francis Group, London, ISBN 978-0-415-44083-7

Biodegraded characteristics of ground soil for crude oil contaminants

Xiaoyan Liu, Chun Li, Jungang Li, Chunlei Dai, Jing Wen & Yingli Li
Geochemistry laboratory, Daqing Petroleum Institute, Daqing, P R China

ABSTRACT: In order to prevent soil environments from contamination during petroleum production, storage, and transportation in the oilfield, degradation of microbes for oily soil was studied with the Geofina Hydrocarbon Meter (GHM) by experimental analysis qualitatively and quantitatively in the paper. Analytical results showed that the crude oil could be considerably degraded by oil-eating microbes in oily soil and the number of oil-eating microbes increases when the working hours of oil-wells rise. As a result, contaminated oil could be degraded more quickly by a lot of oil-eating microbes in the soil. At the same time, the degradation rate of contaminated oil increases gradually as the time goes on. In addition, the amount of gaseous components in the oily soil samples increases over time and microbes could selectively consume contaminated oil, so biodegradation might alleviate the degree of contamination and destruction to the soil and environments in the process of oil production in the oilfield. The biodegradation laws for oily soil are studied and some significant conclusions are gained in the investigation as well.

Keywords: soil; crude oil; contaminant; biodegradation; experiment.

1 INTRODUCTION

The phenomena of crude oil falling and leakage often occur in the process of oil exploitation, storage, transportation, processing and production. The crude oil soaking into the soil of the earth's surface may result in the problem of environment pollution (Zhang Xingru and Zhang Shiquan, 1998; Al-Hassan, 2000). Oily soil can not only give rise to local pollution but also destroy ecological environments and function of the ecosystem, which may make the comprehensive fertility of soil decline and influence the growing of plants (Xia Lijiang and Wang Hongkang, 1997). It is important that organic contamination in soil may harm human health by biological accumulation and enlargement action of the natural food chain (Li Tianjie, 1996; Ren Lei and Huang Yanlin, 2000). Many organic contaminants in petroleum can result in various diseases to human beings, for instance, cancer, malformation and gene mutation, so the research of degradation for crude oil contaminants is of great significance (Li Peijun, 2003; Genouw, 1994). At present, the method of biological degradation is considered as a better way to remedy oily soils in the world owing to microbes ability to degrade organic contaminants in soil. The process of biological remediation is simpler than other ones and the expense is lower. In addition, its effect is better and it can't produce additional pollution. So some scholars begin to study the method of biological degradation according to the characteristics of the oilfield soil in different areas (Scherrer 1990; Scott, 1993; Chang, 2002). Biological methods to resolve soil contamination will be a significant direction of technology for environmental protection in the future. In China, the soil environments of most oilfields are suitable for applying this biological method. Contaminated soil samples of the First Production Factory at Daqing Oilfield were taken to perform the experiment on oily soil biodegradation and the characteristics of microbial degradation for crude oils were analyzed qualitatively and quantitatively by utilizing technology of thermo-evaporation/gas chromatograph by Geofina Hydrocarbon Meter (GHM) made in Norway Geolab. Moreover, the biodegradation law for oily soil was studied by the data analysis of experiment results.

2 METHOD

2.1 *Experimental samples*

In this paper, experimental samples were obtained from a well area far away from the site of human activities. Sample I was collected from the place where wells had being worked for 15 years. There were more oil-eating microbes selected and domesticated by the nature in the oily soil near wells with longer working history. Sample II was gained from the area where wells had being worked for two years. The basic parameters of the oily soil samples are listed in Table 1.

Table 1. Basic parameters of the oily soils in Daqing Oilfield.

Samples	Density (g/cm^3)	Porosity (%)	pH	Water content (%)	Weight loss (%)	Electric conductivity ($\mu s/cm$)
I	2.014	3.3	8.75	1.38	4.92	91.6
II	2.054	3.7	9.40	3.06	2.34	94.9

Microbes conform well to natural environments. The degree of degradation of petroleum hydrocarbons is related to the species and amount of microbes, petroleum composition, environmental parameters influencing biochemical action, and so on. Temperatures between 10 ~ 40°C provide the optimal condition for microbial reproduction, so the experiment was conducted at room temperature (about 25°C) and in a natural moisture environment. At first, the sample (about 100 g) was triturated into particle size (1mm) in a clean transparent glass column, and then adequate warm boiled water was add (about 25°C) to the sample to make it just immersed in water. The amount of water was decided by sample absorption and soakage degree. This experiment should be carried on in the laboratory without direct sunlight and wind to avoid artificial disturbance in the process of the experiment. Experimental time was set for different intervals, such as 5 days, 10 days, 15 days and 360 days respectively.

2.2 *Analytical instrument and method*

The main instrument used in the analytical experiment was large-scale GHM apparatus. The three GHM detectors were all FID. The exit pressure of high pure hydrogen was 0.26 MPa, and the constant pressure of air compressor was set to 0.6 MPa. The carrier gas was helium whose exit pressure was 0.4 MPa and the pressure ahead of column was 0.06 MPa. The chromatographic column was OV-1032d quartz capillary column with the length of 25 meters. The velocity of helium going through the chromatographic column was 0.85 ml/min. Sample injection, detector, and auxiliary parts of the instrument were set at the same temperature (300°C).

The initial temperature of quantitative analysis was set at 220°C, then the temperature increased to 300°C at the rate of 40°C/min and was kept for two minutes. Organic components of petroleum in soil were volatilized out of the samples. The products of thermo-evaporation were carried by carrier gas to FID C and FID A respectively, then they were detected quantitatively and qualitatively. The ratio of gas distribution between FID C and FID A was 30:1. The temperature of chromatographic oven was initially set at 30°C. Then it was raised at the rate of 4°C/min to 300°C and kept for 20 minutes until the experiment was finished.

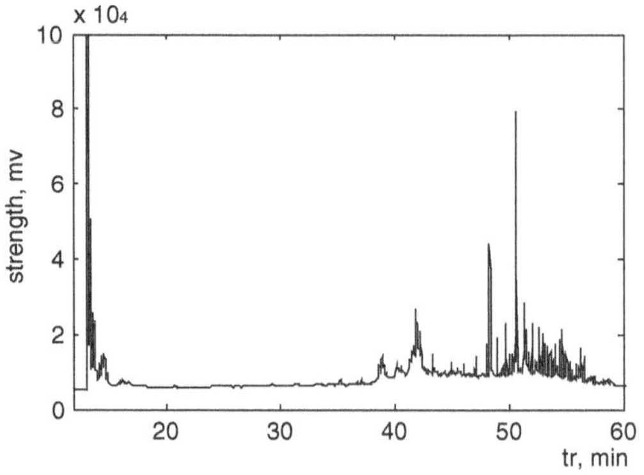

Figure 1. Initial detecting diagram of biodegradation in the oily soil sample II for 360 days. Some compounds in crude oils were utilized selectively, such as the hydrocarbons with lower to medium molecular weight, and the remaining compounds, were relatively enriched after bacteria had consumed certain parts of the oil components.

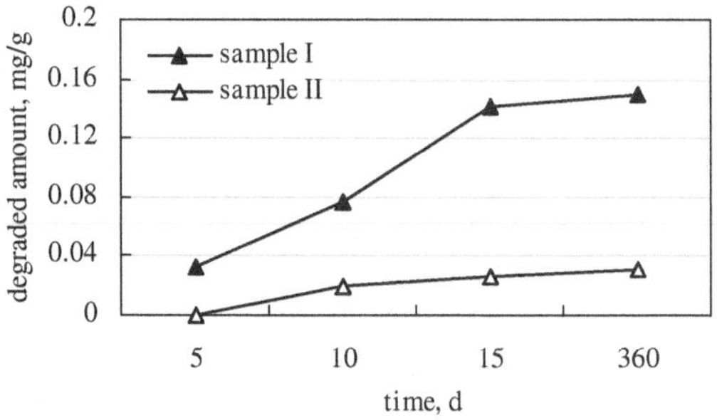

Figure 2. Relative degradation rate of petroleum-pollutants in oily soils at different times.

By the above experiment, the quantitative and qualitative data of biodegradation for oily soil were obtained in order to calculate the amount and composition of oily organic matter in the samples. The initial qualitative detecting diagram is shown in Figure1. Some compounds in crude oils were utilized selectively, such as the hydrocarbons with lower to medium molecular weight, and the remaining compounds were enriched after bacteria had consumed certain parts of the oil components.

3 RESULTS AND DISCUSSION

3.1 *Quantitative analytic results*

Degraded rates and amounts were gotten by experimental data in the process of microbial degradation for oily soil. The characteristics of relative degraded rate for samples were shown in Figure 2, Figure 3

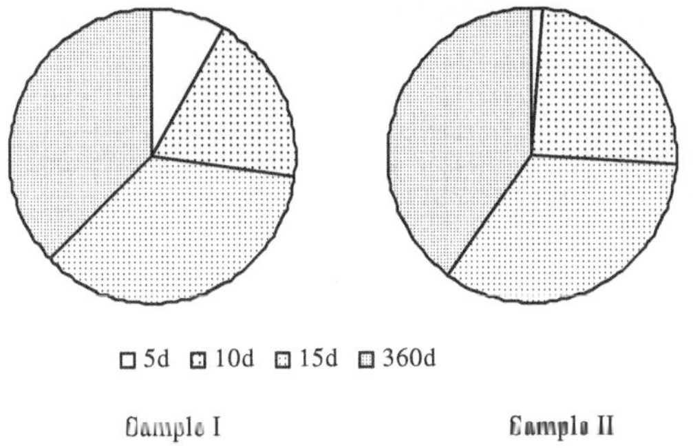

Figure 3. Biodegraded amount of soil organic matter at different times. The absolute degraded amount was higher in the sample I than sample II, and the degraded amount increases continuously over time longer, whereas the change rate was falling gradually.

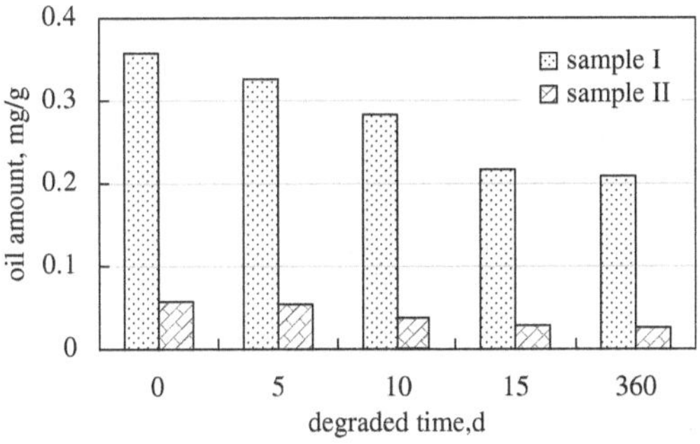

Figure 4. Quantitative biodegraded characteristics of soil organic matter. The quantity of the contaminated oil is different in two samples, and the amount of soil organic matter in Sample I is clearly higher than that in Sample II every time. In addition, the amount of organic matter in two samples decreases as microbial degradation progresses.

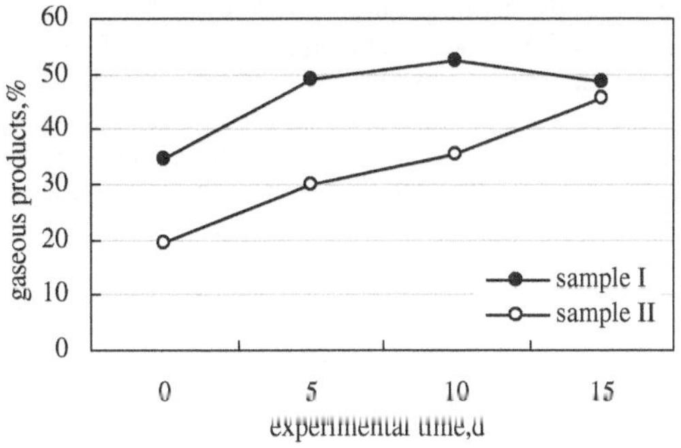

Figure 5. Qualitative biodegraded characteristics of soil organic matter for the products percentage. The component amount of gaseous hydrocarbons in sample I was always higher than sample II during the process of microbial degradation for 15 days. Moreover, the gaseous amount of the samples increased as the time of degradation progressed.

and Figure 4. The numerical value in Figure 2 has been normalized. Because there are more oil-eating microbes in Sample I than those in Sample II, the relative degraded rate for organic matter for 5 days is much higher than that of Sample II. In addition, with microbial degradation continuing, the velocity of degradation of the two samples becomes faster and faster and their relative degraded rate also becomes higher and higher. Experimental result shows that the more the oil-eating microbes, the faster the microbial degradation for oily soils. In other words, oily soils around oil wells with longer production history have more oil-eating microbes, so the degraded degree of contaminated oil increases obviously. The absolutely degraded amount of soil organic matter in different times by microbial action is shown in Figure 3. The data confirms that the absolutely degraded amount is higher in Sample I than Sample II, and the degraded amount increases over time, but the rate of change is falling gradually.

The quantitative analytical result of the oil amount with time changing in soil samples is shown in Figure 4, which shows that the quantity of contaminated oil is very different in two samples. The initial amount of organic matter in Sample I (0.358 mg/g) is obviously higher than that in Sample II (0.056 mg/g). Sample I comes from the district where oil wells has longer production history, so the contaminated oil has accumulated longer, resulting in a higher amount of organic matter in Sample I. In contract, Sample II comes from the district where oil wells has shorter production history, therefore the amount of organic matter in Sample II is lower. Figure 4 also shows that the amount of organic matter in both samples decreases with the advancement of microbial degradation, indicating microbial degradation for organic matter becomes stronger as time goes on. There really exist oil-eating microbes in the soil around oil wells and they could degrade petroleum in the soils. Certain components of petroleum can be selectively consumed by microorganisms, therefore, the degree of destruction of contaminated oil in soil and environments is alleviated gradually and naturally.

3.2 *Qualitative analytic results*

The qualitative analytic result for organic matter in the samples can be demonstrated by the percentage content of gaseous and non-gaseous components at normal temperature. The analytical characteristics qualitatively for microbial degradation of organic matter in soils are shown in Figure 5.

The component amount of gaseous hydrocarbons in Sample I is always higher than that in Sample II during the process of microbial degradation over 15 days (Figure 5). The gaseous amount of the samples increases as the time goes on, however the gaseous amount of Sample I decreases over the 15 days. The main reason could be as follows: on the one hand, microbial degradation for soil organic matter is of selectivity. Hydrocarbon compounds with medium long carbon chain are preferentially degraded; however, gaseous hydrocarbon molecules and heavy components remain. As a result, the amount of light hydrocarbons increases in samples. On the other hand, the phenomenon might be related to the methane produced in the biodegradation process. It is also demonstrated by the qualitative analysis that microbes have a really stronger ability of degradation and consumption for crude oils.

4 CONCLUSIONS

(1) Oil-eating microbes exist in oily soils around oil-wells and microbes could considerably degrade petroleum which has dropped or leaked into the soils; these could decrease petroleum destruction for soils and pollution for environments.
(2) The longer the well history is, the more the oil-eating microbes are in the soil near wells. As a result, contaminated oil could be degraded quickly. More contaminated oil is accumulated with older wells, so the amount of organic matter in oily soil increases obviously.
(3) The relative degraded rate of the contaminated oil increases gradually as the time goes on. The amount of gaseous components increases with degradation in the oily samples. Microbes could selectively and markedly consume contaminated oils.
(4) This study on microbial degradation for oily soils is of practical value for environmental protection and great significance in the sustainable development of society and economy in oilfield areas.

ACKNOWLEDGEMENTS

The financial support from the Nature Science Foundation of Heilongjiang Province (B2010) and Scientific & Technological Foundation (GZ05A601, 10541005) is gratefully acknowledged.

REFERENCES

Al-Hassan J. M., Afzal M & Rao C. V. N., 2000. Petroleum Hydrocarbon Pollution in Sharks in the Arabian Gulf. *Bull. Environ.Contam.Toxicol.* 65:391–398.

Chang B.V., Shiung L.C & Yuan S.Y., 2002. Anaerobic biodegradation of polycyclic aromatic hydrocarbon in soil. *Chemosphere* 48:717–724.

Genouw G., Naeyer F & Meenen P., 1994. Degradation of oil sludge by land farming a case study at the Ghent harbour. *Biodegradation* 5:37–46.

Li Peijun, Tai Peidong & Guo Shuhai, 2003. Two phases bioremediation of oil contaminated soil from Liaohe oilfield. *Environmental science* 24(3):74–78(in Chinese).

Li Tianjie, Gong Shiguo & Pan Genxing, 1996. *Soil environment-precaution-treatment and entironmental protection of soil.* Beijing:*Higher education press* 92–101 (in Chinese).

Ren Lei & Huang Yanlin, 2000. Petroleum pollution of soil. *Environment protection in agriculture* 19(6): 360–363 (in Chinese).

Scherrer D., 1990. Biodegradation of crude oil in experimentally polluted clayey and sandy mangrove soil. *Oil &Chem. Pollu.* 6: 163–176.

Scott T. W., Brker G. W & Cook R. C., 1993. Biodegradation of hydrocarbon contaminated soils at crude oil production sites in west. *Texas and New Mexico. SPE.* 2598.

Xia Lijiang & Wang Hongkang, 1997. *Soil pollution and its precaution-treatment.* Shanghai: *East China university of science and technology press* 135–143(in Chinese).

Zhang Xingru & Zhang Shiquan, 1998. *Exploiting construction and environmental influence of petroleum field.* Beijing: *petroleum industry press* 129–148 (in Chinese).

Petroleum Geochemistry and Exploration in the Afro-Asian Region – Digang Liang, Darui Wang & Zhenxi Li (eds)

Application of fluid inclusion to understand oil charging in Palogue Field, Melut Basin, Sudan

Dingsheng Cheng, Lirong Dou, Zhi Li & Yingying Li
China National Oil and Gas Exploration and Development Corporation, Beijing, China

ABSTRACT: Fluid inclusions have been increasingly applied to petroleum exploration over the last two decades. In this paper, fluid inclusion and biomarker analyses have been done; the research results indicate there are two oil accumulation periods for the reservoir of Palogue oilfield, Melut Basin, Sudan. The primary oil pool had been degraded by structure uplift in Late Oligocene. Up to Neogene Period, source rock of Palogue Field was reburied deeply enough to generate hydrocarbons again, and the later oil is light oil that has the same maturity as the preliminary one. The original Palogue Field oil generated in Late Cretaceous, accumulated in Upper Cretaceous and Tertiary; the erosion in Late Oligocene Period interrupted the formation of the original oil pool. With the reburial of source rock, oil generation restarted, the secondary Tertiary reservoirs were formed. The secondary oil generation potential is much more than the preliminary one. The latest Palogue Field reservoirs had come into being in the Middle – Late Oligocene to Late Miocene.

1 INTRODUCTION

The trends in the application of fluid inclusions in recent years may be divided into three categories: (1) the increasing application to additional branches of earth science, (2) the continued dominant application in ore-deposit research, and (3) the rapidly increasing application in petroleum geology. Hydrocarbon fluid inclusions serve as direct records of hydrocarbon migration and charge history. The frequency of quartz grains containing oil inclusions in sandstone reservoirs has been used to study the charge history and evolution of oil-water contact. The simple observation of oil inclusions in core or cutting samples could be the first real evidence for a hydrocarbon system.

Fluid inclusions are micron-scale volumes of ambient fluid entrapped during mineral growth and rehealing of micro-fractures. Hence, analysis of the inclusions can provide direct evidence of the fluid composition and physicochemical conditions pertaining to cementation and fluid migration, including oil charge. They are identified and studied in doubly polished wafers (thick sections, typically 100–130 μm thick) of well cuttings or cores. Inclusions of aqueous fluid are subjected to heating and cooling in order to determine temperatures of homogenization into a single fluid and temperature of ice-melting (which reflects the fluid salinity). The theory of fluid-inclusion analysis is described in detail by Roedder (1984). Inclusions of oil are recognized by fluorescence in ultraviolet light (McLimans, 1987). Many samples contain multiple populations of inclusions, which may yield different homogenization temperatures and salinities or involve different fluid types (aqueous, oil, gas). Study of their paragenesis allows reconstruction of an evolving fluid-migration history and links it to an evolving thermal history.

2 GENERAL INFORMATION OF PALOGUE FIELD

As Petrodar drilled the discovery well Palogue-1 in the Melut Basin of Sudan during 2002, a world-class giant Palogue field was confirmed by subsequent exploration and appraisal wells (Figure.1, 2). So far, this field is the biggest oil field in Sudan, even in the Central-Western Africa Shear Zone. The proven reservoir rock of Yabus and Samma mainly consists of the Paleogene delta sandstone (Figure 3). The structure of the field is a giant drape anticline belt located in a relay slope (Figure 2). The effective hydrocarbon source rocks are believed to be the Lower Cretaceous shales deposited during the syn-rift stage of basin evolution (Figure 3). Crude oils are mainly normal oil with API of 20° ~ 34°, some of them are heavy oil with API < 20°. In general, the crude oils in Palogue field have high asphaltene content, high wax content, high pour point, high total acid number (TAN) and low sulphur content (Table 1).

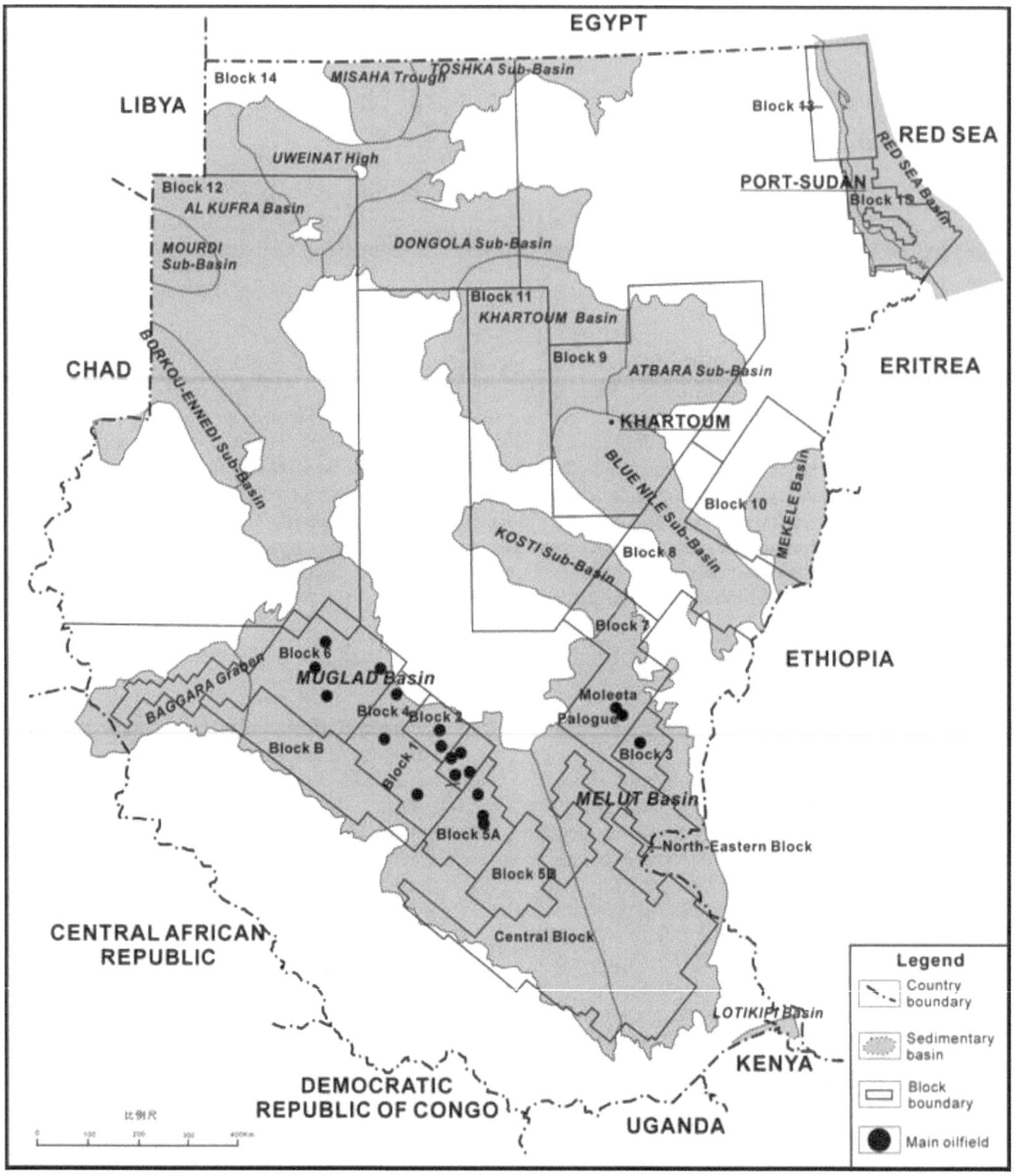

Figure 1. Location map of Melut Basin.

3 SAMPLES AND ANALYTICAL METHODS

3.1 *Samples*

The samples used are 12.99 m core from Yabus Formation in Well Fal-2 (1208.2–1226.56 m) for fluid inclusion analysis and 7 oil samples are selected from Yabus and Samma Formations in Well Fal-1 for biomarker analysis.

3.2 *Experimental methods*

All of the data used in this study had been reported by CNPC Key Laboratory based on analyses on the oil and core samples. Fluid inclusion analyses were performed using GP600 MOTOR-DRIVED GEOLOGY HEATING/FREEZING STAGE according to the criterion of SY/T 6010-94. JY-1000 Laser Raman Spectrum was used to analyze the component of fluid inclusion using

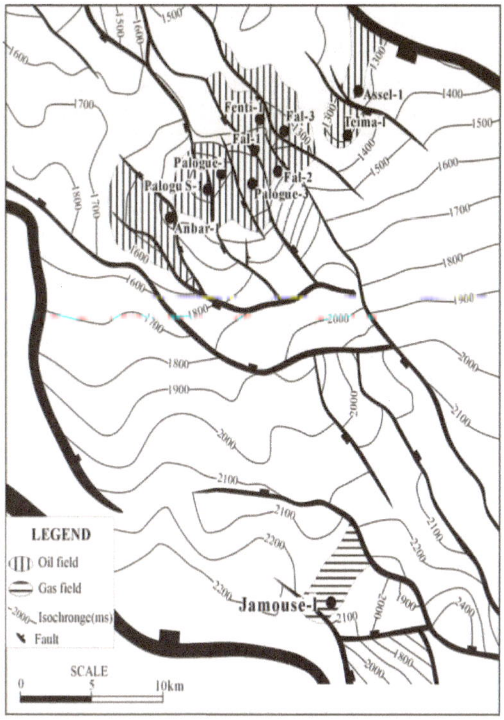

Figure 2. Top Yabus structure map of Palogue Oilfield.

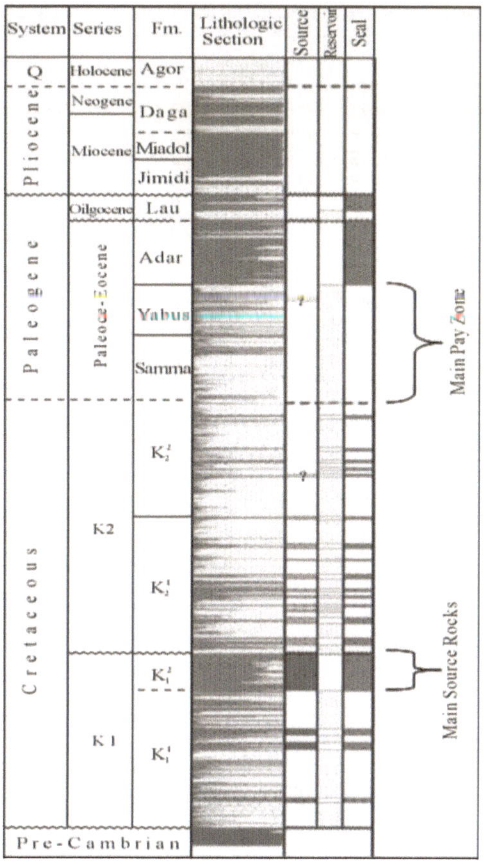

Figure 3. General column of Melut Basin.

Labran-infinity-010 Laser with Excitation Line: 514. 543 μm, Objective: 100.

Oil fluorescence was determined by Oil Fluorescence Analyzer (OFA) with 150 W Mercury Lamp, interference filter with 254 nm exciting wavelength.

The GC/MS is fitted with a 60 m DB5 ms (J & W) fused silica capillary column, 0.25 mm I.D., 0.25 um film thick. Analytical conditions are as follows: 100°C held for 10 minutes, then programmed at 2°C/min. to 320°C, Helium as carrier gas, head pressure 20 psi, split ratio 30:1, injector temperature 300°C. transfer line temperature 300°C, EI source 200°C, electron energy 70 V, MID scan mode was cycled at 2 seconds. Identification of the major biomarker families was carried out by comparison with published retention characteristics and by co-elution with reference compounds. Corrected peak areas obtained by baseline subtraction were used to calculate biomarker ratios.

4 RESULTS

Inclusion homogenization temperature data of Well Fal-2 (Table 2) indicate the Yabus reservoir is formed by two oil charge periods, the primary Yabus oil charge has been destroyed into heavy oil, with the light brown fluorescence and low homogenization temperature (average 78°C); the secondary charge oil is normal oil, with no fluorescence but high homogenization temperature (average 227.5°C) caused by CO_2 intrusion from the deep earth crust. The CO_2 component can be easily proven by Roman spectroscopy data (Table 3). The Fal-2 well oil sand has three peaks (main peak of 360 nm, medium peak from 380 nm to 390 nm, weak peak of 310 nm) by OFA data. Under fluorescent microscope, some dissemination grains and grain margins show brown fluorescence, which is typical for a heavy oil. While, blue and white fluorescence in pores indicates the existence of light hydrocarbons in the Yabus crude oil. Both heavy oil and light oil tested in Well Fal-2 prove the Yabus reservoir is formed by two oil charging times.

Geochemical data of the Yabus and Samma oils, in Well Fal-1 (Table 4), show that the oil pool may be formed in two oil charge periods, which can be seen from the data as follows: (1) The wax content and API value of these oils decrease with the depth increasing while TAN and asphaltene content of them increase;

Table 1. Basic geochemical data for oils in Palogue Field.

Well	Depth (m)	Fm.	API°	TAN (mgKOH/g) By Crude	TAN (mgKOH/g) By PVT	Pour point (°C)	C_{6+} Asphaltene (%m/m)	Wax (%m/m)	Sulphur (%)
Anbar-1	1392–1397	Yabus	30.9		0.8	39		33.2	
	1415–1430		30.2		1.3	36.1	14.94	29.97	
	1445–1459		28.4		2	39		30.67	
Fal-1	1145–1148		23.8	6.4				5.73	0.16
	1203–1213		23.3	3.4	3.2	42	22.68	26.6	0.13
	1243–1247		22.7	3.5	3.5	42	23.2	30.23	0.13
	1260–1291		21.8	4.2	4.1	39		25.75	0.14
	1311–1332	Samma	19.2	6.2	6.8	27.2	32.47	2.62	0.19
	1343–1348		18.2		10.4	39		3.03	0.17
	1366–1382		14.5	8.9	8.4	15		2.99	0.19
Fal-2	1183–1202	Yabus	23.8		3.5	42	19.4	30.35	
	1206–1240		21.5		3.1	39		29.4	
	1282–1302	Samma	21.8		5.3	42	26.76	16.57	
Palogue-1	1265–1270	Yabus	25.3		2.6	42		15	
	1287–1301		24	3.1	2.9	45	20.7		0.13
	1312–1333		26	3.6	3.64	42		24.9	0.15
	2249–2253	K_2^1	30.6	0.53				43.66	0.12

Table 2. Fluid inclusion data of the Well Fal-2 oil sand in 1225.84 m.

No	Type	Shape	Mineral	Size (μ)	Gas/liquid (%)	Feathers	Color of gas	Color of liquid	Homogenization temperature (°C)
1	Secondary inclusions	Irregular line	Quartz	5 × 12	6	Distributed at enlarged quartz margin in group	Light brown	Colorless	77.8
2		Irregular ellipse		9 × 24	6	Distributed in quartz in group			71.8
3		Curve line		6 × 15	8		Colorless		235
4		Irregular block		10 × 15	10	Distributed at quartz margin scatteredly	Light brown		73.6
5		Irregular line		5 × 10	6	Distributed in suture			89.7
6				6 × 15	8	Distributed in quartz in group	Colorless		220

Table 3. Composition (percent) of fluid inclusions from Yabus Formation of Well Fal-2 by Laser Lamma Spectrum.

Depth (m)	CO_2	SO_2	H_2O	CH_4	C_2H_6	C_2H_4	C_2H_2	C_3H_8	C_3H_6	C_4H_6	C_6H_6
1213.79	85.12				1.46		2.97	3.15	4.5	1.24	1.58
1216.58		50.19		25.67	5.51			5.42		7.07	6.13

(2) Based on GC data of crude oils, light component content of the deeper oil is low, showing "bump" and double peaks; (3) C_{25} nor-hopanes (Figure 4) were detected in saturated hydrocarbons, indicating serious biodegradation of the crude oil, yet the normal paraffin content is intact. In general, the ratio of normal oil content of Palogue field decreases with the depth increasing.

Table 4. Geochemical profile of Well Fal-1.

Depth (m)	Fm.	API°	TAN	C6+ Asphaltene (%)	Wax (%)	Whole oil GC	m/z 191	m/z 217	Maturity LH (%)	Rc (%)	C29-Steranes 22S/(22S+22R)
1145–1148	Yabus		6.4	22.68	5.73	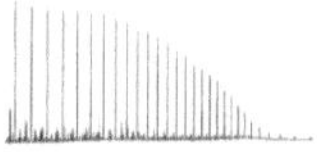			0.91		0.56
1203–1213		23.3	3.4		26.6	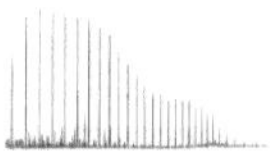	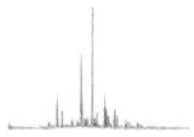	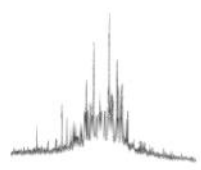	0.93	0.66	0.55
1243–1247		22.7	3.5	23.2	30.23						
1260–1291		21.8	4.2		27.7						
1311–1332	Samma	19.2	6.8		4.84	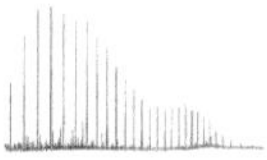	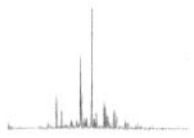	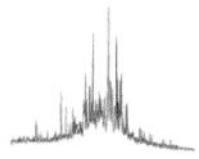	0.93	0.67	0.57
1343–1348		18.2	10.4	32.47	3.03	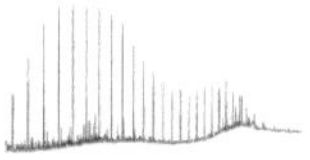					
1366–1382		14.5	8.9		2.99				0.95		0.57

Note: TAN: Total Acid Number; Rc: Calculated Ro by MPI1; LH: Light hydrocarbon

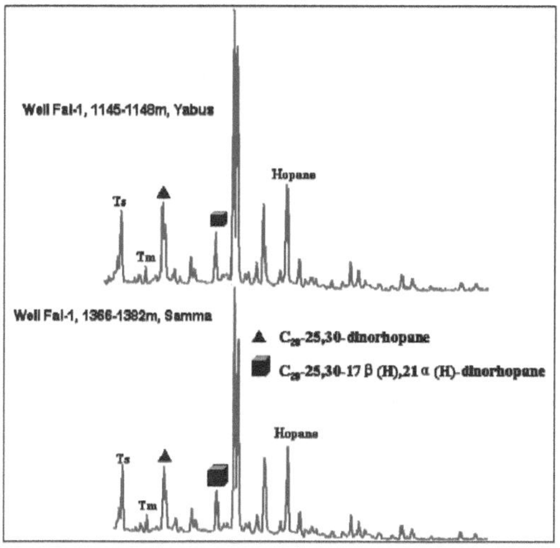

Figure 4. 25-norhopanes were detected in mass fragmentograms of terpanes (m/z177) in saturates of oils from Well Fal-1.

5 CONCLUSIONS

In one word, the intact normal paraffin content and the C_{25} nor-hopane detected in saturates of oils from the Yabus and Samma formations, Palogue oil field, indicate there are two oil accumulation periods for the reservoir. The primary oil pool had been degraded by structure uplift in Late Oligocene. The tectonic change also suspended the oil generation in this area. Up to Neogene Period, source rock of Palogue Field was reburied deeply enough to generate hydrocarbons again, and the later oil is light oil that has the same maturity as the preliminary one. Thermal evolution analysis and oil – source rock correlation showed that source rock is K12 shale. So we conclude that the original Palogue Field oil generated in Late Cretaceous, accumulated in Upper Cretaceous and Tertiary; the erosion in Late Oligocene Period interrupted the original oil pool; With the re-burial of source rock, oil generation restarted, the secondary Tertiary reservoirs were formed. Based on data analysis, the secondary oil generation potential is much more than the preliminary one. The latest Palogue Field reservoirs had come into being in the Middle – Late Oligocene to Late Miocene.

REFERENCES

Chi G, Chou I-M and Lu H-Z, 2003. An Overview on Current Fluid-Inclusion Research and Applications. *Acta Petrologica Sinica*, 19(2): 201–212.1.

Marjrie Wilson and Rene Guiraud, 1992, Magmatism and rifting in Western and Central Africa, from Late Jurassic to Recent times. *Tectonophysics*, 213: 203–225.

Peters K. E., and Moldowan J. M., 1993. *The Biomarker Guide: Interpreting Molecular Fossils in Petroleum and Ancient Sediments* [M]. Englewood Cliffs, NJ: Prentice Hall Inc, 1993. 252–265.

R. Guiraud, R.M. Binks, J.D. Fairhead and M. Wilson, 1992, Chronology and geodynamic setting of Cretaceous-Cenozoic Rifting in West and Central Africa. *Tectonophysics*, 213: 227–234.

Roedder, E., 1984. Fluid Inclusions. *Mineralogical Society of America, Reviews in Mineralogy*, 12. 6. S. E. Browne and J. D. Fairhead, 1983, Gravity study of the Central Africa rift System: A medel of continental disruption, 1.The Ngaoundere and Abu Gabra rifts. *Tectonophysics*, 94: 187–203.

Ten Haven H L, 1996, Application and limitation of Mango light hydrocarbon parameters in petroleum correlation studies. *Organic Geochemistry*, 24: 957–976.

Thomas J. Schull, Rift Basins of Interior Sudan: Petroleum Exploration and Discovery. *Bulletin of American Association of Petroleum Geologist*, 72(10): 1128–1142.

Tim R. Mchargue, Tom L. Heidrick and Jack E. Livingston, 1992, Tectonostratigraphic development of the Interior Sudan rifts, Central Africa. *Tectonophysics*, 213: 187–202.

Petroleum Geochemistry and Exploration in the Afro-Asian Region –
Digang Liang, Darui Wang & Zhenxi Li (eds)
© 2008 Taylor & Francis Group, London, ISBN 978-0-415-44083-7

Geochemical monitoring technique in reservoir performance

Xiaohui Jin
SINOPEC Exploration production Research Institute, Beijing, P.R. of China

Quan Shi, Renzi Lin & Dan Zhu
University of Petroleum, Beijing, P.R. of China

ABSTRACT: Due to hetergenities of reservoir rocks and the fluid distribution, the contribution of a single formation to liquid production is different from others in the production. By the chromatographic fingerprint correlation analysis technique developed in the 1990s, the geochemical monitoring of reservoir performance is applied to estimate the variation of bulk liquid production of individual pay formation. Consequently the behavior of fluid in the reservoir can be monitored, the reservoir management can be carried out, and the elevation of recovery efficiency can be measured. According to the process of commingled production of different pay bed, physical and mathematical models based on the HQR algorithm have been made. By the allocation experimental technique, the database of chromatographic fingerprint parameters can be established so as to carry out the research on chemical and mathematical simulations. The simulation results show that the constraint conditions of chromatographic fingerprint parameters in calculation are: (1) the parameter reflecting each single pay formation remains constant with the variation of contribution; (2) for one fingerprint compound, the content of this compound in the commingled samples is under the control of the relative contribution of single pay bed; (3) a certain discrete degree must exist between the array of the parameters of each bed. Therefore, the mathematical simulation on dynamic monitoring commingled production of three pay beds has been resolved by choosing new parameters.

1 INTRODUCTION

In order to optimize management and enhance oil recovery, commingled production of multiple reservoirs is taken in most oil fields. But subsurface fluid will move when the reservoir has been developed, simultaneously, the contribution of a single formation to liquid production is different from one to another in the production due to heterogeneities of reservoir rocks and the fluid distribution. At present, production logging is the most popular method to measure the production allocation of commingled production wells and it usually determines the fluid volume produced from each reservoir in an oil well by a turbine flowmeter. All above methods are time-consuming, labor-intensive and expensive. In addition, they can not monitor producing status of a well at any moment and have to make the production halt.

Dynamic geochemical monitoring of oilfield development is to estimate the variation of bulk liquid production, which is based on the correlation analysis on the composition of crude oils in reservoirs by using chromatographic fingerprint technique developed in the 1990s. To assign oil production allocation by means of gas chromatographic fingerprints of crude oils was firstly introduced by Kaufman in 1987, which calculated production allocation of various wells in two oilfields in San Joaquin Valley by constraint least squares. Later the method was applied in the Liaohe oilfield, Daqing oilfield and Tarim oilfield (Renzi Lin, 1996;Bowen Mei, 1997; Peirong Wang, 1997; Tieguan Wang, 1998, et al). However, the method can only figure out the commingled production of no more than two pay beds presently. So it can't be applied widely *in situ* of oilfields. The main problem is the constraint condition of chromatographic fingerprint parameters in calculation is vague, at the same time, the simply constraint least squares applied will result in collinearity problem and will lead to the failure of solution. The dual aims of this paper are to determine the constraint condition of chromatographic fingerprint parameters and to resolve the commingled production allocation of three pay beds by using allocation experimental technique and mathematical simulation. Particularly, it should be pointed out that its application is of importance to petroleum industry but can not be more feasible without improvements in basic understanding and/or analytical techniques.

2 METHODS TO CHARACTERIZE RESERVOIR FLUID HETEROGENEITY

Gas chromatographic system capable of obtaining high-quality gas chromatographic fingerprints of crude oils has revolutionized the application of molecular geochemistry to the study of petroleum reservoirs. In this paper, samples were collected from Gangxi oilfield in Tianjing, China. Agligent 6890 gas chromatograph equipped with FID(flame ionization detection) was used for the crude oil analysis. The GC column was a J & W fused silica capillary DB-5 column (60 m × 0.25 mm i.d., 0.25 μm methy1 silicone film). The temperature was programmed to start with 80°C, be isothermal for 1 min, then increase at 5°C/min to 300°C and be isothermal at 300°C for 20min. The injector temperature was 300°C, and the carrier gas was helium. Most samples were analyzed twice, and the standard deviation was ±1.

3 CHEMICAL AND MATHEMATICAL MODEL

3.1 *Chemical model*

It is supposed that the commingled production oil of discrete three pay beds labeled as II III and I corresponds to three vessels that hold the same oil samples, respectively (Fig.1). The commingled production oil IV is in another vessel. Therefore, the process of commingled production of different pay formations during the reservoir production can be simplified in the laboratory as the process to blend oilfrom different vessels having different features of chromatographic fingerprint in unlike proportion to form a new kind of oil. Chemical and mathematical models based on the HQR

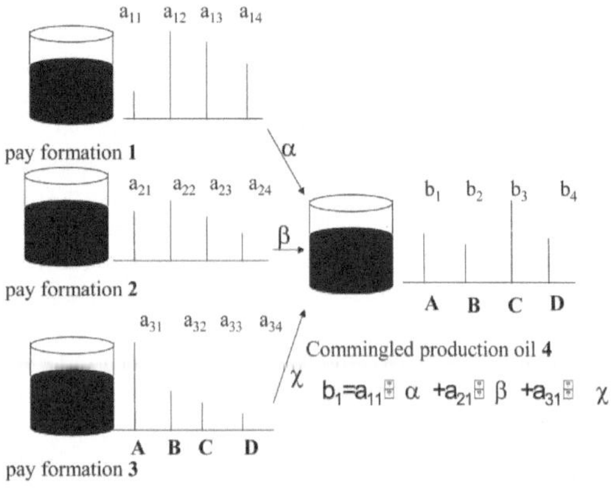

Figure 1. Chemical model of a commingled production reservoir (α, β, χ refer to the percentage of each formation; a_{11}, a_{21} ... to gas chromatographic parameter; A, B, C and D to gas chromatographic compound, respectively).

calculation can be made according to the process. Primary, basic and specific chromatographic fingerprint parameter data can be constructed using allocation experimental technique to carry out the research of chemical, mathematical simulation and practicability of allocation of compounds.

3.2 *Mathematical model and algorithm*

Nearly all the gas chromatographic compounds like A, B, C, D... can be always detected according to the fluid hetergenities. The fingerprint parameter of that compound can be described as aji, and the fingerprint parameter of commingled production oil can be calculated as

$$b_i=\sum_{j=1}^{m} a_{ij}x_j+\varepsilon_i \qquad (i=1,2,3,\cdots n) \tag{1}$$

where **i** is the number of characteristic parameters of Compound **i** in each pay formation, **j** is the number of individual reservoirs (j = 1,2,3 ... m), $\mathbf{b_i}$ is characteristic value of Compound **i**, ε_i is a statistical error, $\boldsymbol{a_{ij}}$ is a calibration value($a_{ji} \neq a_{j+1,I}$), $\boldsymbol{\chi_j}$ is the percentage of the pay formation **j** in the commingled production.

Bold face capital letters are used as matrices, e.g., ***A*** and ***b***. Equation 1 can be still described as (2) by using matrix algebra:

$$Ax = b \tag{2}$$

The matrix A (n × m) is calibration matrix in which samples are prepared by using oils from M individual reservoirs and N compounds can be measured, if the same compounds are measured in a produced oil, then b matrix could be used to express the composition of the commingled production oil.

Usually the numbers of characteristic compounds are more than those in the composition of the commingled production oil (so n > m), It can be expressed as the equation (3) below.

$$rankA \neq rank\,[A,b] \tag{3}$$

The standard equation (3) can be shown as

$$A^T Ax = A^T b \tag{4}$$

If $rank\,A = m(m \leq n)$, the solution of Equation 4 can be calculated as

$$\eta = (A^T A)^{-1} A^T b \tag{5}$$

where η is the column matrix of contributions of individual reservoirs.

If the variables a_{ij} are independent from each other (i.e. little correlated) and the number of Compounds ***n***, is large, compared to the number of samples (individual reservoirs) ***m***, the estimation of η is best made by ordinary least squares (*OLS*) with multiple regression (*MR*). However, oils within an oilfield often have the same source and consequently have very similar geochemical nature, which leads to the similarities of their gas chromatographic fingerprints In the meantime, if the end members of individual reservoirs are more than two, the a_{ij} variables are strongly correlated. If ordinary least squares is applied, the collinearity problem will appear, which will result in solution failure. Therefore, the transformation of *Householder* and the method of *Arranging-primary-member* in matrix *QR* analysis have been named *HQR* algorithm based *QR* analysis. This makes HQR solution have equal or better predictive properties for η, especially in the case when the end members of individual reservoir are more than two. Meanwhile, no explicit regression step is given in the method and it is also much faster than *OLS*.

4 CHEMICAL SIMULATION RESULTS OF PARTITION EXPERIMENTS FOR GC FINGERPRINT COMPOUNDS WITH DIFFERENT CARBON NUMBERS

An example of a binary stratum commingled production well is given as follows. Based on the distribution characteristics of GC fingerprints and the quantitative principle of fingerprint parameters, the 47 characteristic fingerprint parameters were first divided into 3 different stages for the mathematical simulation and all the parameters were then calculated according to the principle of the selected fingerprint parameters whose carbon numbers are possibly neighboring and to the physical properties of different carbon number compounds (i.e., the boiling points and chemical stability of the compounds with different C number). There were 14 parameters in the first parameter segment whose carbon numbers were between $C_7 \sim C_{12}$, and also 14 parameters in the second parameter segment whose carbon numbers were between $C_{13} \sim C_{16}$, and 11 parameters in the third parameter segment whose carbon numbers were between $C_{17} \sim C_{20}$. In addition, the fourth parameter segment is all the partial characteristic fingerprint parameters whose carbon numbers were between $C_7 \sim C_{20}$.

The real values of the 5 partition samples based on the chemical simulation process of every single stratum GC fingerprints and the concentration of every single stratum were calculated and compared with the mathematical simulation results. Then an error analysis of partition extents (Table 2) was also conducted. In the first parameter segment, the average relative error of the 5 samples is 4.82%, with the minimum error of 0.24%, the maximum error of 13.25%, and mid-error of 3.11%. In the second parameter segment, the average relative error is 2.13%, with the minimum error of 0.46%, the maximum error of 5.63% and mid-error of 1.15%. In the third parameter segment, the average relative error is 2.89%, with the minimum error of 0.4%, the maximum error of 5.43% and mid-error of 2.38%. In the fourth parameter segment, the average relative error is 2.05%, with the minimum error of 0.00%, the maximum error of 4.77% and mid-error of 1.83%. Thus, all relative errors in above cases are smaller than 4.93%, and the precision is also in the allowable range (<10%). The absolute errors of different partition samples are also smaller than 2.3%. Based on the chemical simulation results of three stratum partition experiments, the fingerprints in $C_{12} \sim C_{20}$ at the whole hydrocarbons of GC analysis have good partition characteristics, and these are the optimum parameter segments for the chemical monitoring of oilfields. The reason for this is as follows. A) The boiling point of this carbon segment is always higher than 110°C, which is nonvolatile, non-dispersive and has good distribution in a specified time. B) The GC retention time of these fingerprints is in accordance with that of the standards, leading to small errors for quantitative calculation. C.) There are very few isomers of the compounds in this segment.

5 SIMULATION RESULTS OF EXPERIMENTAL SAMPLES

The reliability of the HQR- chromatographic fingerprint method has been evaluated in the following case. A set of samples with known commingled proportion for calibration was prepared with the oils in different wells of II, III and I reservoirs in the Gangxi oilfield. The good consistency between actual and calculated values is showen in Table I. The relative errors are less than 10% except for the No. 6 sample.

6 RESULTS AND DISCUSSION

6.1 *Simulation results in the Gangxi oilfield*

The oil production allocation for all the commingled production wells in the researched area is assigned using the calibration sample sets by *HQR*-chromatographic fingerprint method. Fig.2 shows the variation of oil production of $NmII^4$, $NmII^6$ and $NmII^7$ formations during the three-time monitoring work (about 90 days, the interval is three month). It can be seen that the percentage of production allocation of $NmII^4$ is gradually declining from 25.19% to 14.89% during the three-time monitoring work. As time goes on, the percentage of $NmII^6$ is 30.10% then 29.94% and 14.17%, but that of $NmII^7$ is gradually changeable from 70.94% to 44.71% and then to 44.06%. Actually,

Table 1. Analytical results of the commingled production oil of three simulated end members.

Sample no.	Actual content (%) I: II: III	Calculated value (%) I: II: III	Absolute error (%) I: II: III	Relative error (%) I: II: III
1	20.23: 20.69: 59.08	17.91: 17.01: 54.84	2.32: 3.68: 4.25	6.08: 9.76: 3.73
2	20.53: 38.80: 40.66	17.92: 32.31: 44.30	2.61: 6.49: 3.64	6.97: 9.13: 4.28
3	40.98: 20.14: 38.88	40.17: 15.41: 36.11	0.81: 4.73: 2.77	1.00: 13.31: 3.69
4	39.56: 29.04: 31.94	42.37: 24.11: 27.43	2.81: 4.93: 4.51	3.43: 9.28: 7.60
5	0.00: 38.72: 61.28	1.34: 29.22: 56.27	1.34: 9.50: 5.01	/: 13.98: 4.26
6	59.47: 0.00: 40.53	81.07: −8.7: 29.64	21.60: 8.70: 10.89	15.37: /: 15.52

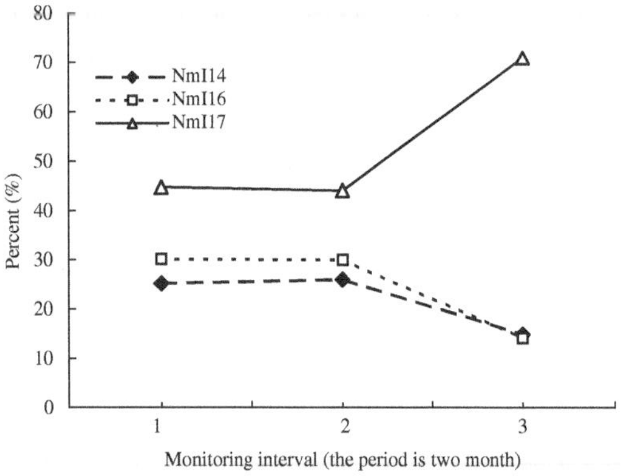

Figure 2. Assignment of $NmII^4$, $NmII^6$ and $NmII^7$ oil production in the 6-9-1 well of the Gangxi oilfield.

the same result has been proven by the logging data of the well.

6.2 *Geochemistry significance of model parameters*

The constraint conditions of chromatography fingerprint parameters in calculation have been concluded from the process of establishing mathematical models and obtaining simulation results. They include: (1) the parameter reflecting each single pay formation remains constant with the variation of contribution; (2) for one fingerprint compound, the content of this compound in the commingled samples is under the control of the relative contribution of single pay bed; (3) a certain discrete degree must exist in the array of the parameters of each bed; (4) negative contribution shows that a discrete pay formation is not produced.

The constraint conditions of (1) and (2) are explanation for geochemistry significance of model parameters. And (3) is an explanation for mathematical significance of model parameters.

Therefore, the mathematical simulation on the commingled production dynamic monitoring of three pay beds has been resolved by choosing new methods and a new index system. The comprehensive researches have developed the theoretical models of geochemical monitoring and analysis of reservoir performance. The stabilization of petroleum fingerprints is influenced by content, properties and resolution of chromatographic analysis. Petroleum chromatographic fingerprints are steady in both the underground and subsurface temperature. At the same time, they show similarities during different producing periods. The result means a new thought for the quantitative simulation of reservoir performance.

As a test tool to determine contributions of the oil production allocation of commingled production, the method of crud oil gas chromatographic fingerprints is more economic, convenient and accurate than the methods used before. The new method needs only for samples at wellhead and indoor analysis. It provides a new way for performing monitoring of oil-producing section, which can be done at any time by using the method. Thereby, the application of the new method in practical researches on various oilfields and various types of oils (especially to heavy oil) will be theoretically important and economically valuable. It is of significance in theory and practice in carrying out stratification management and making measure to raise recovery efficiency.

REFERENCES

England W.A., 1990. The organic geochemistry of petroleum reservoirs. Org. Geochem.. Vol16, No.1–3, pp.41–425.

Gelad, P. and Kowalski, B.R., 1985. Partial least-squares regression: a tutorial: Analytical Chimica Acta 185, 1–17.

Kaufman, R.L. Ahmed A.S.and Moldowan, J.M., 1987. A new technique for the analysis of commingled oils and its application to production allocation calculations: in Proceedings of the Sixteenth Annual Convention of Indonesian Petroleum Association, Indonesian Petroleum Association 247–268.

Kaufman, R.L., Ahmed, A.S. and Elsinger, R.L. 1990. Gas chromatography as a development and production tool for fingerprinting oils form individual reservoirs: application in the Gulf of Mexico: Gulf Coast Section of the Society of Economic Paleontologists and Mineralogists Foundation Ninth Annual Research Conference Proceedings 263–282.

Larter, S.R. and Aplin, A.C., 1995. Reservoir geochemistry: methods, application and opportunities, In: The Geochemistry of Reservoir, Eds. J. N. Cubitt and W.A. England, Geological Society Special Publication No. 86, 5–32.

Mark A. McCaffrey, Henry A. Legarre, and Scott J. Johnson, 1996. Using biomarkers to improve heavy oil reservoir management: An example from the Cymeic field, Kern County, California. AAPG Bulletin, V. 80, No. 6, P. 898–912.

Petroleum Geochemistry and Exploration in the Afro-Asian Region –
Digang Liang, Darui Wang & Zhenxi Li (eds)
© 2008 Taylor & Francis Group, London, ISBN 978-0-415-44083-7

The formation and recognition of thermo-evaporative condensate oil and gas

Jianian Shen, Shuangfang Lu & Min Wang
Daqing Petroleum Institute, Daqing, P R China

Wei Fang
Exploration and development research institute of Daqing oilfield, Daqing, P R of China

ABSTRACT: Till now only Well Gu109 and Well Ying51 located in the Gulong sag show a condensate gas reservoir in the Songliao basin. And under the condensate gas reservoir, there exist oil layers with dissolved gas. The oil-source correlation shows that all the oils come from the high-maturity source rock of Qingshankou Formation in the sag. According to the analysis of the geological background, the formation of the condensate gas cannot be interpreted by using the phase state origin mechanism of the condensate oil and gas reported. Based on the experimental study of thermo-evaporation, the calculation of the fluid equation of state and the correlation of ratio relationships of main hydrocarbon compounds between the condensate gas reservoir and the underlying oil layer, the condensate gas in the Gulong sag is formed as a result of thermal evaporation effect. The relative volatility (α) is defined as a link that restricts the ratio relationship between the condensate oil gas formed by thermo-evaporation and the underlying residual oil-gas components. Therefore, different components have different evaporating capabilities when thermo-evaporative fractionation occurs, and the difference can be denoted by the relative volatility (α). The relationship between the ratios of main hydrocarbon components in the condensate gas and in the underlying oil is consistent with the calculated α under the stratum condition. Basic geochemical characteristics of condensate formed by thermo-evaporation are both oil and gas in condensate gas and in residual oil coming from the same source rock and have the same maturity; ratios in molar concentration of n-alkanes (nC_m/nC_{m+1}) in condensate gas layer is always bigger than those in residual oily layers.

Keywords: Gulong sag; condensate oil-gas; relative volatility; thermo-evaporation; phrase state origin.

1 INTRODUCTION

The discovery of condensate gas in the Gu109 well in 1987 and in the Ying 51 well in 2002 in the Gulong sag of the Songliao basin attracted wide attention to the exploration prospect of the condensate oil-gas in this area. As the phase state origin directly determines the distribution and the resource assessment of the condensate gas, and affects the direction of exploration and decision-making, it was widely investigated when it was firstly found in the Gu109 well. Xu (1989) believes that it is the hydrocarbon compound that is derived from the overmature source rock of Qingshankou Formation in the Gulong sag, and then accumulates directly, i.e. original type (Huang, 1997) or initial type (Zhou, 1998). This view is also supported by some latter scholars (Li, 1997; Chen, 1997), and it is taken as the representative of the original condensate gas that generates from the sapropel type source rock during the overmature stage (Huang, 1997). But the relationship of main hydrocarbon components in the condensate gas and in the underlying oil is quite similar to the relative volatility obtained by thermo-evaporative experiment and theoretical calculation, so this phenomenon cannot be explained by original type. Thus it is thermo-evaporative condensate gas rather than the original type.

2 GEOLOGICAL BACKGROUNDS

The Gulong sag is located to the west of the Daqing placanticline, the north of the Qijia sag and the west of the Da'anlonghupao terrace. It is one of the two major oil producing sags in the north of the Songliao basin, and also the main oil producing sag in the Daqing placanticline. The main source rocks are Sections 1 (K1qn1) and 23 (K1qn23) of Qingshankou Formation and Section 1 (K1n1) of Nenjiang Formation in this sag. The main organic matter belongs to types I and II_1 in lacustrine sediments. It is also the classic sedimentary and oil generating formation in China. The values of Ro for

Table 1. Basic data of the Ying 51 and Gu 109 wells.

Well no.	Oil layer no.	Deep (m)	Formation	Oil per-day (t)	Gas per-day (m^3)	Gas-oil ratio (m^3/m^3)	Temperature (°C)	Pressure (MPa)	Gas $\delta^{13}C$(‰, PDB) C_1	C_2	C_3	C_4
Ying 51	P	1818.0–1822.6	K_1yl(Yiaojia)	7.08	88522	7399.4	85.0	18.45	−47.8	−34.8	−31.0	−30.0
	G1	1923.8–2021.0	K_1qn23 (Qingshankou)	0.346	0		87.9	/				
	G2	2187.2–2198.6	K_1qn23 (Qingshankou)	26.50	16546	505.4	104.6	/	−47.7	−35.6	−30.7	−30.1
	qn1	2270.0–2276.0	K_1qn1 (Qingshankou)	3.523	0		108.4	/				
	F	2306.8–2349.6	K_1q4 (Quantou)	3.353	0		/	/				
Gu 109	P1	1645.6–1648.6	K_1yl(Yiaojia)	12.39	110565	7109	75	23.47	−47.8			
	P2	1678.4–1704	K_1yl(Yiaojia)	10.83	685	60	80.5		−48.6			

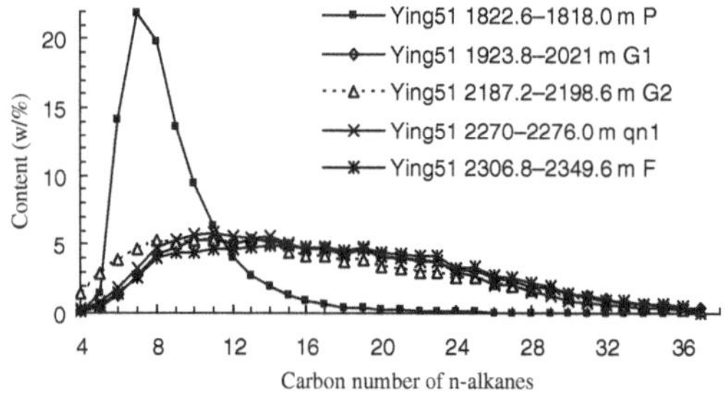

Figure 1. Comparison of n-alkanes from different oil-gas layers (Well Ying 51).The major components of the oil (P layer) are lower carbon number hydrocarbons, and different from that of the underling oil layers.

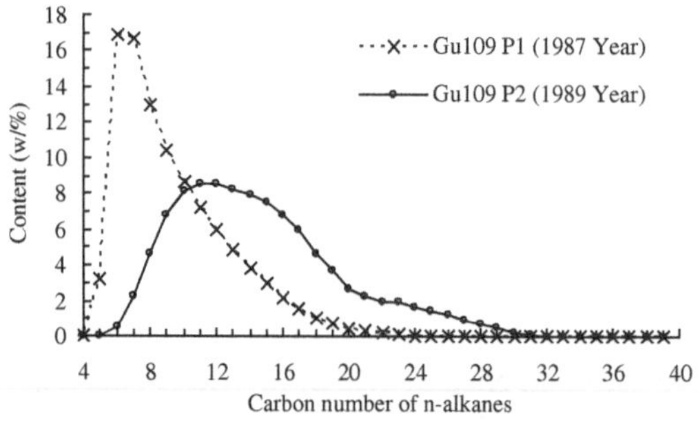

Figure 2. Comparison of n-alkanes in different oil-gas layers (Well Gu 109).

the K1qn1, K1qn23 and K1n1 sections are 1.1 ~ 1.8%, 0.9 ~ 1.7%, 0.5 ~ 1.0%, respectively, in the mature to overmature stage. The Ying51 well is located in the west slope of the Gulong sag and has five oil producing layers altogether (Table 1). The top Putaohua oil layer is a condensate gas reservoir in the lithologic trap (with a subsurface crude oil density of 0.6974 g/cm^3), the other four oil layers are fault traps, and the G2 oil contains lots of dissolved gases (with a surface crude oil density of 0.8334 g/cm^3). Fig. 1 shows the n-alkane distribution of five oil samples. The Gu109 well is located in the abdomen of the Gulong sag, and the drilling displays three oil-gas layers, the Heidimiao oil layer is a low producing oil layer, and the Putaohua oil layer has two oil-gas layers. The upside of the Putaohua oil layer is an anticline condensate gas reservoir (with a surface crude oil density of 0.7450 g/cm^3) and the lower part is a lithologic oil-gas reservoir (with a surface crude oil density of 0.8194 g/cm^3), which contains dissolved gases. Fig. 2 shows the n-alkane distribution of P1 and P2, which were analyzed in 1987 and 1989, respectively.

3 THERMO-EVAPORATIVE FRACTIONATION EXPERIMENT AND ANALYSIS RESULTS

3.1 *Experimental*

3.1.1 *Equipment and samples*

Fig. 3 shows the experimental equipment, the oil sample used comes from the G2 formation (2187.2–2198.0 m) of the Ying 51 well.

3.1.2 *Experimental steps*

① First weigh the sample tube, fill the tube with some crude oil and then weigh the filled tube, and the difference between the above two is the oil weight (16 ~ 43 g), then put the tube into the heating device. ② Quickly vacuumize the device. ③ Use the temperature controlling apparatus to heat the sample tube in order to achieve the fluid-gas balance, then keep the expected temperature for two hours, collect condensate and gas in the condensate liquid collecting tube and gas collecting tube, respectively, under the normal temperature (approximately 20°C) and atmospheric pressure conditions. ④ Weigh the residual oil in the

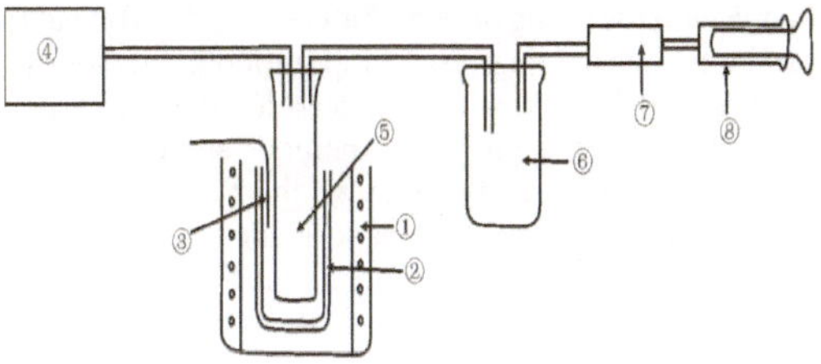

① Intermediate frequency heating up power source
② Heated steel tube ③ Temperature sensor
④ Vacuum pump ⑤ Sample installing tube
⑥ Condensate liquid collecting tube ⑦ Thick wall emulsion tube ⑧ Gas collecting tube

Figure 3. Apparatus of thermal-evaporation.

sample tube and the condensate in the condensate liquid collecting tube, collect gas products in the injector and perform GC (HP6890) analysis on the condensate and residual oil. ⑤ Repeat ① ~ ④ steps and start the next temperature-point test at 150°C, 200°C, 250°C, 300°C, respectively.

3.2 *Elementary theory and general characteristics of thermo-evaporation*

In the process of thermal evaporation of the hydrocarbon system, different components turn into vapor phase with different capabilities, which can be represented by relative volatility (Zhan, 1994).

$$y_A/y_B=\alpha^N x_A/x_B \tag{1}$$

Where, A and B represent two kinds of components; y and x represent the mole fraction of A or B in the vapor phrase (maybe one part is gas phase and the other is condensate phase) and residual liquid phase, respectively, α represents the relative volatility, N(>0) represents the number of plates. It is obvious that if α is larger than 1, the volatile capability of A is heavier than that of B, then A is enriched in the vapor phase more easily than B ($y_A/y_B > x_A/x_B$), the degree of fractionation depends on α and N when thermo-evaporative fractionation occurs in A and B. And if $\alpha > 1$, the bigger the values of α and N are, the richer A is than B in gas phase.

There are lots of complicated factors that affect the relative volatilities between different components, such as temperature, pressure, molecular weight, chemical structure and composition, and there has been no simple equation to denote the relative volatility up to now, so only the equation of state can be employed to calculate it. Generally speaking, among the hydrocarbon homolog (the same temperature and pressure) the less the molecular weight of hydrocarbons is, the greater the volatility is, such as the n-alkane homolog. That is, the relative volatility ratio of low carbon number to high carbon number is always larger than 1; compounds with low carbon number is usually easier to volatilize than those with high carbon number. For compounds with the same carbon number, the volatility will be evidently affected by molecular structure, and it is also affected by the temperature, pressure and composition of the hydrocarbon system. The high temperature and low pressure is beneficial to volatilization. The condensation temperature can also influence the value of α in theory, and N is the function of evaporation and it is shown in the condensation process of chemical industry (Zhan, 1994).

In this paper, the PR state equation is used to calculate the component relative volatility of special hydrocarbon system at the given temperature and pressure (Peng and Robinson, 1976). The concrete calculating method and details will not be included due to the limited space of the paper and readers can consult relevant literature (e.g. Reed et al., 1994; Danesh, 2000; Guo, 2002).

3.3 *Result analysis*

3.3.1 *GC characteristics of thermo-evaporation*

Table 2 and Fig. 4 show the chromatograph parameters of the residual. It can be seen from parameters that the main carbon numbers of the residual move backwards with the temperature rise, and the ratio of Pr/Ph also drops regularly. It indicates that, because the molecular weight of pristane is smaller than that of phytane, the quantity of pristane turning into gas phase is larger than that of the residual, so the quantity of pristane in the residual is reduced and so is the ratio of C_{21}^{-}/C_{22}^{+}. Ratios of Pr/nC_{17} and Ph/nC_{18} maintain invariable, which is obviously caused by the similar saturated steam pressure between Pr and nC_{17} or Ph and nC_{18}. The $(nC_{21}+nC_{22})/(nC_{28}+nC_{29})$ ratio increasing with the temperature rise is the result of the relative enrichment. The relative quantities of nC_{21} and nC_{22} increase greatly with the temperature rise, however, those of nC_{28} and nC_{29} increase slowly. The reason is that the evaporation effect can make both higher and lower carbon number hydrocarbons turn into evaporative gas phase. The quantity turning into evaporative gas phase is closely associated with many factors, such as temperature, pressure, single component content in the liquid phase and the saturated steam pressure. The $(nC_{21}+nC_{22})/(nC_{28}+nC_{29})$ ratio increasing with the temperature rise is the result of combined actions. If high-molecular compound like nC_{28} and nC_{29} cannot turn into evaporative gas phase, only nC_{21} and nC_{22} can (with the vaporization temperature rise, the quantity turning into gas phase increases inevitably), the $(nC_{21}+nC_{22})/(nC_{28}+nC_{29})$ ratio will not increase with temperature increasing, on the contrary, it will decrease. The $(nC_{21}+nC_{22})/(nC_{28}+nC_{29})$ ratio of the initial oil is 1.92, and that of the residual is 2.04 at 150°C. It means that when the vaporization temperature reaches 150°C, the thermo-evaporation will affect the high-molecular compounds like nC_{28} and nC_{29}.

The carbon number of condensate also increases with temperature increasing, but the relative content of alkanes (corresponding to main carbon) decreases and the envelope curve becomes broad (Fig. 5). It reveals that the carbon number scope of hydrocarbon compounds in the condensate becomes broad with the rising temperature. It can be seen from the comparison of n-alkane distributions of the residual and initial oils (Fig. 4) that the main loss of hydrocarbon components is the light hydrocarbons ($<C_8$) at the vaporization temperature of 150°C. Moreover, the smaller the carbon number is, the larger the loss is; and the more the components are turning into liquid condensate. All the components ($<C_9$) of the initial oil turn into liquid condensate at 300°C, so does a part of $C_{12} \sim C_{15}$. It is obvious that the higher the evaporation temperature, the higher the degree of evaporation, if viewed from physical evaporation. The relative content of the main carbon and the width of the envelope curve change with the rising of temperature (Fig. 5), it reveals that the higher the degree of evaporation is, the lower the main peak carbon is, the broader the envelope curve is. When the initial oil includes some components ($<C_6$), it is ready to cause them to evaporate into the condensate oil at the vaporization temperature of 150°C. Because of the limited vaporization period in the laboratory conditions; to achieve the same transpiration rate, the extension of vaporization time may cause the decrease of temperature. Another important factor is the vaporization space, so the bigger it is, the lower the temperature is needed. Both points can be physically proved. At a certain temperature, long-time and large-space can increase the content of heavy components in the liquid condensate. For example, when the temperature reaches 150°C, time and space being suitable, the components ($>C_9$) will increase and form a trail.

3.3.2 *The characteristics of the n-alkane ratios*

Fig. 6 shows the relative volatility (α) obtained by applying the PR state equation to the oil sample (G2)

Table 2. Parameters of liquid products from the oil evaporation of the Ying 51 well.

		Main C No.	Pr/Ph	Pr/nC_{17}	Ph/nC_{18}	OEP	C_{21}^-/C_{22}^+	$C_{21}+C_{22}/C_{28}+C_{29}$	Scope Of C No.
Condensate	150°C	nC_6							$3\sim14$
	200°C	nC_6							$3\sim14$
	250°C	nC_7							$3\sim12$
	300°C	nC_8							$3\sim17$
Residual	150°C	nC_{10}	1.68	0.12	0.08	0.99	2.65	2.04	$4\sim34$
	200°C	nC_{10}	1.63	0.13	0.09	1.00	2.50	2.09	$3\sim33$
	250°C	nC_{11}	1.58	0.13	0.09	1.01	2.39	2.13	$8\sim34$
	300°C	nC_{14}	1.54	0.13	0.09	1.00	2.17	2.43	$9\sim34$
Initial		$nC_{10}C$	1.57	0.13	0.09	0.97	3.19	1.92	$4\sim37$

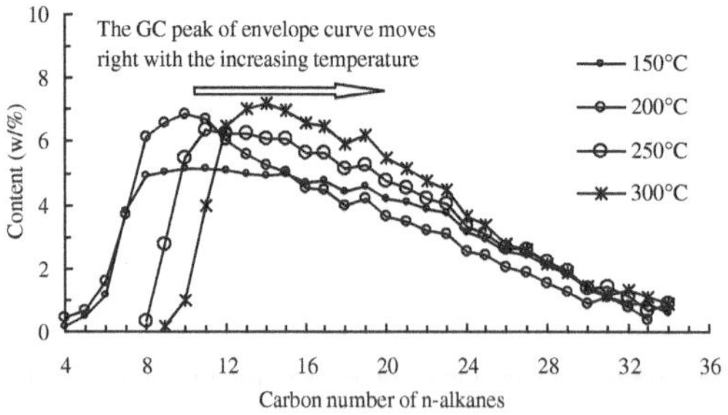

Figure 4. Distribution of n-alkanes in the residual liquid. (The residual liquid is obtained from the thermal evaporation experiment, and the experimental oil sample comes from the G 2 oil layer of the Ying 51 well. The higher the evaporative temperature is, the more the light hydrocarbons enter into evaporation phase, so the content of hydrocarbons with high carbon number is high in the residual).

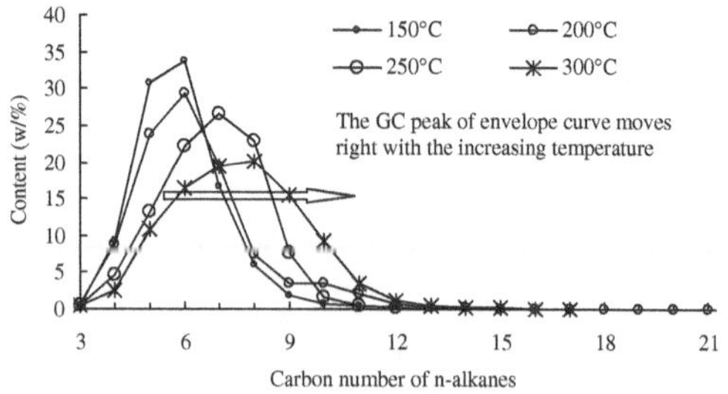

Figure 5. Distribution of n-alkanes in the condensate. (The condensate is obtained from the thermal evaporation experiment, and the experimental oil sample comes from the G 2 oil layer of the Ying 51 well).

and the ratios of n-alkanes based on experimental data. It can be seen that as to the given neighboring component pairs, the calculated relative volatility is larger than 1 under experimental conditions. According to the definition of relative volatility, the concentration ratio of those component pairs in the liquid condensate is larger than that in the residual. Fig. 6 also shows that the results are in agreement with the theoretical results except for nC_4/nC_5. The reason why nC_4/nC_5 is different from the theoretical results is that liquid nitrogen is not used in the process of condensate collection, which leads to the loss of nC_4 in gaseous form. In fact, this also applies to nC_5, nC_6 and others, which also affect their liquid condensate density. There is no doubt that the ratio in the condensate oil is greater than that in the residual, which is in agreement with the relative volatility ($\alpha > 1$).

3.3.3 *The ratios of C_6 and C_7*

N-hexane, methylcyclopentane (MCP), cyclohexane (CH), n-heptane, methylcyclohexane (MCH) and toluene (Tol) are the major components of light hydrocarbons, and they are characterized by their easy determination and importance to the light hydrocarbons. Table 3 presents some concentration ratios of $C_6$and C_7 hydrocarbons in the condensate oil, the

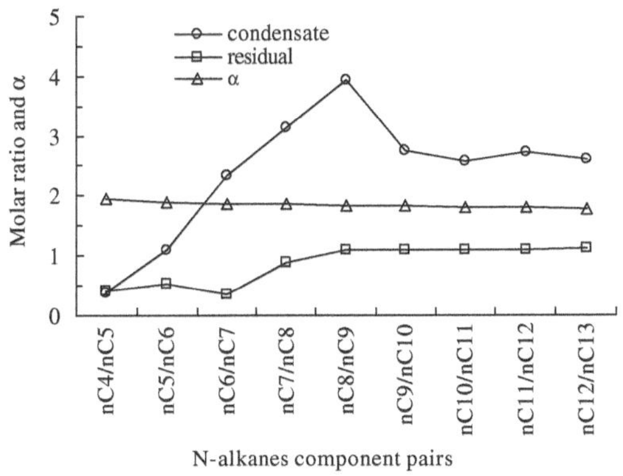

Figure 6. Molarity ratios and α of the thermo-evaporation experiment (at 150°C). ($\alpha > 1$, and the nC_m/nC_{m+1} ratio in the condensate liquid is greater than that in the residual).

residual and the initial oil, and the calculated relative volatility is also given in Table 3. It can be seen from which that the volatility of nC_6 is larger than that of MCP and CH, the volatility of MCP is larger than that of CH., the nC_6/MCP, and nC_6/CH and MCP/CH ratios (5.00, 5.09 and 1.17, respectively) in the condensate oil are much larger than those in the residual oil (3.55, 1.66 and 0.47, respectively) and also larger than those in the initial oil (4.11, 3.39 and 0.83, respectively). Thermo-evaporation enriches nC_6 in the condensate oil compared with MCP and CH, which is in agreement with the fact that α, the calculated relative volatility, is larger than 1. The concentration ratios of C_7 compound pairs in the condensate oil, the residual oil and the initial oil are consistent with the calculated α.

Because the experimental installation is simple, the number of plate N value cannot be estimated. But from Equation (1), the relative value between y_A/y_B and x_A/x_B is controlled by α, in a qualitative or a semi-quantitative analysis, the concentration ratio between compound pairs from the condensate oil and the residual oil produced by thermal evaporation is not subject to the value of N.

The experimental results clearly show that the changing regulation of components between the condensate oil and the residual oil can be well explained by using the relative volatility caused by the thermal-evaporation. Relative volatility is a bond that determines the relative value of concentration ratio of any compound pairs in both the condensate oil and residual oil.

4 THE FORMATION OF CONDENSATE GAS AND THERMAL EVAPORATION

4.1 *Oil source of the condensate oil-gas*

Both the condensate oil-gas and the underlying reservoir oil come from the same source rock. Only the Gu 109 well and Ying 51 well produce the condensate gas from the Putaohua oil layer in the Gulong sag, and underlying them are oil reservoirs with dissolved gas. If the condensate gas comes from the lower part reservoir by thermo-evaporation, the condensate gas from the same source of the underlying oil reservoir is the prerequisite. The source of the condensate gas

Table 3. Molar concentration ratio of light hydrocarbons and the calculated α in the experiment (150°C).

	nC_6/MCP	nC_6/CH	MCP/CH	nC_7/MCH	nC_7/Tol
Condensate	5.00	5.09	1.17	2.02	11.28
Residual	3.55	1.66	0.47	1.52	7.32
α	1.04	1.21	1.16	1.08	1.11
Initial	4.11	3.39	0.83	1.63	8.48

Note: MCP-methylcyclopentane; CH-cyclohexane; MCH-methylcyclohexane; Tol-toluene.

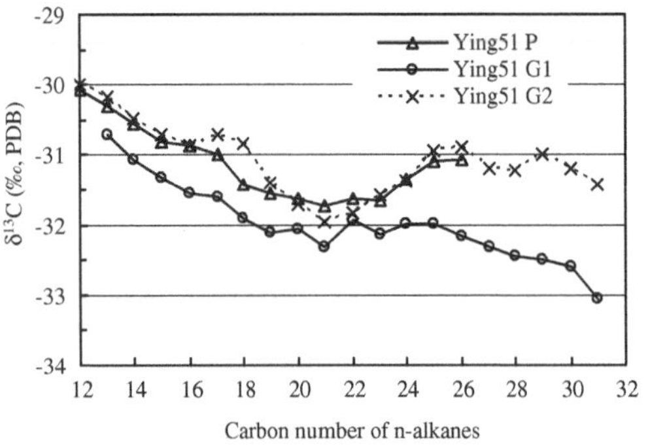

Figure 7. Comparison of carbon isotopes of monomeric compounds (Well Ying 51) (The mode of P is similar to that of G 2, so this similarity means that they come from the same source rock, and the P oil may be one part of G 2.).

in the Gu 109 well has already been studied by some scholars (Xu, 1989; Li, 1997). They arrived at the conclusion that it comes from Qingshankou Formation and shares the same source rock with the underling oil. In view of the condensate gas reservoir resulting from thermo-evaporation and the systematic variation in the hydrocarbon composition in the underlying oil reservoir, and the effect on the hydrocarbon parameters, we also agree that the monomeric hydrocarbon isotope with the tracer character, especially carbon ($>C_{14}$) isotope data which do not have isotopic fractionation in the fluid-gas phase state (Carpentier, et al., 1996; Bjoroy, et al., 1994). In order to prove this view, the monomeric hydrocarbon isotope (Fig. 7) and the gas isotope (Table 1) of the Ying 51 well were studied. The data showed that the uniformity of the monomeric hydrocarbon isotopes of the condensate gas in the layer P and oil in the layer G 2 is extremely good, the maximum deviation is only 0.58‰, most are less than 0.2‰. Such a good uniformity not only shows that the condensate in the layer P and the oil in the layer G 2 come from the same source rock, but also indicates that the oil in the layer P possibly comes from the layer G 2, thus the gas isotope data of the two layers with a high similarity also further explains this possibility.

4.2 *Relationship between the component characteristic of condensate oil-gas and thermo-evaporative mechanism*

4.2.1 *Well Ying 51*

The component relationship of the oil and gas in the two oil layers (P and G 2) in the Ying 51 well coincides with the thermo-evaporative fractionation mechanism. According to the thermal evaporation theory and the experimental results, if the condensate oil-gas in the layer P comes from the underling oil layer G 2 by thermo-evaporation, there should be a relationship based on α between component pair concentrations from both layers. In Fig. 8, the conditions for calculating relative volatility are as follows: according to the geological reserves of P and G 2, through the weight average method, the initial composition of components can be obtained. Temperature and pressure are similar to the geological conditions of the layer G 2 (104°C and 20 MPa). The component pair ratio of the layers P and G 2 is calculated by applying the mass conservation principle to the wellhead oil-gas from the both layers. Fig. 8 shows that the component pair concentration ratio of oil-gas from the layer P is larger than that from the layer G 2, except for $nC_5/nC_6$and nC_6/nC_7 ratios. Among 24 pairs of ratios, 22 pairs are well consistent with the value of α. Therefore, from the view of n-alkane series, the corresponding relationship between the layer P and G 2 oil and gas of the Ying 51 agrees with the mechanism of thermo-evaporative fractionation.

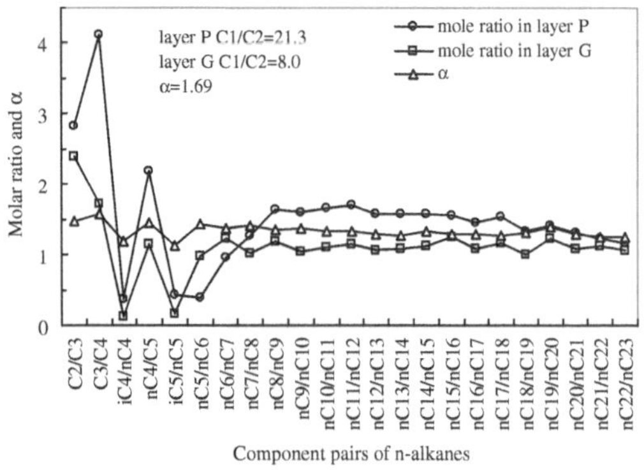

Figure 8. Component pair molarity ratio in the layers P and G. (All the ratios of P are greater than those of G, except for nC_5/nC_6 and nC_6/nC_7. This is because C_5, C_6 and C_7 components are easy to losing and very sensitive to preserving conditions.)

The variations of Pr, Ph, C_6 and C_7 compounds also support the theory of thermal evaporation. The ration valves of seven compound pairs from the layer P are greater than those from the layer G2, respectively, and this is also consistent with the fact that the relative values of volatility for the seven compound pairs are greater than 1 (Table 4). There is such a range of compound pair ratios between the layers P and G 2 that they accord with the mechanism of thermal evaporation, so the above two have a genetic relationship in thermo-evaporative fractionation.

4.2.2 *Well Gu 109*

For the oil sample, only the limited information of the layer P1 in Gu109 well can be obtained from the database, including the relative contents of n-alkanes, Pr and Ph. Fig. 9 shows the mole ratios of n-alkanes. These ratios and Pr/nC_{17}, Ph/nC_{18} and Pr/Ph ratios (Table 5) of P1 are all greater than those of P 2, and this

Table 4. Molar concentration ratio of hydrocarbon compounds from the P, G2 oil-gas layers of Well Ying 51 and α.

	nC_6/MCP	nC_6/CH	nC_7/MCH	nC_7/Tol	Pr/nC_{17}	Ph/nC_{18}	Pr/Ph
P	4.50	3.44	2.17	19.66	0.16	0.14	1.73
G2	4.12	3.39	1.63	8.48	0.13	0.11	1.31
α	1.105	1.231	1.153	1.295	1.017	1.122	1.154

Note: MCP-methylcyclopentane; CH-cyclohexane; MCH-methylcyclohexane; Tol-toluene.

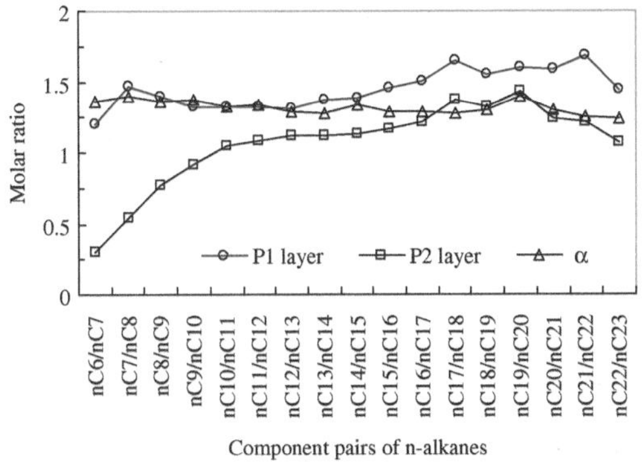

Figure 9. Ratio relationship of n-alkanes in the condensate gas reservoir and the underling oil reservoir (Well Gu 109).

agrees with α calculated by PR state equation according to the stratum liquid composition, the subsurface temperature and pressure of P 2. Therefore, these data retain to support the thermal-evaporative mechanism.

For the gas sample, Table 5 shows the molar ratios of main gas components in the layers P1 and P2. From which it can be seen that these ratios of P1 are also greater than those of P 2, this is also consistent with their relative volatilities ($\alpha > 1$). In fact, there are some components ($>nC_4$) in the analysis process, and their ratios (for example nC_4/nC_5, iC_5/nC_5) have not been shown in the table because a part of these components is assigned in the oil under the surface condition. Therefore, the ratio of the gas sample cannot represent the real relative volatility under the stratum situations.

4.3 *Possibility of the other phase states of the condensate gas in the Gulong sag*

Presently the condensate gas origin falls into four types: (1) original type (Zhou, 1998) also named initial type; (2) gas washing type (Meulbroek, 1998), including the enrichment type (Zhou, 1998) in connotation; (3) gas invasion type; and (4) separate type (Zhou, 1998). Moreover, there are other types, such as migration fractionation type (Zhang, 2000) and Thompson evaporation fractionation type (Thompson, 1987).

What should be added is that the thermal-evaporative effect in this paper is different from the Thompson evaporation fractionation in mechanism. Based on the Thompson experiment, Thompson's type should be exactly called the gas washing, and extremely tallies with the result of gas washing by theoretical calculation (Shen, 2005). In fact, Meulbroek (1998) also named Thompson's type as gas washing type. There is little possibility for the existence of original type in the Gulong sag. Original type means that the hydrocarbon expulsion takes place in the form of the condensate gas phase and still remains in this kind of phase during the process of migration. It is generally considered that, at the subsurface temperature and pressure, if hydrocarbons can be transformed into condensate gas, the gas-oil ratio needs to reach a certain value, generally 500–800 m^3/m^3. If the temperature and pressure of the oil layer are high enough, then the oil-gas ratio (500 m^3/m^3) can be used properly, in view of the low reservoir temperature and pressure (see Table 1), the minimum ratio is selected (800 m^3/m^3) in the Gulong sag. It also means that if there is original type condensate gas in the Gulong sag, the oil-gas ratio must achieve 800 m^3/m^3. Based on the half quantitative thermal evolution model for different types of organic matter occurring in Daqing Oilfield for Songliao Basin (Wang, 1993), only the gas-oil ratio of the products from type II organic matter of the deeper Qn1 formation is close to 800 m^3/m^3. The quantification result of chemical kinetics shows that the gas-oil ratio of hydrocarbon produced by Qn1 is 435 m^3/m^3, and the ratio of Qn23 is only 109 m^3/m^3 in the Gulong Sag. These values are far different from the gas-oil ratio of the gas condensate reservoir in the Gu109 and Ying 51 wells ($>$7000 m^3/m^3) (see Table 1). Even if for hydrocarbons expelled from local source rocks with high oil-gas ratio, the oil-gas ratio cannot reach 7000 m^3/m^3, let alone the effect of the initial migration and secondary migration, which may cause the oil-gas ratio to decrease. The above analysis shows that the possibility of the existence of original type in the Gulong sag is very low, at least for the condensate gas in the Gu109 and Ying 51 wells there is not.

Even by an extensive exploration, massive overmature source rocks and natural gas with over maturity

Table 5. Molar concentration ratio of hydrocarbon compounds from the P1, P2 oil-gas layers of Well Ying 51.

	N_2/C_1	C_1/C_2	C_2/C_3	iC_4/nC_4	C_3/C_4	Pr/nC_{17}	Ph/nC_{18}	Pr/Ph
P1	0.13	18.19	2.41	0.43	5.42	0.18	0.18	1.58
P2	0.02	7.10	1.79	0.33	3.63	0.13	0.12	1.26
α	1.443	1.577	1.389	1.163	1.286	1.077	1.190	1.101

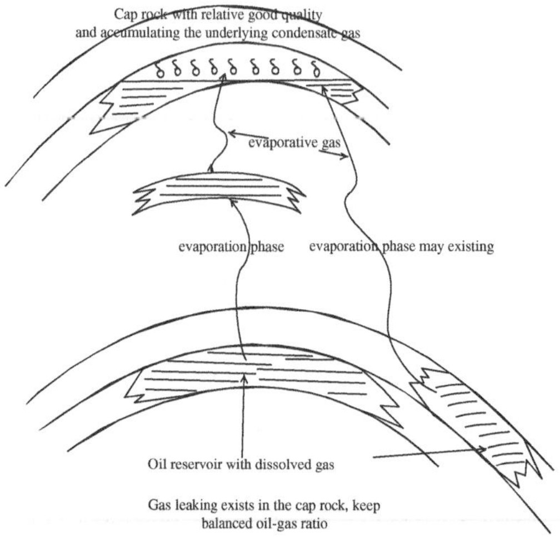

Figure 10. Mode of thermal evaporation (The thermal evaporation will occur in the underling oil reservoir under a certain thermal driving force. Oil and gas exist in the evaporation phase, and the higher the evaporative temperature is, the more the oil in the phase is. The oil and gas in the evaporation phase move above through the cap rock, if cap rock is in good quality, they will accumulate in it. In the cap the condensate gas may be formed, and a little condensate may exist in the gas reservoir. If there is gas leaking in the reservoir, a light-oil reservoir may be formed.).

have not been found in the Gulong sag. Therefore, there are no basic material conditions for the formation of gas invasion and gas washing types, which are related to the process of gas re-injection.

The whole source rocks of the Gulong sag are in the mature to high-maturity stage, and there is no condition for far migration of oil-gas, so the separate type can not be generated here.

The above analysis indicates that using the condensate gas genesis mechanisms mentioned above can not explain the condensate gas origin of the Gulong sag.

5 DISCUSSIONS

Thermal evaporation caused by the thermal movement of molecules occurs not only at normal temperature and atmospheric pressure, but also at subsurface temperature and pressure. Fractionation caused by the thermal evaporation is called thermo-evaporative fractionation. Obviously, the result of thermal evaporation is that more light components accumulate in the vapor phase while more heavy components are retained in the residual liquid phase, which is shown apparently in Fig. 10.

Obviously, the differentiation caused by thermal evaporation has the special demand for cap rock of a residual oil reservoir, i.e. it can neither completely seal up the reservoir, nor can it completely open it widely. If the cap rock completely seals it up, the evaporation discharges can not go upward, reversely, the reservoir fluids will migrate upward entirely, which causes the oil reservoir destroyed. Therefore, the sealing characteristics of cap rocks become an important condition whether the evaporation occurs or not. There are many micro-fractures (Wang, 1993; Ding, 2003) in the Qingshankou, Nenjiang and Yaojia formations of the Gulong sag, and some of the micro-fractures even form the fractured oil-gas pool (This phenomenon hardly occurs in other areas of Songliao Basin). As for the growth and distribution of fractures, most fractures in Qingshakou Formation grow vertically, and those in Nenjiang and Yaojia Formations are less developed; the distribution of fractures in the cross section is not even, which is related to structure spots. The existence of fractures in the Gulong sag is possibly related to the particularity of the sedimentary process and the evolvement of structures. Because the micro-fractures have low penetrability and the intermittence, they play a leading role in both incompletely sealing up and being incompletely unobstructed. This kind of cap rocks is precisely necessary to thermal evaporation. A good sealing capacity of the upside cap rock is needed for the formation of a condensate gas reservoir. The cap rock with a good sealing capacity can preserve most of vapor, and make it achieve a certain gas-oil ratio, and then only at proper temperature and pressure can the condensate gas phase be formed. Obviously, the forming condition of a thermo-evaporative condensate gas reservoir is too hard to be formed, and this is possibly why there exist only a few condensate gas reservoirs in the Gulong sag.

The covariant relation caused by gas washing is similar to that caused by thermal evaporation (Shen, 2005), but there is an important difference between them. The characteristic of gas washing is that the oil and gas are not homologous or homologous but with different maturities in the condense gas and oil, and the oil and gas are associated products of the condensate gas reservoir which forms by thermo-evaporation. The condensate oil and gas from the Gu109 and Ying 51 wells not only have the same maturity but also come from the same source rock, which is in the high-maturity stage.

From another point of view, the formation of condensate gas by thermal evaporation hints important information. Because the thermal evaporation possibly exists universally under the geological condition, the process of evaporation will occur if the condition is proper. The condensate gas reservoir is only one of its results, so maybe there is a light oil reservoir or gas reservoir that contains condensate. The authors believe that the research method applied in this paper will serve as the guidance to the explanation for the special oil-gas distribution.

The condensate gas reservoir found in the Ying 51 well in Songliao Basin cannot be reasonably explained by genesis mechanism of the current phase state. The relative volatilities of a series of compound pairs of the Ying 51 and Gu 109 wells are calculated by the state equation under subsurface conditions. Through comparing the ratios of condensate gas with those of the underlying oil, we find that the relationship decided by relative volatility is in accordance with the ratios mentioned above. Therefore, the condensate gas in the Gulong sag comes from the underlying oil through thermal evaporation, and the micro-fractures in the cap rock must be the migration pathway. The covariant rule of oil-gas components is also revealed and exists in the condensate oil and the residual caused by thermal evaporation.

REFERENCES

Bjoroy, M., Hall, P.B., Moe, R.P., 1994. Variation in the isotopic of single components in the C4–C20 fraction of oils and condensates. Org. Geochem. 21(6/7):761–776.

Carpentier, B., Ungerer, P., Kowalewski, I., Magnier, C., Courcy, J.P., Huc, A.Y., 1996. Molecular and isotopic fractionation of light hydrocarbons between oil and gas phase. Org. Geochem. 24 (12). 1115–1139.

Chen, Y.C., Li, Y.J., Yang, Y.C., Wang, T.D., Li, Y.K., Wen, H. F., Kong, Q.Y., 1997. Discrimination of condensate gas reservoir and its favorable area in Gulong sag, Songliao basin. Xinjiang Petroleum geology, 18(4):313–318. (in Chinese)

Danesh, A., 2000. PVT and phase behavior of petroleum reservoir fluids, Tr. by Shen, P.P., and Han,D., Beijing: Petroleum Industry Press. 263–264. (in chinese)

Ding, W.L., Zhang, B.W., Li, T.M., 2003. Formation of non-tectonic fractures in mudstones in Gulong Sag, Oil & Gas Geology, 24(1), 50–54. (in Chinese)

Guo, T.M., 2002. Multiple unit steam – fluid equilibrium and rectification. Beijing: Petroleum Industry Press. 81–161.(in Chinese)

Huang, R.C., 1997. Generation and occurrence of immature oil and condensate oil in China. Beijing: Petroleum Industry Press. 165–169. (in Chinese)

Li, Y.J., Chen, Y.C., Zhu, J., 1997. Genesis of oil and gas, and distribution of gas condensate reservoirs in Putaohua formation in Gulong area, Songliao basin. Journal of South west Petroleum institute, 19(3):14–19. (in chinese)

Meulbroek, P., Cathles, L., Whelan, J., 1998. Phase fractionation at South Eugene Island Black 330[J]. Org. Geochem. 29(1–3):223–239.

Peng D.Y., Robinson D.B., 1976. A new two-constant equation of state. Ind Engchem Fund. 15(1), 59–64.

Reed R.C., Prausnitz J.M., Poulin B.E., 1994. The properties of gases and liquids. Tr. by Li Z.F., and Yang Y.S., Beijing: Petroleum Industry Press. 11–26. (in Chinese)

Shen, J.N., Lu S.F., 2005. Influence of gas washing on oil-gas composition. Geochimica, 34(2), 161–172. (in Chinese)

Thompson K.F.M., 1987. Fractionated aromatic petroleum and the generation of gas-condensates. Org. Geochem. 11(6), 573–590.

Wang, Z.W., Yang, J.L., Gao, R.Q., 1993. China petroleum geology records (volume II), Daqing Oil Field, Beijing: Petroleum Industry Press. 451–452. (in Chinese)

Xu, Y.X., 1989. Features and formation condition of condensate gas in Gu109 well. Petroleum Geology & Oilfield Development in DaQing, 8 (4), 29–34. (in Chinese)

Zhan, S.L., 1994. Petrochemical industry separation engineering. Beijing: Petroleum Industry Press, 128–130. (in Chinese)

Zhang, S.C., 2000. Migration fractionation: an important mechanism for condensate and wax oil. Chinese Science Bulletin, 45(1), 667–670. (in Chinese)

Zhou, X.X., 1998. Generation and occurrence of natural gas in Tarim basin. Beijing: Petroleum Industry Press. 129–130. (in Chinese)

Petroleum Geochemistry and Exploration in the Afro-Asian Region – Digang Liang, Darui Wang & Zhenxi Li (eds)
© 2008 Taylor & Francis Group, London, ISBN 978-0-415-44083-7

Characteristics of crude oil components for different current recoveries

Xiaoyan Liu, Jungang Li, Chunlei Dai, Chun Li, Jinlai Feng & Jing Wen
Geochemistry laboratory, Daqing Petroleum Institute, Daqing, P R China

ABSTRACT: In order to investigate the petroleum exploitation and enhance oil recovery, the expelled oil characteristics of experimental flooding should be detected accurately. By improved analytical instrumentation (Geofina Hydrocarbon Meter–GHM) with thermal-extraction method, the component characteristics of crude oils at different production stages were analyzed directly without complex pretreatment in the process of a water flooding simulation experiment. The internal relation between crude oil components and current recoveries was studied in detail. Results showed that there is a regular relation between current recovery of crude oils and the PV number of fluid. Furthermore, the carbon-number range and the main-peak carbon number of normal alkanes shift backward and the ratio of light/heavy hydrocarbons decreases when the current recovery rate increases. Moreover, some molecules of isoprenoids are more easily flooded than the normal alkane molecules with similar carbon number because of their column structure in space. This investigation is of practical significance for the research of flooding mechanism and enhancement of recovery efficiency in crude oil production.

Keywords: crude oil; current recovery; oil production; simulation experiment; oil component.

1 INTRODUCTION

In the water flooding simulation experiment, the percolating law of fluid is usually studied with the help of the physical models that concern the factors such as water cut, the output of oils and the recovery efficiency of crude oils (Waston, 1964; Zhang et al. 1997). Some new analytical models have been developed for predicting the swept area of water flooding and for the oil recovery efficiency behind the waterfront in a naturally-occurring reservoir (Fourar & Enormand, 2001; Granet et al. 2001). The effects of wettability and residual oil spreading characteristics in the process of petroleum production during water-flooding have been studied (Carlos & Richard, 2003; He et al. 2002; John, 2002; Shehadeh, 2003; Zhuo et al. 2002). Geochemical technique of oil exploiting is being developed quickly in the field of reservoir geochemistry (Genetti et al. 2003; Whelan et al. 1994; Chen et al. 2002; Hou, 2002). The study on the change characteristics of the inner composition of the produced oil is beneficial to petroleum production. As we know, some crude oils in the strata can be extracted under geological conditions, whereas the others can not. In theory, this is not only related to the porosity, permeability and other parameters of strata, but also related to the inner composition and chemical property of crude oils. With the help of Geofina hydrocarbon meter (GHM) made in Norway Geolab, the composition of crude oils which come from different stages in the process of water flooding simulation experiment is effectively studied by the thermal evaporation method. Therefore, the change characteristics of the crude oil components can be found and the internal relation between the crude oil components and the current recoveries may be sought by the simulation experiment. This investigation is of practical significance, especially in the research of flooding mechanism from the points of microcosm and the enhancement of recovery efficiency of crude oils.

2 METHOD

2.1 *Experimental samples*

The experimental samples came from the crude oils produced by the water flooding core simulation experiment. The geological conditions of the actual stratum were simulated in the experiment. The core used in the experiment is a natural cylinder core that is 52 cm long and 2.5 cm in diameter. The average permeability of the core is 0.665 μm^2. The water and oil samples in the experiment came from the flooding water of the Second Production Factory in Daqing Oilfield. The experimental temperature was the same as the actual strata temperature (45°). The concrete process of the water flooding in the simulation experiment is as follows: firstly, the experiment core was extracted for 72 hours with an organic solvent – chloroform under

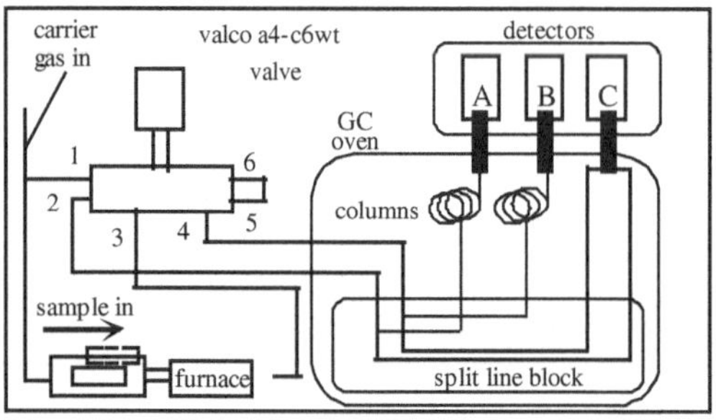

1, 2, 3, 4, 5, 6 – the exit of a 6-port valve

Figure 1. Flow sketch of the analytic instrument.

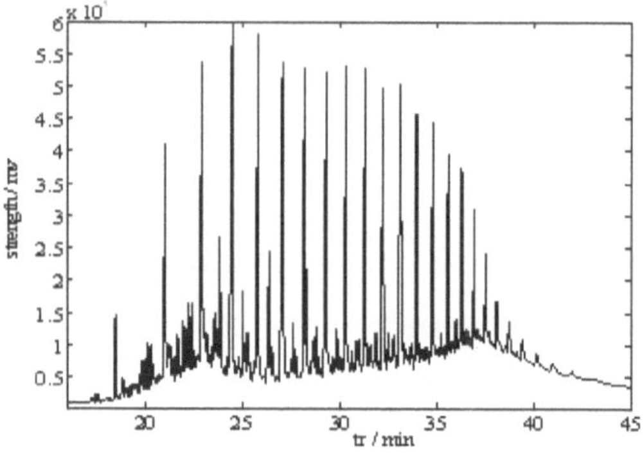

Figure 2. GHM gas chromatogram of the sample with a current recovery rate of 68.7% in the flooding process.

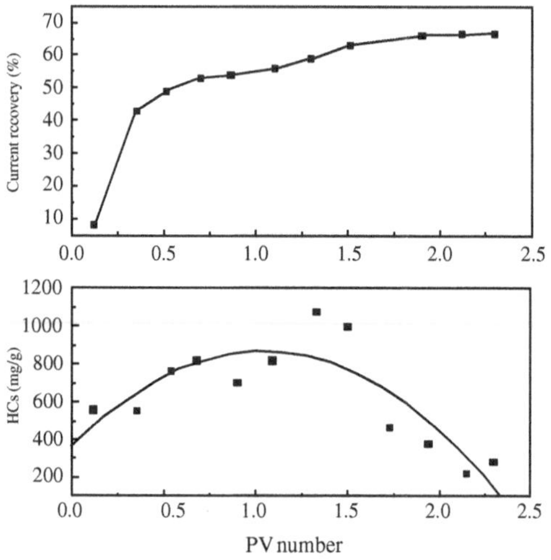

Figure 3. Relation of oil current recovery rate and hydrocarbon content or PV number, showing a good corresponding relation between the current recovery rate and the ratio of the produced fluid to all the fluid in the core.

common pressure conditions; secondly, the core was saturated with the strata water after drying at under 100°C for 1 hour, and the porosity of the core was determined; thirdly, the strata water in the core was expelled until there was no more water coming out from the exit of the core; fourthly, the core was saturated with the crude oil, then the original oil saturation of the core was determined. Afterwards, the oil was expelled by the water until the water saturation of the core was more than 98%. The expelled samples were taken every 20 minutes. Therefore, the various oil samples could be gotten under the different current recoveries in the process of oil expulsion.

2.2 *Analytical instrument*

The analytical instrument used in the investigation was GHM hydrocarbon analyzer. The GHM was mainly composed of five parts: the main engine of GHM, the temperature controller of GHM, the GC separating system, the data-collecting system, and the computer managing system (Fig.1). GHM can be used to analyze organic compounds qualitatively and quantitatively (Liu et al. 1998; Liu & Lü, 1999; Liu et al. 2001).

2.3 *Analytical procedure*

The main procedure of GHM analysis is as follows: the experiment sample was weighed precisely with approximately about 0.2~0.6 mg in the small sample cup, and the cup was put on the sample shelf after programing the pyrolysis and the GC separation. At the beginning of experiment, the temperature was constantly kept at 200°C for one minute, then the temperature increased to 300°C at the rate of 25°C/min and this temperature was kept for 3 minutes. The pyrolysis hydrocarbons were detected in the quantitative detector with a dividing ratio of 30:1. Then the qualitative and quantitative analysis charts (Fig.2) of the sample of current recovery 68.67% could be gotten from the experiment.

3 RESULTS AND DISCUSSION

3.1 *Current recovery of oil*

In the process of the experiment, there is a good corresponding relation between the current recovery rate and the ratio of the produced fluid to all the fluid in the core (PV number). By quantitatively analyzing the peak area and the amount of the sample in the analysis chart, the changing character of the PV number for different evaporable hydrocarbon contents in the sample can be obtained in different current recoveries at 300°C. These evaporable hydrocarbons have light molecular weight and low polarity (Fig.3).

It can be seen that the relation of the amount of the evaporable hydrocarbons to the PV number has

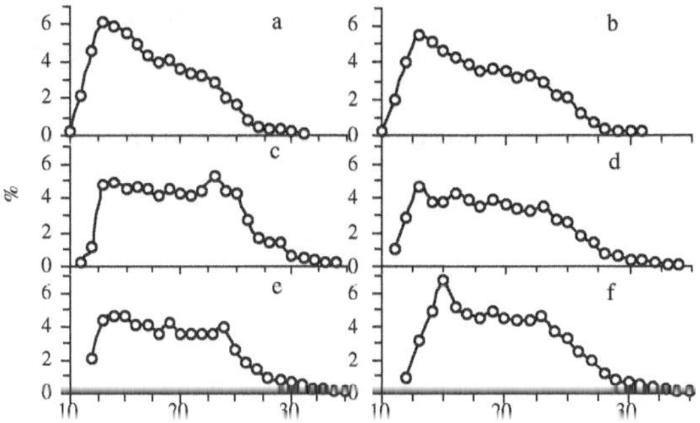

Figure 4. Distribution of n-alkanes in crude oil samples with carbon number ranging mainly between nC_{10} and nC_{35}, the curent recovery rates of Samples a, b, c, d, e and f are 17.1%, 43.02%, 56.75%, 62.97%, 67.37% and 68.67% respectively.

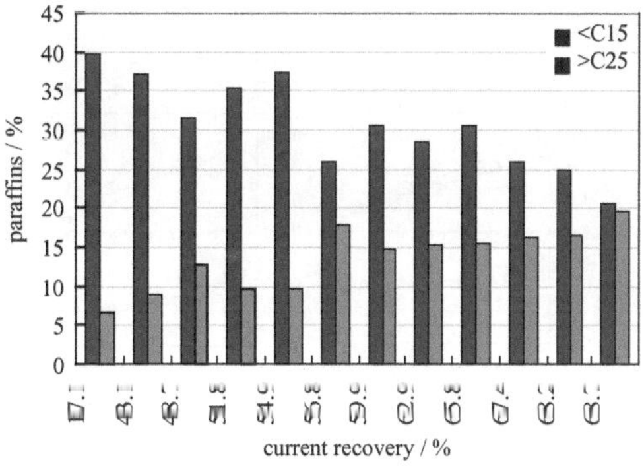

Figure 5. Column diagram of lighter & heavier components under various current recovery rates.

a good correlativity. With the improvement of the current recovery rate, the amount of the evaporable hydrocarbons in the expelled oil also increases gradually. A peak value appears when the PV number of the produced fluid is about 1.3, and then it declines gradually.

3.2 *Composition characteristics of oils*

Through the experiment, the changing character of the oil components in the oil expulsion process can be obtained, and the initial diagrams in flooding process are shown in Figure 4. According to the pristane, phytane and other compounds of biomarkers, the oil components may be analyzed qualitatively, and the resulting data can be gained; for instance, the carbon number range of normal alkanes is mainly between nC_{10} and nC_{35}.

The distribution characteristics of n-alkanes in the oil samples at different current recoveries are shown in Figure 4, and the column diagram of lighter and heavier components in the crude oil is shown in Figure 5.

In Figure 4, as the current recovery rates of Samples a, b, c, d, e and f increase gradually, the carbon number range of n-alkanes in the crude oil moves backward from $C_{10}{\sim}C_{31}$ to $C_{12}{\sim}C_{35}$. When the current recovery reaches about 43%, the components of the produced oil change little. Its main peak carbon number is 13. However, when the current recovery rises again, the components of the crude oil change gradually. The component with medium molecular weight in the crude oil increases gradually and the peak carbon number of n-alkanes changes from C_{13} at low current recovery to C_{15} at high current recovery (Fig.4). In addition, there is no clear odd-even predominance of n-alkanes in both the original oil and the produced oil samples. Because of the high oil content of the rock in its original state, some heavy components are brought out when the light components with low viscosity and low gravity are expelled synchronously. The content of light components in the core decreases with the current recovery rate increasing. So the light components in the produced crude oil reduce gradually, while the heavy components increase relatively. The oil saturation of the core decreases with the constant advancement of the expulsion process. The rock surface has a certain absorption ability due to the existence of the polar molecules in the rock (Fig.5).

The absorption degree between the rock and the heavy components with strong polarity is high. The heavy components are difficult to be expelled due to their high viscosity and high gravity. So both the light components and the heavy components cannot ultimately be expelled out of the rocks.

3.3 *Special indicator change*

In the oil samples, the peak area ratios of some characteristic components change with the PV number in the oil expulsion process. The change curves are shown in Figure 6. A certain rule can be found between the oil components and the change of PV number. Pr/Ph, iC_{16}/iC_{17}, nC_{21}^-/nC_{22}^+, nC_{14}^-/nC_{15}^+, Pr/nC_{17}, Ph/nC_{18} and other parameters decrease while the current recovery increases. Because of the similar structural arrangements between Pr, iC_{16} and Ph, iC_{17}, all the regular isoprenoids composed of isoprene units are joined by "head-to-tail" linkage; therefore, paraffins with low molecular weight and low carbon number are expelled out first in the oil expulsion process while the ones with high molecular weight and high carbon number are difficult to do so. Moreover, nC_{21}^-, nC_{22}^+, nC_{14}^- and nC_{15}^+ all belong to n-alkanes, and they have similar structure type, so the light n-alkanes are also relatively expelled first. With the current recovery increasing, the ratio of light and heavy hydrocarbons decreases at the same time.

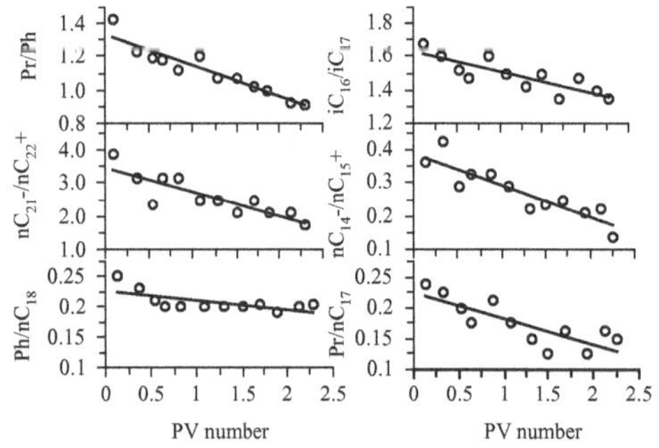

Figure 6. Variation of indicator ratios in the oil samples for different current recoveries, showing the decrease of parameters (Pr/Ph, iC_{16}/iC_{17}, nC_{21}^{-}/nC_{22}^{+}, nC_{14}^{-}/nC_{15}^{+}, Pr/nC_{17} and Ph/nC_{18}) with the increase of the current recovery.

However, Pr, Ph and nC_{17}, nC_{18} have different structures with the difference of two $-CH_2$ groups in the molecule. Pr and Ph have regular branches while nC_{17} and nC_{18} are straight chain alkanes. According to the traditional viewpoint, n-alkanes should be expelled earlier than isoalkanes, but the analysis result of the experiment showed that isoprenoid hydrocarbons are expelled earlier in the experiment. A logical explanation to the experimental phenomenon is as follows: there are four regular branches in Pr and Ph molecules to make the repelling strength among the atoms in the molecule the smallest and let it exist in the space with the most stable structure, while the space direction of the 4 branches should take the distance the longest. Therefore, Pr and Ph molecules are like a cylinder and their cross section areas are bigger than those of nC_{17}, nC_{18}. For this reason, they are more easily drained through big pores and more difficult to enter small or invalid pores in the migration process. Thus, Pr and Ph molecules are more easily expelled than nC_{17} and nC_{18} molecules. In addition, there are many light components ($nC_{10}\sim nC_{14}$) in the earlier expelled oil sample. With the current recovery rate increasing, the light components in the oil sample decrease while the heavy ones ($nC_{30}\sim nC_{35}$) increase.

Because of the low molecular weight, low viscosity, and low polarity of the light components, the absorption ability of the rock for the light components is low, and the light components flow more easily and are expelled more readily than heavy ones. Therefore, the amount of the light components in the later produced oil sample decreases step by step and the heavy ones take on a trend of increasing.

4 CONCLUSIONS

The characteristics of the oil expulsion can be studied by the experiment and analysis not only in a macroscopic view but also in detail from a micro-molecule point. Based on the study result in this paper, the following conclusions can be drawn:

(1) There is a good correlation between the PV number and the current recovery rate. The amount of the evaporable hydrocarbons in the produced oil increases with PV number increasing gradually, and then decreases step by step.
(2) With the current recovery rate increasing gradually, the carbon number range of n-alkanes in the crude oil and the carbon number of the main peak move backward.
(3) The light components in the crude oil are more easily expelled than heavy components. The ratio of the light and heavy components in the crude oil decreases when the current recovery rate increases.
(4) Pr and Ph are more easily expelled than nC_{17} and nC_{18} due to their cylinder-like structural characteristics.

The change law of the crude oil components in the oil expelling process is analyzed and the mechanism of oil expulsion is discussed from a micro-molecule view in the paper. They may be used in the research of oil expulsion process and increasing recovery efficiency of crude oils. With the research going deeper, the oil saturation of strata under geological conditions can also be determined by the change law of the molecular characteristic parameters in crude oils.

ACKNOWLEDGEMENTS

The financial supports from the Nature Science Fundation of Heilongjiang Province (B2010) and Scientific & Technological Fundation (GZ05A601, 10541005) are gratefully acknowledged.

REFERENCES

Carlos A.G., Richard A. D., 2003. Gas and oil production from waterflood residual oil: effects of wettability and oil spreading characteristics. *Journal of Petroleum Science & Engineering* 39(3-4):297–308.

Chen Zulin, Zhu Yangming, Chen Qi, 2002.The change of crude oil components in different production stages. *Acta Sedimentologica Sinica.* 20(1):69–173(in Chinese).

Fourar M. and Lenormand R., 2001. A new model for two-phase flows at high velocities through porous media and fractures. *Journal of Petroleum Science & Engineering* 30(2):121–127.

Genetti D.B., Whitaker C.A., Smith D.P., 2003. Applying Improved Recovery Processes and Effective Reservoir Management to Maximize Oil Recovery at Salt Creek. *SPE 13th Middle East Oil Show & Conference, Bahrain* 6:9–12.

Granet S., Fabrie P., Lemonnier P., 2001. A two-phase flow simulation of a fractured reservoir using a new fissure element method. *Journal of Petroleum Science & Engineering* 32(1):35–52.

He Fengyun, Yu Tianzhong, Zhang Jifen, 2002. Effect of water drive on crude properties in reservoirs and formations. *Journal of Daqing Petroleum Institute* 26(2):21–23(in Chinese).

Hou Dujie, 2002. New developments on geochemical technique of oil exploit: geochemistry of oil production. *Oil & gas geology* 23(1):8–11(in Chinese).

John C.R., 2002. Water advance and oil production rate in a naturally fractured reservoir during waterflooding. *Journal of Petroleum Science & Engineering* 36(1–2):19–32.

Liu Xiaoyan, Chen Xin, Lü Jing, 1998. Improvement of the standard experimental program on GHM hydrocarbon analysis. *Petroleum Instrument* 12(6):10–11(in Chinese).

Liu Xiaoyan, Lü Jing, 1999. Experimental methods of pyrolytic parameter measurement using GHM instrument. *Petroleum Instrument* 13(1):30–31(in Chinese).

Liu Xiaoyan, Lu Shuangfang, Feng Yali, 2001.Generation characteristics of later-mature hydrocarbons in Songliao Basin by GHM experiments. *Petroleum Instrument* 15(2):35–37(in Chinese).

Shehadeh K.M., 2003. The effect of wettability heterogeneity on capillary pressure and relative permeability. *Journal of Petroleum Science & Engineering* 39(3–4):399–408.

Waston R.E., 1964. Model studies on the inverted nine-spot injection pattern. *JPT.* 16:801–804.

Whelan J.K., Kennicutt M.C.K.K., Brooks J.M., 1994. Organic geochemical indicators of dynamic fluid flow processes in petroleum basins. *Org.Geochem.* 22: 587–615.

Zhang Jifen, Zhao Mingguo, Liu Zhongchun, 1997. The principle of enhanced oil recovery. *Petroleum industry press* 145–153(in Chinese).

Zhuo Xingjia, Guo Xianguang, Liu Hong, 2002. Development effect of the low-permeability reservoirs by fracture at different drive manners. *Journal of Daqing Petroleum Institute* 26(2):15–17(in Chinese).

Petroleum Geochemistry and Exploration in the Afro-Asian Region – Digang Liang, Darui Wang & Zhenxi Li (eds)
© 2008 Taylor & Francis Group, London, ISBN 978-0-415-44083-7

Author Index

Banerjee, A. 21
Bian, Lizeng 103

Chandra, K. 3
Cheng, Dingcheng 213
Cheng, Keming 95

Da, Jiang 95
Dai, Chunlei 209, 235
Dai, Jinxing 47, 59
Dou, Lirong 213

Fadul, O. Ibrahim, Mohamed 37
Fang, Wei 113, 225
Feng, Jinlai 235
Fu, Xiaochun 187

Hou, Dujie 201
Hu, Kai 103
Hu, Mingxia 103
Hu, Yong 95
Huang, Baojia 201

Jia, Ping 169
Jiang, Lin 169
Jin, Qiang 47, 59, 127
Jin, Xiaohui 219

Kumar, B. 69

Li, Chun 209, 235
Li, Jungang 209, 235
Li, Yanxia 75
Li, Yingli 209
Li, Yingying 213
Li, Zhiming 103, 153
Li, Zhi 213
Li, Zhenguang 119
Liang, Yingbo 47
Lin, Renzi 219
Liu, Baoquan 153
Liu, Chongxi 137
Liu, Shaobo 27
Liu, Wenbin 103
Liu, Xiaoyan 187, 209, 235
Long, Zulie 201
Love, G.D. 145
Lu, Shuangfang 113, 187, 225
Lu, Xiancai 103

Meredith, W. 145
Mohamed, A. Mahgoub 37
Murae, T. 87
Mustafa, A.A. 177

Nwachukwu, J.I. 177

Pahari, S. 21
Panicker, Smitha, K. 69
Pisutha-Arnond, S. 161
Pisutha-Arnond, V. 161
Prasad, I.V.S.V. 21

Qin, Jianzhong 103, 153
Qin, Shenfei 27

Shen, Jianian 225
Shi, Quan 219
Sirimongkolkitti, A. 161
Sinha, A.K. 3
Snape, C.E. 145
Song, Yan 27
Sun, Xiaodong 11

Tang, Youjun 201
Toh, K. 87

Uguna, C. 145

Vardhan, C. Vishnu 69
Varshney, Meenu 21

Wan, Congli 127
Wang, Bo 187
Wang, Guanmin 59
Wang, Jianwei 11
Wang, Min 225
Wang, Qinhua 27
Wang, Xibin 169
Wang, Yuewen 113
Wang, Zhaoyun 11
Wang, Zhaoming 27
Wen, Jing 209, 235
Wu, Chuanzhi 137
Wu, Liyan 195
Wu, Xiaozhou 11

Xong, Ying 95
Xu, Guanjun 11
Xu, Li 127
Xu, Xinde 201
Xue, Haitao 187

Zeng, Yi 127
Zhang, Dajiang 95
Zhang, Juhe 113
Zhang, Jinchuan 201
Zhang, Linye 47
Zhang, Shuichang 27, 47, 59
Zhang, Zhihuan 75
Zhang, Zhenling 195
Zhao, Kebin 137
Zhao, Menjun 27
Zhong, Jianhua 169
Zhong, Ningning 75
Zhu, Dan 219
Zhu, Guangyou 47, 59
Zhu, Junzhang 201
Zou, Caineng 11